D0559339

Selection in Natural Populations

Common killifish, *Fundulus heteroclitus*, swimming over a bed of blue mussels, *Mytilus edulis*

Selection in Natural Populations

JEFFRY B. MITTON

Oxford New York • Oxford University Press 1997

Oxford University Press

Oxford New York
Athens Auckland Bangkok Bogota Bombay Buenos Aires
Calcutta Cape Town Dar es Salaam Delhi Florence Hong Kong
Istanbul Karachi Kuala Lumpur Madras Madrid Melbourne
Mexico City Nairobi Paris Singapore Taipei Tokyo Toronto Warsaw

and associated companies in
Berlin Ibadan

Published by Oxford University Press, Inc.
198 Madison Avenue, New York, New York 10016

Library of Congress Cataloging-in-Publication Data
Mitton, Jeffry B.
Selection in natural populations / by Jeffry B. Mitton.
p. cm.
Includes bibliographical references and index.
ISBN 0-19-506352-x
1. Molecular evolution. 2. Natural selection.
3. Enzymes—Evolution. 4. Variation (Biology)
I. Title.
AH371.M68 1997
572.8'38—dc21 96-49688

9 8 7 6 5 4 3 2 1

Printed in the United States of America
on acid-free paper

Things Eternal

What seems more stable than some mighty mountain range?
And what more frail than wing of butterfly?
Yet mountain ranges rise and wear away
Through eon after eon of unending change.

While patterns of the insect wing, unchanging
In the species, though each butterfly must die,
Persist through countless generations,
Outlasting, in the end, the very granite hills.

Such constancy, in fragile organism—
Its physiology in equilibrium
With all the forces of environment,
Through geologic time—is worth our awe
Like constellation spangled skies at night,
Or like the Andes rising from Pampean plain.

Karl P. Schmidt, *Scientific Monthly*, October 1944, 260.

Preface

Both the beginning and the completion of this writing project were prompted by methodological innovations that profoundly influenced the questions asked by evolutionary biologists. I began the book to document the exciting developments in evolutionary biology fueled by studies of protein genetic variation. I was stimulated to finish the book, or at least to bring it to an end, by the realization that biologists were turning away from the use of protein polymorphisms, favoring the use of the rapidly expanding set of techniques to survey genetic variation of DNA. The common desire to use the most modern techniques has caused biologists to put aside some important issues, for the newest techniques are most appropriately applied to a different set of questions.

The use of electrophoresis and histochemical staining to survey genetic variation freed evolutionary biologists from the constraints that had limited classical geneticists to laboratory species. Electrophoretic techniques allowed evolutionary biologists to examine a wide variety of plants and animals, including such exotic species as coast redwoods, elephant seals, blue mussels, and American eels. The levels of genetic variation revealed by the electrophoretic surveys could not be accommodated by the existing models of fitness determination, stimulating some theoretical evolutionary geneticists to first propose and later assert that the protein genetic variation was adaptively neutral. At the same time, empirical geneticists found evidence of natural selection in their data. The neutralist-selectionist debate flared. But with more data and some thoughtful consideration, the neutralist and selectionist hypotheses were recognized as diametrically opposed and equally unrealistic alternatives. The natural world must lie between these extremes. Before the controversy was abandoned, laboratory and field studies inadvertently wedded the fields of physiological ecology and population biology, and revealed insights into the strength and variability of selection. The evolutionary insights from these interdisciplinary studies are the focus of this book.

Several strategies were employed to test whether protein polymorphisms were influenced by natural selection. Some biologists chose to sample a variety of samples and to test whether genetic variation changed over space in a pattern consistent with environmentally mediated selection. Others imposed selection in the lab or watched it in the field, while still others estimated selection coefficients associated with mating, survival, or fecundity. A few exemplary research programs include studies of enzyme kinetics, physiology, demography, and behavior, and these give us the fullest impression of how profoundly the genotypes segregating at a single locus can produce differences among individuals. Truly comprehensive studies, such as the studies of sickle-cell hemoglobin polymorphism in humans, the Lap polymorphism in blue mussels, the LDH polymorphism in killifish, and the PGI polymorphism in *Colias* butterflies, reveal two important insights. First, natural selection measured with enzyme polymorphisms can be intense, with selection coefficients in the range of 0.1 to 0.8. Second, and more generally, these studies reveal dynamic interactions among genotypes, fitness differentials, and fluctuating environmental conditions.

This book is a product of interactions with many friends. Richard Koehn, Dennis Powers, Jeffrey Powell, James Hamrick, and Ward Watt read various stages of the manuscript and suggested numerous modifications. Discussions or correspondence with John Avise, Ronald Burton, Pat Carter, Neil Cobb, Douglas Crawford, John De Fries, Michael Grant, Günther Hartl, Richard Hoffmann, Tom Kocher, Brian Kreiser, Robert Latta, William Lewis Jr., Yan Linhart, Hsiu-Ping Liu, Janusz Markowski, Anders Møller, Susan Mopper, Eviatar Nevo, Ben Pierce, Andrew Pomiankowski, Daphne Rainey-Foreman, Gerald Rehfeldt, Peter Smouse, Lee Snyder, Michael Soulé, Robert Vrijenhoek, Tom Whitham, and Eleftherios Zouros made their way into the text. Jarle Mork, Rebecca Kimball, and Astrid Kodric-Brown provided unpublished data. Theodosius Dobzhansky and George Williams wrote much that I appreciated, and some of their ideas are reflected in this work. Ken and Eric Mitton produced most of the figures, and Ken Mitton formatted the simulation program in appendix 2. Sara Mitton provided valuable support throughout this project.

Rather than adopt a common convention for genes and alleles, I followed the conventions used by the various authors in the primary literature. Thus, lactate dehydrogenase in killifish is *LDH-B*, leucine aminopeptidase in mussels is *Lap*, and alleles are designated either with numbers or letters.

I produced the first outline of this book while on sabbatical at the Marine Biological Laboratory at Woods Hole (academic year 1980–81), and made substantial progress on it while on sabbatical at the Hopkins Marine Station (academic year 1989–1990). Most of the book was written at the University of Colorado, in a lively environment created by my students and colleagues.

Boulder, Colorado J. B. M.
March 1997

Contents

Selection in Natural Populations

1

Natural Selection, Fitness Determination, and Molecular Variation

The principal unit process in evolution is the substitution of one gene for another at the same locus.

Haldane,
"The Cost of Natural Selection" (1957)

Although evolutionary biologists are known for their diversity of opinions, a majority would probably accept the two following generalizations. First, most natural populations contain abundant genetic variation. Morphological (Endler 1986) and physiological (Feder et al. 1987) variation is common, and variability for life-history variation also is apparent in natural populations (Clutton-Brock 1988; Dingle and Hegmann 1982). Surveys of allozyme variation, mitochondrial DNA, and moderately repetitive DNA all reveal abundant genetic variation. Second, selection is common in natural populations. Antonovics (1971) summarized natural selection for tolerance to soil contaminants in plants, and Linhart and Grant (1996) summarized examples of natural selection generating and maintaining microgeographic variation in plants. Endler (1986) compiled studies of selection, which showed that selection was often intense, with selection coefficients exceeding .2 (figure 1.1) in about half the studies that reveal natural selection. We know little about how these two generalizations relate to each other. Furthermore, although selection is common and genetic variation is abundant, we do not know to what degree the abundance of genetic variation is influenced by selection.

In this book I examine the data relating molecular variation to selection in natural populations. After reviewing the methods used to detect genetic variation (chapter 2), I discuss the associations between levels of genetic variation and environmental heterogeneity (chapter 3). Next I present several case studies that demonstrate how genetic variation at a protein locus can influence biochemistry, whole animal physiology, and fitness (chapter 4).

The remaining chapters consider the implications of selection on many loci, the determination of fitness, and several of the evolutionary implications of selection for physiological variation. The following discussion in this chapter describes the focus of the later chapters of the book.

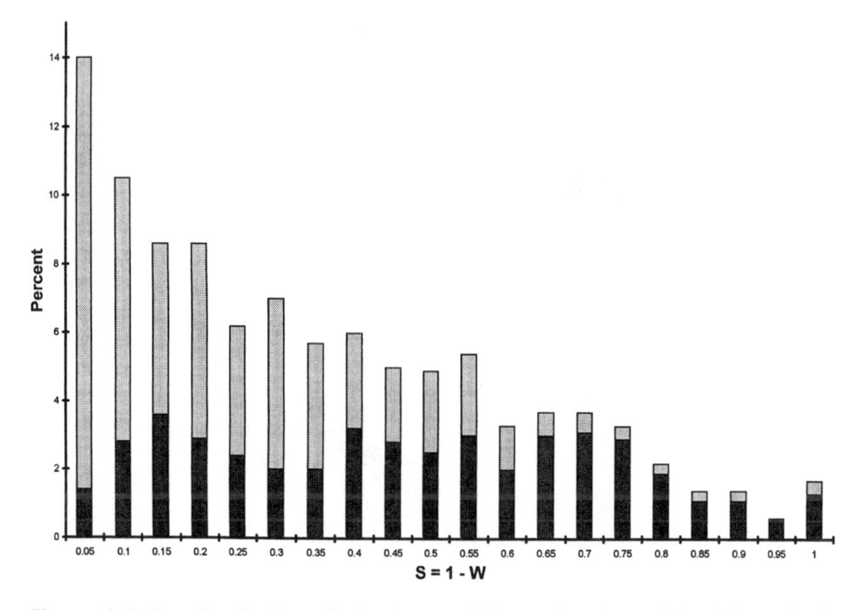

Figure 1.1 The distribution of selection coefficients, S estimated for 36 species in undisturbed natural populations. The height of each bar indicates the percentage of S values in each interval, and the shaded portion marks those that are statistically significant. Of the 566 values of S, 239 were statistically significant. Adapted from figure 7.1 in Endler 1986.

Levels of Genetic Variation in Natural Populations

Nucleotide sequences, amino acid sequences, surveys of intron number and length, surveys of restriction enzyme sites in mitochondrial DNA, DNA fingerprint analyses, and electrophoretic surveys of proteins all indicate that natural populations contain abundant genetic variation.

More than any other molecular technique, surveys of protein polymorphisms have been used to describe levels of genetic variation within and among populations. The range of genetic variation among species is dramatic (chapter 9), but on average, 33 to 50% of the enzymes examined are polymorphic, and a randomly chosen individual is heterozygous at 4 to 15% of the loci surveyed (Brown 1979; Hamrick, Mitton, and Linhart 1979; Nevo, Beiles, and Ben-Shlomo 1984; Powell 1975).

Surveys of mitochondrial DNA (mtDNA) have also been used to estimate levels of genetic variation within populations and to describe patterns of geographic variation (Avise et al. 1987; Moritz, Dowling, and Brown 1987). Most of these surveys have used restriction fragment length polymorphisms to describe the genetic variation of mtDNA. Avise and colleagues (1987) summarized studies of mtDNA in 21 species of animals, finding that the number of mtDNA phenotypes detected in species varied from 2 to 61. Length and restriction site polymorphisms are so common in menhaden (*Brevoortia tyrannus* and *B. patronus*) and chuckwallah (*Sauromalus obesus*) that approximately 98% of the mtDNA phenotypes are seen in only a single individual.

DNA fingerprinting (Jeffreys, Wilson, and Thein 1985a,b) is most useful in forensics and human paternity disputes, but it has also been used to assay variation within natural populations. Probes designed to hybridize with the core segments of intermediately repetitive, tandemly repeating elements detect both variation in restriction enzyme sites and variation in the number of tandem repeats, and consequently they reveal high levels of genetic diversity. Probes can be designed to detect only a single locus or to hybridize with numerous loci sprinkled throughout the genome. Indeed, multilocus probes recognize so much genetic variation that the probability that unrelated human individuals share the same pattern is estimated to be $<3 \times 10^{-11}$ with a single probe and $<5 \times 10^{-19}$ with two probes (Jeffreys et al. 1985b). Probes designed to hybridize to a single locus may identify more than 50 "alleles" (Jeffreys et al. 1988).

A Brief Overview of Neutral Theory

Evolutionary biologists generally agree that natural selection has played a major role in shaping physiological, morphological, and behavioral adaptations, but they are divided over the issue of the primary evolutionary force that influences molecular evolution. Several comprehensive research programs (chapter 4) have clearly demonstrated that selection can act on molecular variation. Nevertheless, important patterns in the data are predicted by neutral theory.

The evolutionary theory treating neutral genetic variation is rich in predictions, and it helps us understand many empirical patterns of genetic variation. Neutral theory can provide the expectations under the null hypothesis, and empirical data may then be tested for their fit to the expected patterns.

The theory of neutral molecular evolution (Crow and Kimura 1970; Kimura 1983; Kimura and Ohta 1971; Nei 1975) is based on an assumption concerning the distribution of selection coefficients for newly arising mutations. Deleterious mutations, which produce alleles with impaired biochemical properties, are assumed to be common, but these mutations are expected to be eliminated by natural selection or kept at very low equilibrium frequencies by the opposing forces of selection and recurrent mutation. The potential for beneficial mutations must exist at many loci, but because the rate of mutation to new, beneficial alleles is assumed to be very low, the impact of these mutations on molecular evolution is assumed to be slight. A major impact on molecular evolution is assumed to result from mutations that have either no influence or only insignificant consequences for the protein's function. Because these mutations do not substantially alter the protein's function, they have no impact on physiological, morphological, or behavioral phenotypes and are thus neutral with respect to adaptation and natural selection.

Because they offer neither advantage nor liability, neutral mutations are either lost or fixed by stochastic changes in allelic frequency from generation to generation. Thus the evolutionary dynamics of neutral mutations are adequately described by equations employing population size, N, effective population size, N_e, neutral mutation rate, u, and migration rate, m (Kimura 1983; Kimura and Ohta 1971; Lewontin 1974; Nei 1975). Neutral theory has had a tremendous impact on population genetics, and many empirical patterns are consistent with predictions arising from neutral theory. The utility of neutral theory can be illustrated with several predictions consistent with empirical data.

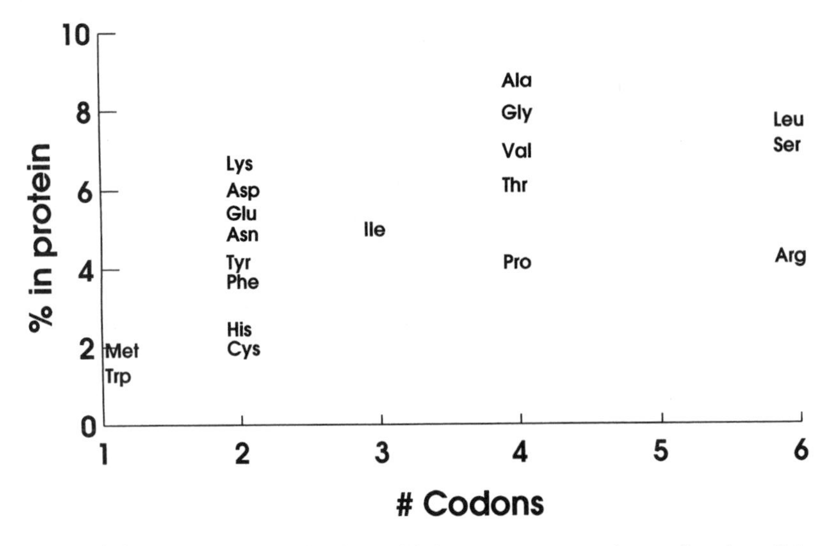

Figure 1.2 The percentage of amino acids in an average protein as a function of the number of codons coding for that amino acid. The correlation between percentage and number of codons is $r = .63$. Adapted from Jukes, Holmquist, and Moise 1975.

Redundancy in the genetic code predicts amino acid frequencies

Let us consider as a null hypothesis the average protein that would result if proteins first evolved unconstrained by natural selection, with their amino acid compositions initially determined by a random permutation of the four nucleotides in the DNA and subsequently evolving only by historical constraint on opportunity. The four nucleotides—adenine, thymine, guanine, and cytosine—exist in approximately equal frequencies in extant species (Ayala and Kiger 1984). Twenty amino acids are inserted into proteins, with some later being modified. The amino acids are coded by one to six codons; methionine and tryptophan are specified by a single codon, whereas arginine, leucine, and serine are specified by six codons. Given the approximately equal abundance of the nucleotides, differences in the abundance of amino acids in proteins should reflect primarily the redundancy of the code. Analyses of the expected and observed frequencies of amino acids in proteins (Jukes, Holmquist, and Moise 1975; Ohta and Kimura 1971) are similar in result if not in interpretation. Although the observed distribution of amino acids deviates from the expected distribution, mainly because of the scarcity of arginine (Jukes, Holmquist, and Moise 1975), the distributions are similar (figure 1.2). The correlation between the observed and the expected frequencies is .63, indicating that approximately 40% of the variation in amino acid frequencies is explained by the redundancy in the genetic code.

The molecular clock and phylogenetic inference

The neutral theory of molecular evolution (Kimura 1968; King and Jukes 1969) predicts the continual molecular evolution that results from mutation pressure (Kimura

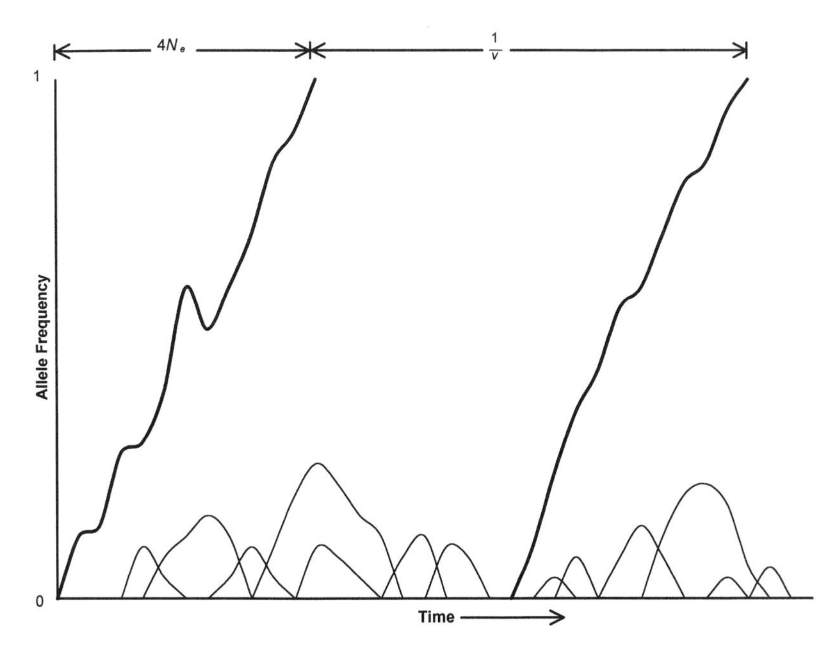

Figure 1.3 A graphical representation of neutral gene substitution. Neutral mutations arise at random, and then their frequencies vary stochastically. Most alleles are lost (*light lines*), but some rise to fixation (*heavy lines*), replacing all the alleles that had been in the population. The time for an allele to drift to fixation is approximately 4 N_e, indicating that the rate of drift decreases with increasing population size. The time between fixations is $1/v$, where v is the mutation rate; thus the rate of fixations is inversely proportional to the mutation rate and independent of population size. Modified from Kimura 1979.

1979, 1983). Consider a mutation rate, u, producing amino acid replacements that result in functionally equivalent proteins. In a population of N individuals, $2Nu$ new mutations should arise in any generation. The chance that any one of these mutations will stochastically rise in frequency to replace all other alleles in the population is equal to its initial frequency, $^1\!/_{2N}$. The rate of allelic fixation (K) is the product of these probabilities, $K = (2Nu)(^1\!/_{2N}) = u$, and is equal to the mutation rate. Thus, for functionally equivalent amino acid replacements, the rate of fixation is independent of population size (figure 1.3). If the rate of mutation remains constant, allelic fixations will occur at a constant rate, and amino acid replacements will provide the data to construct an evolutionary clock. Numerous studies have used molecular data to estimate the times of divergence, but many observations have suggested that the molecular clock is either not perfect or is simply inadequate (Avise and Aquadro 1982; Esteal 1985; Gillespie and Langley 1979; Langley and Fitch 1973, 1974; Scherer 1990; Thorpe, 1982; Wu and Li 1985).

Whether or not amino acid or DNA sequences provide an accurate molecular clock, they can be used to infer phylogenetic relationships. Sequences of cytochrome c were the first to be used to reconstruct the phylogenetic relationships among 20 species of mold, yeasts, plants, and animals (Fitch and Margoliash 1967), and this phylogeny was in remarkable agreement with traditional phylogenies based on cellular structure, development, and gross morphology. An extension of this study to include 29 species (Fitch and Markowitz 1970) emphasized the utility of amino acid sequences in the reconstruction of phylogenies. The phylogeny based on the amino acid sequence of this single protein is similar to the traditional phylogeny discovered in studies of development and comparative anatomy. Protein and DNA sequences contain immense amounts of information for cladistic analyses.

Loci evolve at different rates, thereby suiting them to different phylogenetic studies. Microevolutionary phenomena, such as differentiation associated with recent isolation or the influence of families on population structure, can be analyzed with the rapidly evolving mitochondrial DNA (Avise et al. 1987; Moritz et al. 1987). On a very different scale, sequences of ribosomal DNA, evolving at a much slower rate, provide little information about recent microevolutionary events, but they do record distinctions among such fundamental groups as coelenterates, arthropods, and chordates (Appels and Honeycutt 1986; Field et al. 1988; Halanych et al. 1995).

The increase of genetic variation with population size

Although population size is not a significant consideration regarding the rate of fixation of neutral mutations, population size is important to determining the level of heterozygosity for neutral alleles. At equilibrium, $H = 4N_e u / (4N_e u + 1)$, where H, N_e, and u are heterozygosity, effective population size, and the neutral mutation rate, respectively (Kimura and Ohta 1971). Thus, if populations are at or near equilibrium for neutral alleles, heterozygosity should increase with increasing population size.

Empirical data are consistent with some aspects of this prediction. Soulé (1976) examined estimates of protein heterozygosity and population size in lizards, fish, mammals, marine invertebrates, and *Drosophila* and found a general pattern of increasing heterozygosity with increasing population size (figure 1.4). Yet these data do depart from the expectations generated by the preceding equation in that the observed levels of heterozygosity fall well below the predicted levels. Soulé (1976) pointed out that this discrepancy may be explained either by the fact that electrophoresis does not detect all genetic variation or by the possibility that the populations have not had sufficient time to accumulate the equilibrium number of alleles.

An apparent agreement between data and predictions from neutral theory

Several major patterns in the empirical data are consistent with expectations based on the assumption of functional equivalency of allelic forms. The amino acid composition of proteins is predicted solely by the redundancy of the genetic code. Rates of allelic substitution may not be constant (Langley and Fitch 1974; Scherer 1990) over extended periods of evolutionary time, but amino acid or DNA sequences can be used to reconstruct phylogenetic trees. Among species, average heterozygosity increases with population size. These observations conform to the hypothesis that most of the proteins' molecular evolution is driven by neutral mutations, and they justify the null hypothesis that molecular variation is adaptively neutral.

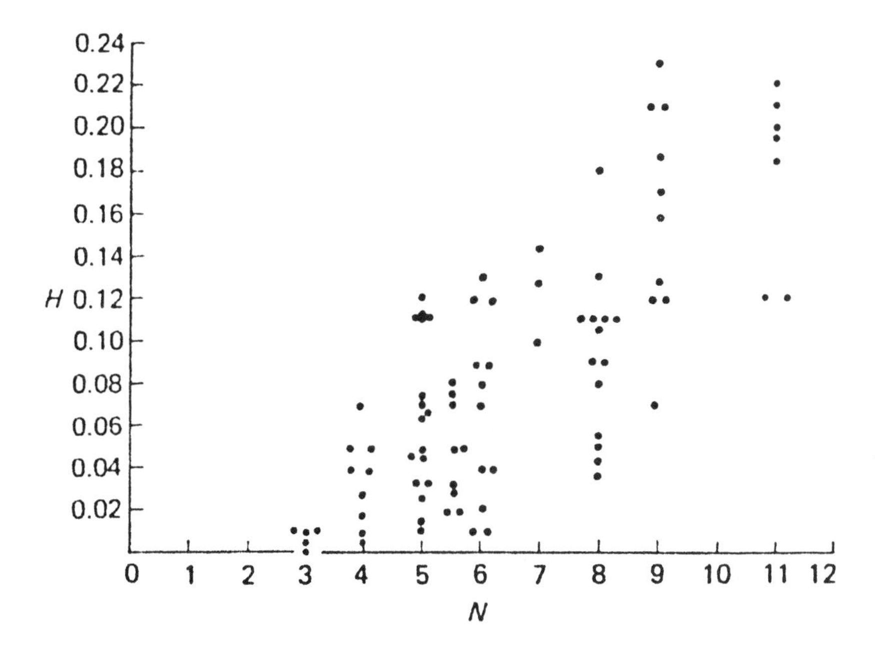

Figure 1.4 Heterozygosity of electrophoretically detectable genetic variation increases with population size in lizards, fish, mammals, marine invertebrates, and *Drosophila*. From Soulé 1976.

Determination of Fitness

Although evolutionary biologists attribute most adaptations to natural selection, they do not agree on the level (figure 1.5) at which selection differentiates among the alternatives to produce these adaptations (Brandon and Burian 1984; Futuyma 1986). Does selection act on either the alternative alleles or the various genotypes at a locus (Williams 1985)? To some biologists, this notion is inconceivable (Sober and Lewontin 1982). After all, genes come packaged in chromosomes, and selection favoring an allele at a locus would affect hundreds or thousands of linked polymorphic loci. If it is not genes or genotypes but some higher level of organization whose reproductive performance drives evolution, it must be phenotypes—morphological, physiological, or behavioral—that really matter. But even the notion of selection of chromosomes or phenotypes seems contrived, for it is individuals—not genes, chromosomes, or phenotypes—that reproduce. Because individuals can enhance their fitness by either producing their own offspring or helping related individuals bear and raise offspring (Hamilton 1964; Michod 1982), it may be more appropriate to measure the reproductive performance of families or demes or perhaps even populations (Wade 1978). What level or levels in this hierarchy are the targets of natural selection? This is not an exercise in semantics but a discussion of justified perspectives on natural selection and fitness determination (Brandon and Burian 1984; Gould 1982).

We have incontrovertible evidence indicating that epistatic (nonadditive on either an arithmetic or a log scale) interactions among loci contribute to the fitness of individuals.

Figure 1.5 Natural selection can act on many levels, from distinguishing genotypes at a locus to favoring one species over another.

These data come from many laboratories and have been observed in many different experimental designs employing various techniques. For example, the pioneering work of Dobzhansky (Dobzhansky and Spassky 1960; Dobzhansky et al. 1963; Spassky et al. 1965) revealed both abundant variation in natural populations and epistatic fitness interactions among loci. Similarly, very strong fitness interactions among chromosomes and segments of chromosome were detected in studies utilizing the parthenogenetic reproduction of *Drosophila mercatorum* (Templeton 1979; Templeton, Singh, and Brokaw 1976). *Drosophila melanogaster* heterozygous for either chromosome 2 or 3 have much higher fitnesses than do homozygous individuals. The genotypes of these two chromosomes are not sufficient, however, to predict fitness; rather, epistatic interactions for fitness are strong (Seager and Ayala 1982; Seager, Ayala, and Marks 1982).

Epistatic interactions among loci or chromosomes in the determination of fitness present a monumental challenge to population geneticists. A thorough contemplation of evolutionary forces must include not only the external environment's physical and biotic components but also the genetic environment. That is, the relative performances of alternative genotypes at a locus are expected to be dependent on both the external environment and the genetic environment. To incorporate this component of biological reality into population genetic theory, multilocus models were examined to determine the extent to which the genetic environment contributed to the evolutionary dynamics of a single locus (Allard 1975; Allard, Jain, and Workman 1968; Hedrick, Jain, and Holden 1978; Lewontin 1964). Multilocus studies suggested that a locus would reflect not just the selection impinging on it but also the forces acting on linked loci (Sober and Lewontin 1982). For example, a model of natural selection presented by Franklin and Lewontin (1970) described the accumulation of linkage disequilibrium along a chromosome containing many overdominant loci. More and more geneticists became convinced that single-locus theory was inadequate to describe evolutionary dynamics.

The theoretical expectation of pervasive linkage disequilibrium in natural populations and the fear of the inadequacy of single-locus models of selection led empirical population geneticists to measure linkage disequilibrium. But when they surveyed natural populations to determine the extent of genetic organization in outcrossing species, they found little or no linkage disequilibrium (Charlesworth and Charlesworth 1976; Langley 1977; Mukai, Mettler, and Chigusa 1971; Mukai and Yamaguchi 1974; but see Epperson and Allard 1987; Schaeffer and Miller 1993). A study of the HLA complex in humans indicated that linkage disequilibrium declined to zero when recombination between markers reached approximately 2% (Klitz et al. 1992).

In contrast to these studies, linkage disequilibrium is common throughout the genomes of selfing species (Allard 1975) and within and directly beside chromosomal inversions (Hedrick et al. 1978); both selfing and inversions effectively restrict the amount of recombination.

An analysis of the fine structure of the *Adh* locus in *Drosophila melanogaster* revealed linkage disequilibrium within the gene and also independent patterns of geographic distribution for the polymorphic sites (Berry and Kreitman 1993). A high-resolution restriction mapping technique was used to survey 113 haplotypes from 44 polymorphic DNA markers in the *Adh* region in 1,533 individuals from 25 populations along the east coast of North America. Linkage disequilibrium was estimated between the variable nucleotide sites in the approximately 3,000 bp coding for Adh. Linkage disequilibrium was strong, almost absolute, and did not decay within the distance measured in this study. However, the geographic patterns varied among the polymorphic sites. Two sites—one coding for the electrophoretic variation detected as the "slow" and "fast" alleles and the other coding for an insertion/deletion polymorphism in an intron—showed clear latitudinal clines, whereas the remainder of the sites did not vary with latitude. Thus despite extensive linkage disequilibrium, this study revealed little or no hitchhiking of closely linked polymorphic sites.

We are left with a fascinating puzzle: although epistatic interactions are commonly reported in studies of components of fitness, linkage disequilibrium is rare in obligately outcrossing species. Apparently we do not yet understand how many loci interact to determine fitness or how selection will influence the genome's architecture.

Studies of molecular variation are directly relevant to the controversy concerning the targets of selection and the determination of fitness. Selection can discriminate among the genotypes at a locus (chapter 4), and selection also appears to act on the levels of multilocus heterozygosity (chapters 6 and 7).

Physiological and Demographic Consequences of Abundant Genetic Variation

For approximately three decades, evolutionary geneticists have studied abundant genetic variation, principally protein variation, to understand the evolutionary processes that influence this variation. It is safe to say, however, that we do not yet know whether the abundant genetic variation is balanced by selection or is adaptively neutral (Lewontin 1974). Neutral theory makes specific predictions concerning evolutionary rates and levels of genetic variation. But these predictions often are not precise because we cannot accurately estimate effective population sizes, rates of migration, and neutral mutation (Lewontin 1974; Stebbins and Lewontin 1972).

Empirical results that agree with a theory may support it, but their agreement does not confirm that theory, and the predictions from competing theories may not be distinctly different. For example, the correspondence of the amino acid content of proteins to predictions arising from the redundancy of the genetic code does not—and cannot—reject the hypothesis that allelic variants of a protein significantly affect physiology and growth and play a major role in adaptation to environmental variation. Furthermore, although an excellent molecular clock would result from the fixation of neutral mutations, an approximate clock would result from a mixture of effectively neutral and adaptive mutations. And although an increase in genetic variation with population size is predicted by neutral theory, it is also predicted by models relying on natural selection to balance genetic variation.

We could reasonably begin a study by assuming that genetic variation is neutral, as this assumption helps us build null hypotheses and formulate predictions. However, if evolutionary biologists are curious about the fitness consequences of abundant genetic variation and want to test the assumption of neutrality, they must address this question directly, by examining the physiological and demographic consequences of the genetic variation (Bennett 1987; Clarke 1975; Clark and Koehn 1992; Koehn, 1978, 1987; Powers 1987, Powers et al. 1994; Watt 1985, 1991). Accordingly, chapter 2 discusses enzyme kinetic studies, and chapter 4 summarizes studies of the biochemical and physiological consequences of molecular variation.

Although the substitution of one allele for another is an important evolutionary event (Haldane 1957) and an event that is especially interesting to molecular systematists, substitutions are of little or no importance to population biologists, for they occur on a time scale much greater than that used by population biologists studying microevolutionary events. New mutations (Kahler, Allard, and Miller 1984) and allelic substitutions occur so infrequently that they cannot play an important role in the daily, seasonal, and yearly dynamics that produce fitness differentials among individuals in populations. For these reasons, in this book I do not treat mutation or gene substitution but instead concentrate on the assortment and rearrangement of genes extant in the gene pool (chapters 7, 9, 10).

I begin with empirical data, as my primary motive is to consider the generality of several contrasting population genetic models of the determination of fitness. In particular, patterns in the empirical data have led me to examine the immediate consequences and the evolutionary implications of truncation or threshold selection acting on levels of heterozygosity in natural populations (chapters 6, 7, 8, 10).

Summary

Decades of studies of natural populations have firmly established that genetic variation is abundant and that selection is common. Oddly enough, we know little about the degree to which selection affects this abundant genetic variation.

New mutations and allelic substitutions occur so infrequently that their significance in the adaptation of populations to the typical range of heterogeneity in their environments must be slight.

Questions concerning the dynamics of abundant genetic variation constitute major intellectual challenges for evolutionary biologists and have profound implications for plant and animal husbandry. Although genetic variation in natural populations of plants and animals is abundant, is fitness variation abundant in those populations? Can we realistically consider and reliably measure selection at a single locus? Answers to these questions evade us because we still do not know the number of loci interacting to determine fitness, the nature of the interactions among loci, the level at which selection acts, and the consequences of natural selection for the architecture of the genome.

2

Classes of Abundant Genetic Variation

> On an absolute temperature scale the earth is extremely cold, and the chemical transformations which comprise metabolism simply cannot occur at life-supporting rates in the absence of biological catalysts.
>
> P. W. Hochachka and G. N. Somero,
> *Strategies of Biochemical Adaptation* (1973)

For more than three decades, population geneticists have been using molecular techniques to estimate the levels of genetic variation in populations. Fisheries biologists (Frydenberg et al. 1965; Sick 1961, 1965a, b) and human geneticists (Allison 1955, 1964; Boyer 1961; Fildes and Parr 1963; Hopkinson, Spencer, and Harris 1964; Pauling et al. 1949; Robson and Harris 1965; Spencer, Hopkinson and Harris 1964) used electrophoresis to describe population structure before Johnson et al. (1966a, b), Harris (1966), and Lewontin and Hubby (1966) first used electrophoretic surveys of proteins to estimate levels of genetic variation. Electrophoretic analyses of more than 3,000 species showed the percentage range of polymorphic loci to be from zero to 100%, with a mean between 33 and 50% (Brown 1979; Hamrick and Godt 1990; Hamrick, Mitton, and Linhart 1979; Johnson 1976; Nevo 1978; Nevo, Beiles, and Ben-Shlomo 1984; Powell 1975; Selander 1976; see chapter 9). Two concerns, however, have haunted the use of electrophoresis to estimate genetic variability—accuracy and generality.

Accuracy

Because electrophoresis in starch and acrylamide separates proteins on the basis of size and charge (Poulik 1957) and, to a smaller degree, conformation (Johnson 1977, 1979), amino acid replacements that do not alter the molecule's size, charge, and conformation will not be detected by electrophoresis (Coyne 1982; Johnson 1977). The assumption that amino acid replacements would be random with respect to the charge of amino acids led to the conclusion that electrophoretic surveys detect only 30% of the actual number of alleles (Nei 1975).

More genetic variation can be detected with electrophoresis if proteins are examined in a sequence of different buffer systems. For example, sequential electrophoresis applied to the xanthine dehydrogenase locus in *Drosophila* revealed much more variation, increasing the number of alleles from 5 to 23 in *D. persimilis* (Coyne 1976) and from 6 to 37 in *D. pseudoobscura* (Singh, Lewontin, and Felton 1976). Similarly, with sequential electrophoresis, the number of alleles at esterase-6 in *D. pseudoobscura* increased from 6 to 23 (Coyne, Felton, and Lewontin 1978). Sequential electrophoresis seems to detect much more variation at loci already known to segregate many alleles, but it detects little or no additional variation at monomorphic loci and at loci with little variation (Coyne 1982).

A novel study of electrophoretic variation used variant human hemoglobins to assess the sensitivity of sequential gel electrophoresis (Ramshaw, Coyne, and Lewontin 1979). Not only did sequential electrophoresis detect all the variants that differed by charge, but it also detected some variants with amino acid substitutions that did not alter charge but produced some conformational change. Sequential electrophoresis detected about 90% of the 20 variant hemoglobins.

Heat denaturation is another method used to detect genetically based biochemical variation at loci coding for enzymes (Bernstein, Throckmorton, and Hubby 1973). A gel that contains samples of individuals can be exposed to high heat before applying a histochemical stain. If enzymes have different rates of heat denaturation, genetic differences will appear as differences in activity. Heat denaturation surveys were applied to laboratory strains of the house mouse, uncovering four new alleles and leading to the conclusion that standard electrophoretic surveys detected only 50% of genetic variation (Bonhomme and Selander 1978). When octanol dehydrogenase from 10 species in the *Drosophila virilis* group was examined with heat denaturation, the number of alleles increased from 3 to 18 (Singh, Hubby, and Throckmorton 1975).

Although electrophoretic surveys typically reveal 33 to 50% of the loci to be polymorphic, and 5 to 15% of the loci of an average individual to be heterozygous, these estimates clearly are low. More variation is detected by both heat denaturation and sequential electrophoresis, and still more would be revealed by sequencing the DNA coding for the protein.

Generality

Lewontin and Hubby (1966) first used the assumption that the genetic variability of soluble proteins is a random sample of the variability of the genome. Because only 1 to 5% of the nuclear genome codes for functional proteins, the assumption that protein variability is representative of the genome deserves scrutiny.

The arsenal of molecular techniques available to population geneticists and systematists is quite impressive. To examine genetic variation at portions of the genome other than that coding for enzymes and soluble proteins, population geneticists use two-dimensional electrophoresis of abundant proteins, restriction analyses of mitochondrial DNA, DNA fingerprinting, and DNA sequencing. We now have enough empirical data to test the generality of estimates of genetic variation obtained with electrophoretic surveys of proteins.

Nonenzymatic proteins exhibit substantially lower levels of genetic variation than do enzymes in humans (McConkey, Taylor, and Phan 1979), *Drosophila* (Brown and Langley 1979), and mice (Racine and Langley 1980). Enzymes are typically detected with stains that rely on enzymatic activity, but the abundant, nonenzymatic proteins, such as tubulin and actin, are visualized with either general protein stains or autoradiography. Several hundred proteins can be examined from a single individual by means of two-dimensional electrophoresis, which first separates proteins with isoelectric focusing, then denatures the proteins in urea, and finally separates the proteins by size in acrylamide (O'Farrell 1975). The average individual heterozygosity of humans, based on 87 enzymes, is approximately 6% (Harris, Hopkinson, and Edwards 1977), but it is less than 1% when the estimate is based on 400 abundant peptides (McConkey et al. 1979). Nonenzymatic proteins may have more stringent steric constraints than do soluble and globular enzymes (McConkey 1982; McConkey et al. 1979), and these constraints may limit their genetic variability.

In contrast to the size of the eukaryotes' nuclear genome, which varies over three orders of magnitude and is packaged into from 1 to 250 chromosomes, the mitochondrial DNA (mtDNA) of higher animals is conservative in size and content. Animal mtDNA is a small and simple loop whose contents and sequence are highly conserved yet whose rate of nucleotide substitution is usually high (Brown 1983; but see Brown 1986; Hoffmann, Boore, and Brown 1992). The mtDNA of most species of animals contains 15,700 to 19,500 base pairs (but see Boyce, Zwick, and Aquadro 1989) and codes for small and large rRNAs, 22 tRNAs, and 13 mRNAs (Brown 1983). Animal mtDNA contains an absolute minimum of intergenic spacers and intervening sequences; more than 90% of the genome is transcribed. Among vertebrates, mtDNA evolves at rates five to ten times as fast as nuclear DNA does (Brown, George, and Wilson 1979), but in *Drosophila*, the evolution rates of nuclear DNA and mtDNA appear to be similar (Powell et al. 1986). MtDNA is very common in most cells and is truly abundant in some cells, such as unfertilized amphibian eggs, which may contain 10^8 mtDNA molecules (Birky, 1978; Dawid and Blackler, 1972). In most species, mtDNA is inherited as a maternal cytoplasmic factor and has little intraindividual heterogeneity (Avise and Lansman, 1983; Birky, Maruyama, and Fuerst 1983). Interesting exceptions to these generalizations have been found. For example, three species of bark weevils have mtDNA with 30,000 to 36,000 base pairs, and both these weevils and two species of crickets are commonly heteroplasmic (Boyce et. al. 1989; Rand and Harrison 1989; Solignac et al. 1984).

Genetic variation in mtDNA is detected most easily by examining restriction fragment length polymorphisms. Restriction endonucleases recognize specific nucleotide sequences 4, 5, 6, or 8 bases long, and they cleave the strands of DNA at specific points either within or beside the recognition sequence. Because mtDNA is a loop, the number of fragments produced is equal to the number of recognition sites. Purified mtDNA is cut with a restriction enzyme and run on a gel that sorts out the segments by their size. The restriction fragments can be detected in several ways. The gel can be stained with ethidium bromide and then viewed under ultraviolet light; the ethidium bromide that intercalates with DNA fluoresces. Alternatively, the DNA can be transferred by a Southern blot to a nylon membrane, which permanently binds the DNA. The DNA can then be hybridized with probes marked with either radioactivity or enzymes and can be detected with autoradiography or an enzymatic reaction.

Studies of mtDNA in deer mice, *Peromyscus*, illustrate the utility of restriction fragment length polymorphisms (RFLP's) for the study of natural populations (Avise, Lansman, and Shade 1979). The phenotypes, or nucleomorphs (Nei and Tajima 1981)—revealed with six restriction enzymes—demonstrated that polymorphism within populations is common, and comprehensive analyses revealed 61 nucleomorphs within *P. maniculatus* (Lansmann et al. 1981). A summary of 13 studies of mtDNA (Avise and Lansman 1983) suggested that sequence divergence of 1 to 4% within species is common. The mtDNA sequences in *P. leucopus, P. maniculatus,* and *P. polionotus* are differentiated from 13 to 20% (Avise et al. 1979). Thus the number of nucleomorphs within species and within populations is quite high, and the differentiation of congeneric species is extensive, offering many opportunities for studying population biology and phylogeny (Avise et al. 1987; Moritz, Dowling, and Brown 1987).

The mtDNA of humans is quite variable and provides new insights into human evolution. A survey of 147 individuals revealed 134 nucleomorphs, and these data were summarized with phylogenetic trees (Cann, Stoneking, and Wilson 1987). This branching diagram led Wilson and his colleagues to propose that all the extant nucleomorphs evolved from the DNA of a woman who lived in the southern portion of Africa approximately 200,000 years ago. This pattern of relationship also led Wilson and his collaborators to propose that modern humans replaced Neanderthals and ancients with virtually no hybridization. This field of study is currently very active (Goldstein et al. 1995; Nei 1995), so it is not possible to provide a summary that will not immediately be out-of-date. For example, the extinction of mtDNA lineages may force a more conservative interpretation of the data, and of Eve's birthday, and nuclear genes might paint a rather different view of human evolution. Finally, estimates of the times of divergence with genetic data are only educated guesses (Avise and Aquadro 1982), for they depend on rates of evolution, which vary among taxonomic groups and through time.

Extreme levels of mtDNA variation are found in menhaden (*Brevoortia tyrannus/patronus* complex) and the chuckwallah (*Sauromalus obesus*) (Avise et al. 1989). These species have sufficient restriction site polymorphism and size polymorphism that mitochondrial haplotypes are almost as distinctive as the phenotypes revealed with DNA fingerprinting (discussed later). In each of these species, more than 98% of the individuals carried unique mitochondrial phenotypes.

DNA fingerprinting (Jeffreys, Wilson, and Thein 1985a,b) reveals the heterogeneity of intermediately repetitive DNA in the nuclear genome and provides a large number of highly variable loci. Some of the moderately repetitive DNA is arranged in tandem repeats scattered throughout the genome. At any site, the number of repeats can vary from individual to individual. Jeffreys and his colleagues found that although the tandem repeats were not identical, they often contained a core, called a *minisatellite*, of about 15 nucleotides whose sequence was highly conserved. Although the repeated segments varied from 16 to 64 base pairs in length, they shared the core sequence GGGCAGGAXG (Jeffreys et al. 1985a). A probe for this sequence reveals the repetitive DNA or minisatellites.

Genetic variation for minisatellites is detected by cutting the nuclear DNA with a restriction enzyme, producing restriction fragment length polymorphisms. These are separated on a gel, and the probe for the core sequence is applied. In humans, the number of loci detected with this technique is approximately 20, and some loci have as many as 20 variants. The probability that two unrelated people would share the same pattern for a single probe is approximately 3×10^{-11}, and the probability that they would have identical patterns for two probes is approximately 5×10^{-19} (Hill 1985; Jeffreys et al. 1985b). This striking diversity is produced by three factors. First, as with other restriction fragment length polymorphisms, restriction sites appear and disappear with mutations. In addition, the core sequences are buried in repeats of different length. But most important, the number of tandem repeats at a locus can vary from 3 to 40 among individuals (Jeffreys et al. 1985a). All these factors alter the lengths of restriction fragments, causing them to migrate at different rates.

Although some probes detect repeated elements scattered throughout the genome, other probes detect single loci (Royle et al. 1987). Alleles at a single locus are inherited in a Mendelian fashion, and some loci segregate more than 50 alleles. New alleles can be produced by unequal crossing over, producing products with novel numbers of repeats. Mutation rates are high at these loci, with some empirical estimates of mutation as high as 0.01 and 0.05 (Jeffreys et al. 1988).

DNA fingerprinting has detected abundant levels of genetic variation in house sparrows (Burke and Bruford 1987; Burke et al. 1989; Wetton et al. 1987), cats and dogs (Jeffreys and Morton 1987), mice (Jeffreys et al. 1987), and trout (Fields, Johnson, and Thorgaard 1989).

Genetic variation is high at loci coding for antigen–antibody systems, but comparisons with other sets of loci are difficult because only polymorphic systems are surveyed and studied. Extremely high levels of variation, particularly at the *HLA-A* and *HLA-B* loci, are associated with resistance to disease (Black and Salzano 1981; Hedrick, Whittam, and Parham 1991; Hill et al. 1991; Hughes and Nei 1988, 1989; Markow et al. 1993).

Ecological geneticists have used protein polymorphisms to estimate the levels of genetic variation in more than 3,000 species (Hamrick and Godt 1990; Nevo et al. 1984), but it is now apparent that those estimates must be interpreted cautiously. Although the number of enzymes and soluble proteins (hemoglobin, myoglobin, haptoglobin, etc.) is large (between 1,000 and 2,000), the portion of the genome coding for soluble proteins has been estimated to be 1 to 5% at most. Other portions of the genome appear to have different levels of variability—abundant proteins are much less variable, whereas minisatellites are much more variable. Furthermore, the patterns of genetic variation may differ between sets of genetic markers. For example, the patterns of geographic variation differ between protein polymorphisms and the mtDNA of both the horseshoe crab, *Limulus polyphemus* (Saunders, Kessler, and Avise 1986; Selander et al. 1970) and the American oyster, *Crassostrea virginica* (Buroker 1983; Karl and Avise 1992; Reeb and Avise 1990). In both species, allelic frequencies of protein polymorphisms are relatively homogeneous, an observation usually interpreted as evidence of high levels of gene flow. The mtDNA in both species exhibits an abrupt change in northeastern Florida. These patterns of geographic variation—homogeneity of allelic frequencies versus a step cline in mitochondrial haplotypes—are so different that they must be reflecting different evolutionary forces. It is not clear which forces are acting on which portions of the genome (see chapter 7 for further discussion).

Because estimates of levels of genetic variability obtained with electrophoresis cannot be generalized to other compartments of the genome, we cannot be sure that this set of loci is representative of the entire genome. For example, in the single study with both protein and DNA markers, the growth rate of the sea scallop, *Placopecten magellanicus*, increased with allozyme heterozygosity but not with the heterozygosity of DNA markers (Pogson and Zouros 1994; Zouros and Pogson 1993; see chapter 7). But on the coarsest level, the impression of the genome glimpsed with protein polymorphisms appears to be general. Whether we look at nuclear DNA sequences, mtDNA restriction fragments, intron number and length, antigen–antibody systems, minisatellites, or protein polymorphisms, we find a diversity of genotypes.

The Assumption of Neutrality: mtDNA

Most ecologists and systematists currently assume that the mtDNA variation used to measure geographic variation, to estimate gene flow, or to infer phylogenetic relationships is neutral. This is the most reasonable assumption, and many plausible arguments suggest that the majority of molecular variation is neutral. Nevertheless, this is just an assumption, so it may not be correct (Ballard and Kreitman 1995). For example, long-term studies of ribosomal DNA spacer length variants in barley revealed that frequencies changed more than expected by neutral theory. Furthermore, the frequency of variants changes regularly among environments. These observations revealed that selection acts on spacer length variants or associated sequences (Allard et al. 1990; Saghai-Maroof et al. 1984, 1990; Zhang et al. 1990). Several observations indicate that selection acts on some aspects of variation of mtDNA.

Drosophila *cage experiments*

Cage experiments have given us valuable insights into the selective forces acting on the inversions of *Drosophila pseudoobscura*, and now they have been used to test the assumption of neutrality of mtDNA variation in *D. pseudoobscura* (MacRae and Anderson 1988). A laboratory population of flies was made polymorphic for mtDNA haplotypes by crossing isofemale stocks from Bogota, Colombia, and Apple Hill, California. A cage initiated with a frequency of 30% Bogota haplotypes increased to 82% Bogota in 32 generations. The pattern of this change suggested that it was driven by selection and had arrived at equilibrium. After this population was perturbed, it rapidly returned to equilibrium. Although some subsequent replicate cages exhibited similar patterns, others did not, making the results difficult to interpret (MacRae and Anderson 1990; Singh and Hale 1990).

Injection experiments

An interaction between mitochondria and the cytoplasm influences the respiration rate and cultures of human cells (King and Attardi 1989). Mitochondria were removed from cultures of HeLa cells by chronic applications of ethidium bromide. These cell cultures could be maintained in a supplemental medium, and the cells' metabolic independence could be restored by injecting purified mitochondria into the cell cultures. Several lines were started by injecting cells devoid of mitochondria with human mitochondria from different genotypes. The performance of the cell lines was measured by means of oxygen consumption and the time required for the number of cells to double. Both measures of performance were heterogeneous among lines; oxygen consumption differed by more than 50% among lines differentiated solely by mtDNA haplotype.

A *Drosophila* cage experiment demonstrated that alien mitochondria can displace resident mitochondria (Miki, Chigusa, and Matsuura 1989). The mitochondria of *Drosophila mauritiana* were injected into the eggs of *D. melanogaster*, and four population cages were started with cytoplasmic hybrids. Within 30 generations, two of the cages were fixed, and a third cage was approaching fixation for the *mauritiana* haplotype. The fourth cage appeared to have arrived at an equilibrium at 70% *mauritiana*.

Selection on mitochondrial genome size

Comparative studies have revealed that the evolution of mtDNA has been conservative with respect to genome size and content (Brown 1983), but there is certainly variation within and among species. For example, although the mtDNA of most animal species analyzed to date lies in the size range of 15,700 to 19,500 base pairs, the mitochondrial genome of three species of weevils of the genus *Pissodes* contains 30,000 to 36,000 base pairs (Boyce et al. 1989). Heteroplasmy, or multiple mitochondrial forms within an individual, reaches an extreme level in the bark weevils *Pissodes nemoresnis*, *P. strobi*, and *P. terminalis*; all 219 individuals examined were heteroplasmic for two to five forms of mtDNA (Boyce et al. 1989).

Heteroplasmy has also been reported in *Drosophila mauritiana* (Solignac et al. 1984). The populations of the crickets *Gryllus firmus* and *G. pennsylvanicus* are, respectively, 60% and 45% heteroplasmic (Rand and Harrison 1989). Comparisons of the heteroplasmy in female crickets and their offspring revealed that the smaller haplotypes are significantly more common in offspring than in their mothers. Mitochondria with smaller genomes are probably able to replicate faster, giving them an advantage as the cells proliferate (Rand and Harrison 1989; Solignac et al. 1984).

Physiological Consequences of Abundant Genetic Variation

Of all the classes of genetic variation, protein polymorphisms currently provide the best resource for examining the physiological, behavioral, and demographic consequences of abundant genetic variation. Whereas we can only speculate about the potential significance of another tandem duplication, or restriction enzyme recognition site, we do know what enzymes do. For hundreds of enzymes, we know the substrates that they use, the products that they produce, the metabolic pathways in which they function, and the tissues that express them. Furthermore, analyses of enzyme kinetics permit the quantitative comparison of the in vitro performance of the products of alternative genotypes. We also have enough data to permit some tentative statements about the consequences of this class of abundant genetic variability (see chapter 4).

Metabolism consists of thousands of biochemical reactions that control digestion, photosynthesis, growth, development, respiration, and excretion. Within the range of temperatures experienced by natural populations, the uncatalyzed rates of most of these reactions are so slow that growth and locomotion would be imperceptible. Enzymatic proteins are the catalysts that allow metabolism to proceed at the temperatures experienced by plants and animals. Enzymes also control the rate of energy flow through metabolic pathways, thereby controlling the rates of growth and development and influencing the proportions of energy allocated to various functions.

The presence of life everywhere on Earth places extant species in a remarkable range of physical environments. Life is found from the top of Mount Everest to the extreme depth of the Marianas trench, a range in pressure of approximately 1,000 atmospheres. Alpine forest trees survive winter temperatures of –70°C, and some algae are restricted to hot springs with average temperatures above 40°C. Life abounds in clear, oligotrophic mountain lakes and in tidal pools with salinities exceeding 40 parts per thousand.

Growth, development, and reproduction rely on metabolic processes that are influenced by many variables, including temperature, pressure, and pH. It is therefore not surprising that when physiologists and biochemists examine the metabolic proteins of species living in very different physical environments, they find differences that allow metabolic processes to occur in the appropriate environmental conditions (Somero 1978). For example, parrot fish have enzymes that regulate metabolism in tropical environments, and ice fish have enzymes that regulate metabolism in Antarctic environments. The differences are most striking when species occupying very different environments are contrasted, but essentially the same type of differentiation of function may be seen among the genotypes segregating at a single locus.

Enzyme Kinetics: Quantitative Analyses of Enzyme Performance

The increasing interest in molecular mechanisms of protein function is focusing on the kinetics of the enzymes produced by alternative genotypes segregating at one locus. Several motives are driving this interest. Although population geneticists now recognize that the percentage of polymorphic protein loci varies from zero to nearly 100% among species, they still know little about both the immediate and long-term consequences of that variation. Dissatisfied with imprecise or nondiscriminating predictions about levels of heterozygosity or distributions of allelic frequencies, geneticists are looking directly at the gene products that interest them. Physiological ecologists, eager to explore the molecular mechanisms of adaptation, examine the biochemical performance of enzymes that are associated with contrasting environments. A third group, perhaps best described as population biologists, seek to better understand the processes of natural selection occurring within a population and within a single generation. At first, these biologists used electrophoretic techniques to identify genetic variation, and now they are studying enzyme kinetics to better understand unexpected dynamics of demography or mating success.

Enzymes catalyze metabolic reactions. If alternative genotypes segregating at a locus are to have differential impacts on fitness, then those genotypes must have different biochemical properties or differential performance (Clarke 1975; Koehn 1978). Thus, studies of the physiological or demographic consequences of enzyme variation may begin with the quantification of biochemical performance, that is, the study of enzyme kinetics.

The products of enzyme genotypes, the enzymes themselves, can differ in various aspects of performance that might have physiological consequences. For instance, enzymes may differ in their interactions with membranes or with other proteins. Some enzymes, such as the esterases and peroxidases, work on not one but a variety of substrates, and the enzymes produced by various genotypes at a locus may differ in their relative specificities for substrates. The half-lives of proteins differ dramatically; depending on the amino-terminal amino acid, the half-lives of proteins vary from a few minutes to more than 20 hours (Bachmair, Finley, and Varshavsky 1986; Rogers, Wells, and Rechsteiner 1986; Tobias et al. 1991). Consequently, genotypes can vary in their enzyme turnover rates (Hawkins, Bayne, and Day 1986; Hawkins et al. 1989). Finally, and most germane to this discussion, the catalytic efficiencies may vary among the genotypes at a locus. Studies of enzyme kinetics typically focus on the measurement of a series of catalytic variables, such as V_{max}, k_{cat}, and K_m, that relate to catalytic efficiency. These variables are introduced with a description of the kinetics of a simple reaction.

Consider the following metabolic reaction, in which E represents the enzyme, A is the substrate, and P stands for product: $E + A \leftrightarrow EA \leftrightarrow E + P$. Imagine mixing a solution of enzyme with a solution of substrate and, with the aid of a spectrophotometer, watching the rate of the product's accumulation. The velocity of this reaction varies with the substrate concentration (figure 2.1). As the solutions meet, the enzyme binds with the substrate to form the enzyme-substrate complex EA. This reaction is reversible, so the reaction may proceed to form a product, or the enzyme may simply release the substrate. When the solutions are initially mixed, the rate of formation of the enzyme-substrate complex is highly dependent on the concentration of the substrate, and therefore the reaction's initial rate is nearly a linear function of substrate concentration. When the solution comes to equilibrium—particularly at higher substrate concentrations—the rate of the reaction is limited more by the formation of product than by the formation of the enzyme-substrate complex. At higher substrate concentrations, then, the reaction rate is less responsive to increases in substrate concentration, producing a hyperbolic-shaped curve. The curve reaches its maximum velocity and a plateau when the substrate concentration exceeds the enzyme concentration and virtually all the enzyme is bound to the substrate. Several enzyme kinetic variables can be estimated from this characteristic hyperbolic curve.

The shape of the curve (figure 2.1) describing reaction velocity as a function of substrate concentration ($[A]$) is described by the following equation: $v = V_{max[A]} / K_m + [A]$. The rate of the reaction represented by v_{max} is the maximum velocity of the reaction, which can also be defined as $v_{max} = k_{cat}[E_0]$, where E_0 is the initial enzyme concentration and k_{cat} is the catalytic efficiency. The Michaelis constant, K_m, is the substrate concentration that brings the reaction velocity to half the maximum value.

At low substrate concentrations, when the rate of the reaction increases linearly with substrate concentration, the reaction velocity is described by the equation $v = [k_{cat} / K_m][E_0][A]$. But at high substrate concentrations, when virtually all the enzyme is complexed with substrate, the reaction velocity is described by $v = V_{max} = k_{cat}[E_0]$. At low substrate concentrations, the reaction velocity is proportional to k_{cat}/K_m, but at high concentrations it is proportional to k_{cat}. Thus there may be no single kinetic variable to be maximized by natural selection. (Further discussions of the interpretation of kinetic variables can be found in Hall and Koehn 1983 and Watt 1985, 1986.)

Evolutionary biologists have analyzed kinetic variation in at least a dozen enzymes, hemoglobin, haptoglobin, and transferrin polymorphisms (Frelinger 1972; Koehn, Zera, and Hall 1983; Templeton 1982; see chapter 4). In most of the studies reported in the literature, kinetic differences were found among the segregating genotypes. Because there probably were biases in the selection of polymorphisms for kinetic analyses and in the publication of results, it would be difficult to determine what proportion of extant polymorphisms have significant kinetic variation.

Just as there may be no single kinetic variable maximized by natural selection, there may be no single protein genotype favored at all times and at all places in fluctuating and heterogeneous environments. For example, consider the three curves of reaction

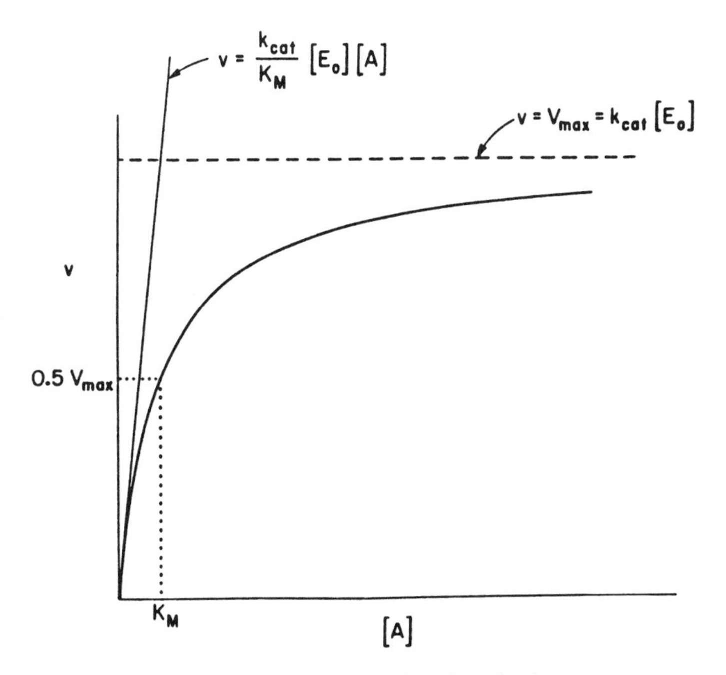

Figure 2.1 Reaction velocity (v) as a function of substrate concentration [A] for a simple metabolic reaction. K_m is the Michaelis constant, k_{cat} is the catalytic efficiency, V_{max} is the maximum velocity, and E_0 is the initial enzyme concentration. From Hall and Koehn 1983.

velocity as a function of the substrate concentration presented in figure 2.2. The three curves result from variation in k_{cat} and K_m among the enzymes produced by three genotypes at a locus. They are highest in the *AA* homozygote, lowest in the *aa* homozygote, and intermediate in heterozygotes. Let us assume that some component of fitness increases with the velocity of this reaction. The species bearing this polymorphism lives in an environment that is heterogeneous in space and fluctuating in time, and as a function of the environment, the substrate levels can be 1, 2, or 3 (figure 2.2).

When the substrate level is 2, the three genotypes perform identically, and so there is no variation among genotypes in the component of fitness associated with reaction rate. When the substrate concentration is 1, the *aa* homozygotes have the highest fitness, followed by the heterozygotes, and then the *AA* homozygotes. At substrate concentration 3, this ranking is reversed, and so the *AA* homozygotes are favored. If this species is locked in a constant environment, three scenarios are possible, depending on the level of substrate concentration. At substrate concentration 1, there would be directional selection for *aa* homozygotes, and if selection prevailed, the population would become monomorphic for the *a* allele. At substrate concentration 3, the other homozygote would be favored, and selection would fix the *A* allele. At substrate concentration 2, the genotypes would have the same kinetics and fitnesses—the genetic variation, stripped of physiological consequences, would become neutral. In this case, allelic frequencies would drift until one of the alleles became fixed in the population.

If the environment occupied by our hypothetical species fluctuates in time, these

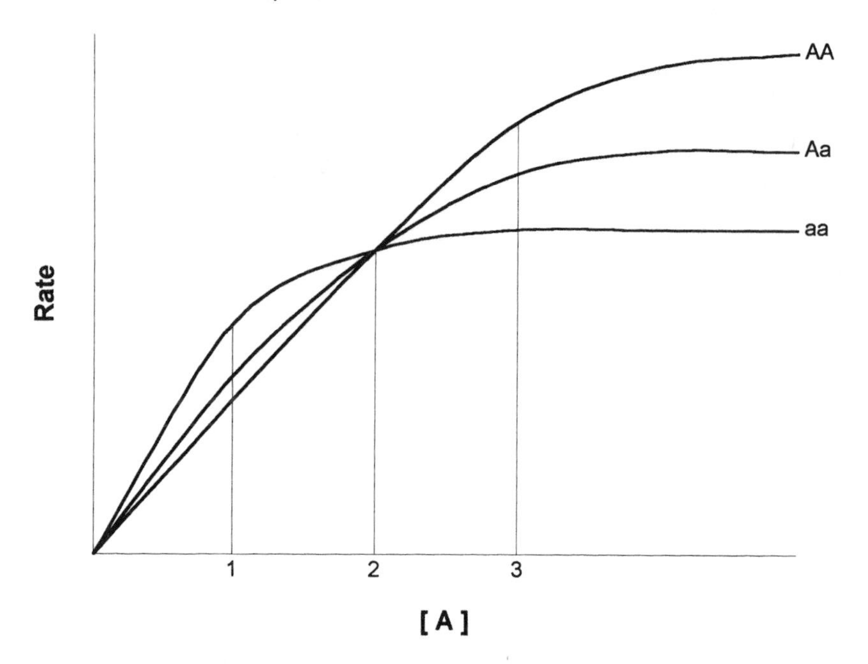

Figure 2.2 Reaction velocities (v) of three enzyme genotypes as a function of substrate concentration [A]. Genotype *aa* has the highest velocity at low substrate concentration (1) but the lowest velocity at high substrate concentration (3), and the genotypes are functionally equivalent at an intermediate concentration (2). This kinetic variation among genotypes might also arise as a function of temperature, pH, and the concentrations of minerals, substrates, and other variables.

three outcomes still are possible, and a fourth is added. A heterogeneous environment, imposing selection that favors first one genotype and then the other, may produce a selection regime that favors heterozygotes over homozygotes, balancing and maintaining the genetic variation (reviewed in Hedrick 1986; Hedrick, Ginevan, and Ewing 1976). This possibility is elaborated for kinetic variation in chapter 6, and empirical examples are presented in chapter 4.

The Adaptive Evolution of Enzyme Kinetic Variation

A fine example of how evolution of enzymes can adapt metabolism to thermal environments is found in a study of skeletal muscle lactate dehydrogenases in barracudas of the eastern Pacific (Graves and Somero 1982). *Sphyraena argentea*, *S. lucasana*, *S. ensis*, and *S. idiastes* occupy north temperate, subtropical, tropical, and south temperate environments, respectively. Genetic studies based on 33 presumptive loci estimated that the times of divergence among these species occurred between 3 million and 15 million years ago. *S. argentea* and *S. idiastes* both currently occupy temperate environments, but those environments are separated by 7000 km, and the species are thought to have diverged 5 million years ago. Nonetheless, their skeletal muscle lactate dehydrogenases are electrophoretically and kinetically indistinguishable. In contrast to this stability, species living in different environments have evolved enzymes that are kinetically dif-

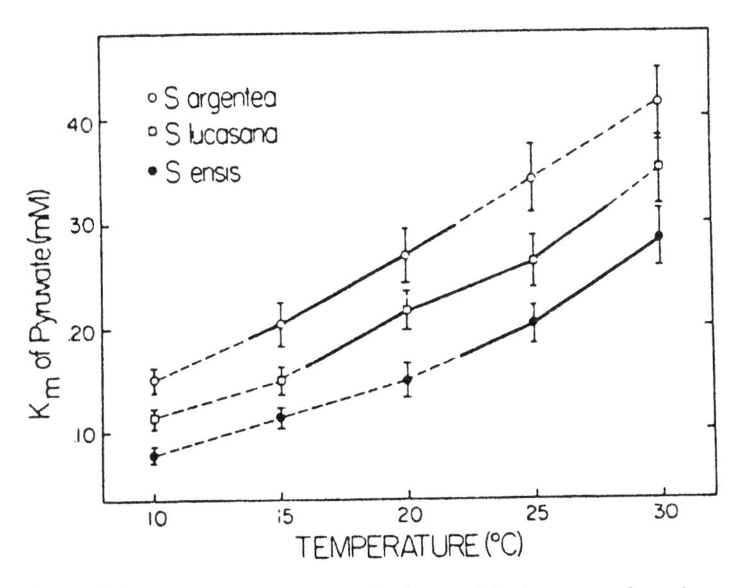

Figure 2.3 The Michaelis constants for lactate dehydrogenases from three species of barracudas as a function of temperature. The K_m for the southern temperate barracuda, *S. idiastes*, not shown here, is indistinguishable from that of the northern temperate species, *S. argentea*. Error bars represent standard deviations. Solid portions of the lines indicate habitat temperature ranges for each species. The lactate dehydrogenases have evolved similar K_m values for their environmental temperatures. From Graves, Rosenblatt, and Somero 1983.

ferentiated and suited to those environments. That is, the lactate dehydrogenases of fishes living in subtropical and tropical environments have Michaelis constants and substrate turnover numbers (k_{cat}s) that differ in interpretable ways from those of fishes living in temperate environments. Within this group of barracudas, Michaelis constants differ among environments and are correlated with temperature, whereas k_{cat}s differ among environments and are inversely correlated with temperature. Although K_m and k_{cat} differ significantly among the three thermal environments, when these parameters are compared at the temperature midranges for their respective environments, they are very similar (figure 2.3), suggesting the accommodation of metabolic rates to thermal environments (Graves and Somero 1982).

When the Isthmus of Panama rose out of the sea 3.1 million years ago, it divided the geographic ranges of many species, isolating populations in the Caribbean Sea from those in the eastern Pacific Ocean. Across this narrow isthmus, the thermal environments differ dramatically. While the eastern Pacific experiences seasonal upwelling—the wind-driven rise of frigid abyssal water to the surface—the Caribbean is tropical. This genetic isolation of populations into contrasting environments produced a unique natural experiment. Since this recent geological upheaval, many of the populations isolated by the land bridge have evolved independently to the degree that they are recognized as cognate species-pairs. The enzymes in these species-pairs have been examined to determine whether separation for this amount of time in contrasting environments has been accompanied by an evolutionary adjustment of enzymes to their respective environments (Graves, Rosenblatt, and Somero 1983). The K_ms and k_{cat}s of skeletal muscle lactate dehydrogenases were compared in pairs of damselfishes, groupers, blennies, and wrasses. In two of the four comparisons (the blennies and the wrasses), adaptive temperature changes in these kinetic variables have evolved. Michaelis constants were lower and substrate turnover numbers were higher in species living in the cooler Pacific waters, but no discernible differences were found in the other pairs of fishes.

In the two studies just reviewed, evolutionary changes in enzymatic proteins occurring in the last few million years can be interpreted as temperature-compensatory modifications in enzyme kinetic parameters (Somero 1978). That is, when examined at the midrange temperature of their respective environments, the changes in the proteins' performance maintained the catalytic rates within a narrow range.

The differences in the catalytic performance of proteins that are conspicuously adaptive among species living in different environments must originally arise within populations. Amid the abundant variation segregating in extant populations resides some variation that will help future populations adapt to currently unforeseen challenges. What is the role of that variation today? Are the differences that will be so important to the future indiscernible in the environment of today, or do they contribute to the differences among individuals in metabolic efficiency, oxygen consumption, growth rate, size, viability, and fecundity? This question cannot be addressed with comparative studies but can be answered only with physiological and demographic studies within populations.

The Randomness of a Sample of Proteins

When electrophoresis was presented to population geneticists as a method of surveying the degree of genetic variation within a population, the loci coding for a sample of proteins was postulated to be a random sample of the genome (Lewontin and Hubby 1966). This postulate is probably not accurate, but the sentiment is still echoed today. I wish to offer a dramatically different view of the nature of the protein polymorphisms used to survey genetic variation. It seems to me that the proteins typically employed in electrophoretic surveys are less a random sample of the genome than a nearly complete sample of the enzymes in central metabolism.

Energy is often stored in glycogen in animals, and glycogen is catalyzed via glycolysis, the pentose shunt, and the citric acid cycle to generate ATP. Plants use the same pathways to release energy stored as starch. Collectively, these pathways constitute respiration, and it is here that the energy utilized in routine metabolic costs—such as circulating blood, ventilating lungs, and maintaining pools of soluble proteins—is generated. These pathways are presented as chains of enzymes in figure 2.4. Although the number of enzymes used by higher plants and animals is clearly in excess of 1,000, approximately 30 enzymes are needed to catabolize glycogen into the end products of CO_2 and H_2O. Furthermore, many of the enzymes used in surveys of genetic variation and in studies of demographic genetics come from these pathways. Certainly, other enzymes—such as esterases, peptidases, and peroxidases—also are employed in surveys of genetic variation, but those readers familiar with studies of protein polymorphisms will recognize the enzymes in these pathways.

Electrophoretic surveys of protein variation among higher plants and animals have demonstrated substantial variation among species in levels of genetic variation. Most

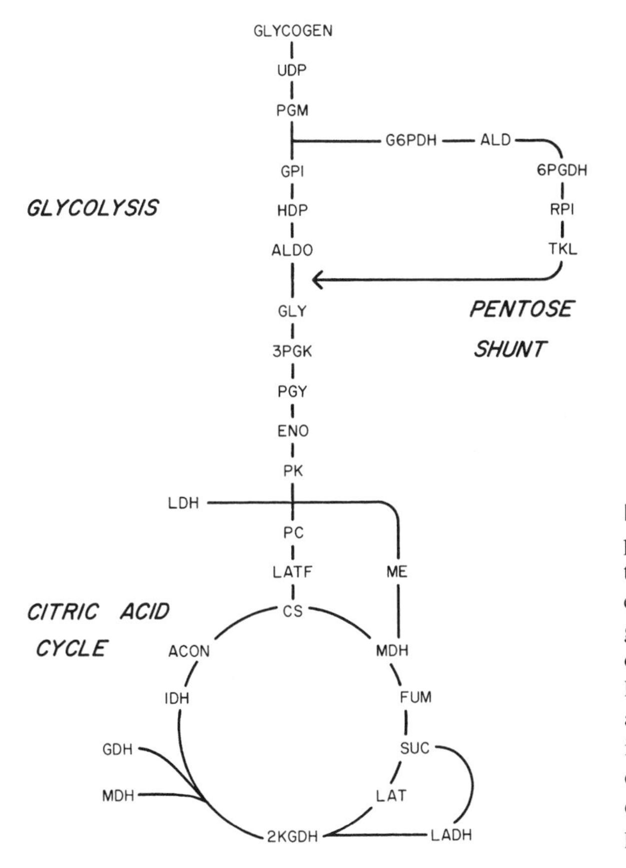

Figure 2.4 The metabolic pathways involved in respiration. Although the number of enzymes in a eukaryote is greater than 1,000, the number of enzymes needed to metabolize a molecule of glycogen to an exhalation of carbon dioxide is on the order of 30. These enzymes are commonly used in electrophoretic surveys of plants and animals.

species are polymorphic at one-third to one-half the enzyme loci that have been surveyed (Brown 1979; Hamrick and Godt 1990; Hamrick et al. 1979; Nevo 1978; Nevo et al. 1984; Powell 1975; Selander 1976). If, on the average, one-third to one-half the enzymes in these pathways are polymorphic within a population, then approximately 12 to 20 of the enzymes in glycolysis, the pentose shunt, and the citric acid cycle should be polymorphic.

In chapter 7 I consider associations between individual heterozygosity and measures of whole animal physiology such as growth rate, oxygen consumption, and routine metabolic costs. In most cases, individual heterozygosity is measured with a set of 4 to 12 polymorphic loci gleaned from the typical array of loci employed in electrophoretic analyses. These loci were initially chosen because of their activity, resolution, and genetic variability, but they need not constitute a random sample of the genome. The perspective offered here is that the 4 to 12 polymorphic loci taken from a survey of 20 to 40 loci supply a very good estimate of the variation at the loci primarily responsible for respiration.

It is futile to rank individuals for heterozygosity of the entire genome (10,000 to 60,000 loci) with a small set of polymorphic enzyme loci (Chakraborty 1981; Mitton and Pierce 1980). For this reason, many biologists have not been inclined to look within populations for associations between protein heterozygosity and either physiological or demographic variation. Yet these associations are common (Mitton 1993a, b, 1994; Mitton and Grant 1984; Zouros and Foltz 1987). The generality of these observations therefore calls into question the assumption that these loci are a random sample of the genome, or the commonly held faith that these loci do not substantially affect physiology or demography.

Conflicting Perspectives from Molecular and Kinetic Data

Comparative studies of amino acid sequences and electrophoretic surveys often lead to the conclusion that most of the substitutions are neutral with respect to adaptation. Yet comparative studies of enzyme kinetics usually reveal an evolution in protein performance, which serves to adapt populations to their current environments. Furthermore, studies within populations of enzyme kinetics, physiology, and demography often disclose variation among those genotypes that are favored during specific environmental conditions and are balanced in fluctuating environments. Is protein genetic variation adaptive in natural populations, or is the variation produced by neutral mutation and influenced primarily by population size? Can a large number of loci contribute to fitness differentials among individuals within populations? This debate and these questions have led population biologists and physiologists to carry out collaborative studies that will profoundly influence our ideas about the role of protein variation and the determination of fitness.

Summary

Population geneticists use surveys of allozyme variation, analyses of mtDNA, analyses of moderately repetitive DNA, and DNA sequencing to estimate levels of genetic variation within populations and the degree of differentiation of populations and species. These techniques probe different compartments of the genome, and the levels of genetic variation that they detect are heterogeneous. Even though the techniques reveal different levels of genetic variation within populations, they all also show substantial levels of genetic variation in natural populations.

It therefore is appropriate to assume that genetic variation is neutral, then to make predictions based on this assumption, and finally to test those predictions. Studies of caged populations, experiments with injected mitochondria, and studies of mtDNA size variation suggest that the variation of mtDNA is not always neutral.

One of the most direct ways to test the assumption of neutrality of enzyme polymorphisms is to study enzyme kinetics of the products of alternative genotypes. Tests of biochemical properties of enzymes will show whether natural selection can discriminate among genotypes.

The enzymes typically used to assay genetic variation are probably not a random sample of the genome and may not be a random sample of enzymes. Many of the enzymes typically employed in surveys are associated with respiration.

3

Environmental Variability and Enzyme Polymorphism

It is clear to even the most casual observer that the environment is in a constant state of flux. It must be the case that fitness differences between genotypes are also in a constant state of flux.

J. H. Gillespie, *The Causes of Molecular Evolution* (1991)

The discovery of abundant protein genetic variation stimulated evolutionary biologists to take a more critical look at their models and to build more biologically realistic models. This reevaluation has included several largely independent goals. Two of the responses were at least partially prompted by the genetic load that would be generated if the variation were maintained by balancing selection. First, a quick response by three papers in the journal *Genetics* (King 1967; Milkman 1967; Sved, Reed, and Bodmer 1967) questioned the traditional calculation of genetic load and devised models of fitness determination that generated little segregational load. Second, concern about an intolerable segregational load spurred the development and elaboration of a theory concerning the evolutionary dynamics of functionally equivalent, or neutral, allelic variation (Crow and Kimura 1970; Kimura 1983; Kimura and Ohta 1971; Nei 1975). Third, emphasis shifted from single-locus models to multilocus models, and biological reality was increased to examine the effect of environmental heterogeneity on the distribution of genotypes.

This chapter begins by reviewing the models of environmental heterogeneity and concludes by examining the empirical studies of the influence of environmental heterogeneity on genetic variability.

Environmental Heterogeneity in Theory

Spatial heterogeneity

Levene (1953) first illustrated the importance of environmental heterogeneity for the maintenance of genetic variation. He modeled a randomly mating population distributed over several environmental patches that imposed different selection regimes on separate portions of the population. This theme of environmental variation inspired the elaboration of numerous related models of the evolutionary dynamics of environmental heterogeneity in space (Gillespie 1991; Hedrick 1986; Hedrick, Ginevan, and Ewing 1976). Several of these models concluded that polymorphism could be maintained if the harmonic mean fitness of homozygotes was less than the harmonic mean fitness of heterozygotes. The conditions for maintaining polymorphisms generally were relaxed when the number of niches increased, when migration among niches decreased, and when individuals selected niches in which they were more fit.

Temporal heterogeneity

Environments are heterogeneous in time as well as in space, and temporal variation can also maintain genetic variation. For example, Dempster (1955) presented a model of natural selection in which heterozygotes were intermediate in fitness between homozygotes, but alternate homozygotes were favored at different times. Although heterozygotes were not superior to all homozygotes at any one time, this model produced cumulative overdominance, or a resultant heterozygous fitness advantage that maintained genetic variation. A similar result was obtained by assigning random fitnesses to homozygous genotypes but stipulating that the heterozygotes' fitness was always intermediate between that of the homozygotes (Gillespie 1973). Temporal variation is most efficient at maintaining genetic variation when the autocorrelation between environments is negative and the cycles of environmental variation are short (Hedrick 1974, 1976). That is, frequent environmental fluctuations are most efficient at maintaining genetic variation. A general result is that for temporal variation to maintain genetic variation, the geometric mean fitness of the heterozygotes must exceed the geometric mean fitness of the homozygotes (Haldane and Jayakar 1963). Because the harmonic mean (the criterion for polymorphism maintained by variation in space) is always less than the geometric mean (the criterion for polymorphism maintained by temporal variation), Hedrick, Ginevan, and Ewing (1976) concluded that variation in space would be more important than variation in time for maintaining genetic variation.

Most models of natural selection that consider temporal variation in the environment focus on environmental variation among generations, but the variation occurring within single life cycles may be important to maintaining genetic variation (Mitton and Grant 1984; Soulé 1976). Consider, for example, a life cycle as a chained series of events among which the relative fitnesses of genotypes change. Assume that in any single event, fitnesses of heterozygotes are intermediate between homozygotes (table 3.1). These events might be embryonic developmental rate, hatching success, growth rate, and fecundity in a fish, or germination, seedling establishment, growth rate, and fecundity in an annual plant. When a series of selective events are chained in this way and alternative homozygotes are favored during successive selective events, the heterozygote may enjoy a substantial fitness advantage over the homozygotes, even though the heterozygotes were not overdominant during any single event. In the example in table 3.1, the fitness of the heterozygote is intermediate during each portion of the life cycle, but the heterozygote has a 17% fitness advantage over the homozygotes for total fitness.

Table 3.1 Fitnesses of genotypes during successive events within a life cycle and resulting cumulative and relative fitnesses

	Life-Cycle Events					
Genotype	1	2	3	4	Cumulative	Relative
AA	1.00	0.50	1.00	0.60	0.30	0.83
Aa	0.75	0.75	0.80	0.80	0.36	1.00
aa	0.50	1.00	0.60	1.00	0.30	0.83

Note: the cumulative fitness is the product of the fitnesses in the selective events, and the relative fitness is the cumulative fitness scaled so that the highest fitness is 1.0. Although the heterozygote had intermediate fitness during each selective event, the heterozygote emerged with the highest relative fitness.

Environmental grain

Levins (1968) used fitness sets to identify the optimal genetic strategy in response to environmental variation. The optimal genetic strategy is not determined solely by the nature of the environmental heterogeneity but partly by the perception of that heterogeneity. For example, it is very unlikely that the spatial heterogeneity in a meadow would be "perceived" or would impinge in the same way on an annual plant, an aphid, and a human. The meadow may be abstracted as a board of environmental patches. A highly mobile species might move from patch to patch, experiencing environments in approximately the frequency in which they occurred. Individuals of a sedentary species or species with severely limited powers of dispersal and mobility, however, might spend most or all of their life in a single patch. The mobile species perceives this meadow as fine grained, whereas the sedentary species perceives it as coarse grained (Levins 1968). The general conclusion of these analyses of fitness sets was that fine-grained species would evolve a strategy of monomorphism, whereas coarse-grained species would maintain polymorphism.

One of the first patterns to be recognized in levels of genetic variability among species was the difference between large and small animals (Selander and Kaufman 1973). Small animals, represented by *Drosophila*, a cricket, snails, and a horseshoe crab, carried substantially more genetic variation than did large animals, represented by a fish, lizards, rodents, a seal, and man. If this study were repeated today, when many more estimates of genetic variability are available, essentially the same pattern would be found. Heterozygosities would overlap more, but the mean levels of genetic variation of small animals would still exceed that of large animals. The difference between these groups was not interpreted as size per se (although it was conceded that size itself could increase homeostasis), but as the way the animals perceived their environments. Large animals were presented as more mobile and more able to make choices concerning their environments and therefore as living in a fine-grained world. Small animals, in contrast, were viewed as more sedentary and less able to choose among alternative environmental states and therefore as living in a coarse-grained world. The pattern of genetic variation between small and large animals is consistent with Levins's predictions (1968), although other interpretations are possible as well (see chapter 9).

When biologists discuss gradients in environmental variability, the subject of latitudinal gradients invariably arises. We homeotherms perceive a dramatic gradient between tropical and polar environments—from relatively constant to dramatically fluctuating, from salubrious to harsh. Ayala and Valentine sampled marine invertebrates from tropical, temperate, and polar environments (Ayala et al. 1973, 1974, 1975a,b; Ayala and Valentine 1974, 1979; Ayala, Valentine, and Zumwalt 1974, 1975; Valentine 1976; Valentine and Ayala 1975, 1976) to test hypotheses concerning the relationship of genetic variability to environmental heterogeneity. Indeed, if tropical and deep-sea environments were relatively invariant and polar environments represented the most variable environments, Levins's work on fitness sets (1968) would lead to the prediction of little genetic variability in the tropics and deep sea and maximal levels of variability in polar environments. In fact, however, exactly the opposite pattern emerged from this extensive series of estimates of genetic variation. Polar species, such as the Antarctic krill, *Euphausia superba* (Ayala et al. 1974; Valentine and Ayala 1976), and the brachiopod *Liothyrella notorcadensis* (Ayala et al. 1975a), have relatively little genetic variation. In contrast, the killer clam, *Tridacna maxima* (Ayala et al. 1973), living on tropical reefs, and deep-sea asteroids (Ayala et al. 1975b) have relatively high levels of genetic variation. Estimates of genetic variation in species of krill were highest in tropical environments, intermediate in temperate environments, and lowest in Antarctic environments (Ayala and Valentine 1979; Valentine and Ayala 1976).

This unexpected global pattern of genetic variation led to a reassessment of the pattern of variability and a shift from a perspective based on temperature to one based on trophic stability (Ayala and Valentine 1979; Ayala, Valentine, and Zumwalt 1975). The consistent pattern of genetic variation, observed in both pelagic and benthic species, led to the hypothesis that the extreme seasonality of polar environments, including the disappearance of phytoplankton during the prolonged winter season, forced species into a strategy of generalism. The researchers proposed that directional selection for flexible alleles in highly seasonal environments would favor monomorphism. Aseasonal environments might be effectively partitioned by genotypes fitting individuals into distinct patches. The alleles of these genotypes were predicted to have narrower functional ranges than did the alleles in the obligate generalists living in polar environments. This hypothesis fits the pattern of genetic variability of a large number of benthic and pelagic species and so deserves further testing. A comparison of the functional ranges of enzymes from tropical and polar species would directly test the postulated mechanism (Ayala and Valentine 1979; Ayala, Valentine, and Zumwalt 1975). It would also be worthwhile to gather estimates of fecundity (see the discussion in chapter 9) for the species in these surveys.

Empirical Studies of Environmental Heterogeneity

Variation in selection coefficients within a life cycle

Some empirical evidence supports the idea that the life cycle can be modeled as a chain of events in which the fitnesses of genotypes switch between successive events. For example, the relative selective values of genotypes at three esterase loci were estimated during two segments of the life cycle of the slender wild oat, *Avena barbata* (Clegg and Allard 1973). Heterozygotes at each of the loci were favored in viability selection, but the most common homozygote at each locus exhibited the highest fecundity. A similar

Table 3.2 Estimates of relative fitnesses in barley for trilocus homozygotes and trilocus genotypes with one heterozygous locus.

Loci	Genotype	Generation 8–9	Generation 19–20	Generation 28–29
ABC	Homozygotes	0.90	1.01	1.76
	Heterozygotes	2.86	5.80	1.23
ABD	Homozygotes	0.94	0.75	1.55
	Heterozygotes	3.09	7.31	0.88
ACD	Homozygotes	0.88	0.83	0.88
	Heterozygotes	4.99	13.15	1.92
BCD	Homozygotes	0.99	0.86	2.08
	Heterozygotes	2.30	7.87	1.08

Note: Data from Clegg, Kahler, and Allard 1978.

pattern was observed in a study of life-cycle components of selection in barley, *Hordeum vulgare* (Clegg, Kahler, and Allard 1978). In this study, three generations were studied, and selective values were estimated for four esterase loci. Selection was intense—selection estimates for trilocus homozygotes typically differed by a factor of 4 or 6. In addition, a bewildering complexity of selective values and changes in selective values was revealed within the life cycle and among generations, so that a single most-fit genotype could not be identified. Finally, a relationship was found between heterozygosity and fitness (table 3.2). The fitness of genotypes heterozygous for one locus generally exceeded the fitness of genotypes homozygous at all three loci.

Heterozygosity and environmental variability: caged populations

The postulated relationships between environmental heterogeneity and genetic heterozygosity prompted experiments with population cages to determine whether more heterogeneous environments would maintain higher levels of genetic variability. The first of these experiments (Powell 1971) contrasted populations of caged *Drosophila willistoni* with different levels of environmental variation. The least variable environments had just one species of yeast, one type of medium, and a constant temperature. The most variable environments had multiple species of yeast, two or more types of media, and a fluctuating temperature. Allozyme heterozygosity was measured with 22 loci, and 50 individuals were scored for each locus. Higher levels of allozyme heterozygosity were maintained in the more variable environments, those cages with multiple sources of yeast and medium and a varying temperature.

This experiment was replicated with *D. pseudoobscura*, and once again higher levels of genetic variability were maintained in the more heterogeneous environments (McDonald and Ayala 1974). There was some concern that environmental heterogeneity was impinging more directly on inversions than on protein polymorphisms (King 1972), so McDonald and Ayala characterized heterozygosity in two ways, first with a complete sample of 20 loci and then with the subset of 17 loci embedded in chromosomes free of inversions. The results of both analyses were essentially the same.

Yet another population cage experiment tested both the effects of environmental heterogeneity and the effects of a competitor on genetic variability (Powell and Wistrand, 1978). *Drosophila pseudoobscura* from 23 isofemale lines lacking inversions founded 36 populations of $N = 500$. The population sizes arrived at equilibrium at approximately $N = 2000$. Population cages differed in the type of food, temperature, and presence or absence of a competitor, *D. persimilis*. Once again, genetic variation increased with environmental heterogeneity, and it was higher in the cages containing the competitor.

An experiment with *D. melanogaster* reported results somewhat different from those from the experiments just discussed (Minawa and Birley 1975). Whereas the preceding experiments were initiated with a large number of gravid females caught in natural environments, the experiment with *D. melanogaster* used a strain that had been kept in the laboratory for more than 200 generations. Three types of cages were established that differed slightly in their degree of environmental variation. The cages were sampled at 27 weeks and 54 weeks after they were established. Variation at 22 protein loci was surveyed, and data from 6 polymorphic enzyme loci and 1 polymorphic locus influencing eye color were monitored. Statistical analyses revealed that selection had influenced the frequencies at every one of the loci in this study. The replicate cages with the lowest level of environmental variation had the lowest levels of heterozygosity, but the differences in levels of heterozygosity among the environments were not statistically significant.

Beardmore (1983) reviewed the laboratory experiments that examined the response of protein heterozygosity (McDonald and Ayala 1974; Minawa and Birley 1975; Powell and Wistrand 1978) or the response of additive genetic variation (Beardmore and Levine 1963; Long 1970; Mackay 1981) to environmental variation. For 15 of the studies, he calculated the ratio of genetic variation in heterogeneous environments relative to their controls. In each case, the ratio was greater than 1.0, and the average ratio was 1.68. This summary of experimental studies supports the hypothesis that more heterogeneous environments maintain more protein genetic variation and more additive genetic variation.

Environmental heterogeneity and single loci

If heterozygosity increases with environmental heterogeneity, then a series of populations or species ranked by their exposure to increasing levels of environmental heterogeneity should exhibit increasing levels of heterozygosity. This expected pattern was first reported for the enzyme glucose-phosphate isomerase (Gpi) in marine mollusks (Levinton 1973). Gpi is polymorphic in a series of mollusks, which spans a wide range of the environments in the western North Atlantic. Marine mollusks can be ranked for the amount of environmental variation that they experience, for intertidal species are exposed to greater variations in temperature and salinity than are subtidal species, and epifaunal species are exposed to a greater range of salinity and temperature than are infaunal species. Gpi and leucine aminopeptidase (Lap) were examined in a large number of population samples of six species of bivalves ranked for the amount of environmental variation that they experienced (table 3.3). For Gpi, both the absolute number of alleles and the effective number of alleles increased with increasing environmental variation.

A second study of the relationship between environmental variation and genetic

Table 3.3 Genetic variation at Gpi and Lap in marine bivalves

		Gpi		Lap	
Species	Life Habit	K	K_e	K	K_e
Mytilus edulis	Epifaunal, I	7	3.9	5	3.0
Modiolus demissus	Semi-infaunal, I	6	2.6	4	2.5
Mercenaria mercenaria	Shallow infaunal, I	6	2.5	4	2.6
Macoma balthica	Medium infaunal, I	3	2.1	4	2.2
Mya arenaria	Deep infaunal, I	3	1.7	3	1.1
Nucula annulata	Infaunal, S	2	1.2	—	—

Note: K is the actual number of alleles, and K_e is the effective number of alleles. I = intertidal, S = subtidal.

variation focused on the marine bivalve genus *Macoma* (Levinton 1975). Gpi and Lap were examined in eight species collected in the San Juan Islands of the state of Washington and Canada. In this study, intertidal species did not exhibit higher levels of genetic variation than did subtidal species. However, estimates of niche breadth were available for the species, and the effective number of alleles increased with niche breadth at the Gpi locus but not at the Lap locus.

The relationship of increasing genetic variation with increasing environmental variability was also reported in a survey of Gpi in population samples of 18 species of marine mollusks collected in Ireland (Wilkins 1975). The average number of *Gpi* alleles was 5.67, 4.40, 3.67, and 2.50 in intertidal epifaunal, intertidal infaunal, subtidal epifaunal, and subtidal infaunal species, respectively. The intertidal species in this study had higher levels of genetic variation at Gpi than did the subtidal species.

The results of these three studies are not perfectly concordant, but that is not surprising, for they involve three different sets of species in three very different environments. Gpi is typically polymorphic in marine mollusks, and Gpi variation increases with environmental variation. The relationships were detected in comparisons of 27 species of marine mollusks, and although linkage and linkage disequilibrium may have contributed substantially to the pattern of Gpi variation in any one species, it is extremely unlikely that the general pattern among the 27 species can be attributed primarily to unseen loci of unknown function in linkage disequilibrium with the locus coding for Gpi.

Environmental heterogeneity and average heterozygosity

If environmental heterogeneity occasionally produces balancing selection and favors higher heterozygosity at individual loci, we may expect to see higher levels of heterozygosity in species occupying more heterogeneous environments. Hidden in this prediction is the assumption that all other factors affecting genetic variation, such as population size and fecundity are approximately equal, and this is clearly not the case in most groups of species. For example, the association between environmental variation and heterozygosity was tested with a group of 18 species from 7 families (Somero and Soulé 1974). Environmental variation was estimated with the range of temperature experienced by each species. No relationship was revealed in this data set. When a similar study was conducted with 22 species of marine fishes from the same family, a positive relationship was found between average heterozygosity and environmental variability (Wallis and Beardmore 1984). If the comparison of Somero and Soulé (1974) had been restricted to related species with similar life histories, it would have enhanced the probability of detecting the effects of environmental heterogeneity on genetic variation.

Indirect tests of the relationship between environmental heterogeneity and genetic variation can be found in reviews of protein variation in animals (Nevo 1978; Nevo, Beiles, and Ben- Shlomo 1984) and plants (Hamrick and Godt 1990; Hamrick, Mitton, and Linhart 1979). Comparisons that approximate axes of environmental heterogeneity are shown in table 3.4. If environmental heterogeneity increases with the geographic range of a species, we could predict that heterozygosity would increase with geographic range.

In a summary of life-history and genetic variation in 480 species of plants, Hamrick and Godt (1990) noted two relationships consistent with the hypothesis that genetic variation increases with the environmental variation experienced by a species. Average heterozygosity was 0.096, 0.137, 0.150, and 0.202 in endemic, narrowly distributed, regionally distributed, and widespread species ($P < .001$), and a similar increase ($P < .001$) was noted for the percentage of polymorphic loci (table 3.4). Early successional environments are influenced primarily by the environment's physical components, whereas late successional environments may experience a richer mixture of physical and biotic variables. Heterozygosity exhibits a nonsignificant tendency to increases from early to late successional species, and the percentage of polymorphic loci is similar in early and mid-successional species but is significantly higher ($P < .01$) in late successional species (table 3.4). But almost all the late successional species in this sample are trees, and they may have high genetic variability for reasons other than their successional status. A principal-components analysis was performed on the genetic, ecological, and life-history data for more than 100 species (Hamrick et al. 1979). The first axis summarized about 30% of the variation in the data and was the axis that explained most of the genetic variation. On this axis, the traits associated with high levels of genetic variation were high fecundity, a long life cycle, wind pollination, late successional status, and large geographic range.

An analysis of protein genetic variation in animals also revealed genetic variation associated with some aspects of environmental heterogeneity (Nevo et al. 1984). Mainland species that were common and widespread and had broad niches were called *habitat generalists*, and rare species or species that were geographically restricted or had narrow niches were called *habitat specialists*. For both invertebrates and invertebrates, habitat generalists had higher levels of genetic variation than did habitat specialists (table 3.4). Heterozygosity increased with geographic range, but in contrast to the pattern in plants, the only significant difference was between endemic species and all the groups with larger geographic ranges.

A summary of surveys of the genetic variation of plants and animals in Israel was analyzed in the context of environmental heterogeneity (Nevo and Beiles 1988). A line drawn from the Mediterranean south and east to the deserts of Israel describes a steep gradient of environmental heterogeneity. Near the Mediterranean, rainfall is approxi-

Table 3.4 Percentage of polymorphic loci (*P*) and average heterozygosity (*H*) in groups of plants and animals differing in levels of environmental heterogeneity

	N	*H*	SE	*P*	SE
Plants					
Geographic range					
Endemic	81	0.096	0.010	40.0	3.2
Narrow	101	0.137	0.011	45.1	2.8
Regional	193	0.150	0.008	52.9	2.1
Widespread	105	0.202	0.015	58.9	3.1
Succession					
Early	226	0.149	0.008	49.0	2.0
Mid	152	0.141	0.010	47.6	2.3
Late	95	0.161	0.011	58.9	3.0
Animals					
Ecological range					
Invertebrate habitat specialists	33	0.064	0.010		
Invertebrate habitat generalists	54	0.149	0.020		
Vertebrate habitat specialists	82	0.037	0.002		
Vertebrate habitat generalists	56	0.071	0.001		
Geographic range					
Endemic and relict	96	0.045			
Narrow	97	0.074			
Regional	267	0.074			
Widespread	211	0.076			

Note: Data for plants from Hamrick and Godt 1990; data for animals from Nevo, Beiles, and Ben-Shlomo 1984. Environmental heterogeneity is not measured directly, but is inferred from the extent of the geographic range, successional status, or the degree of specialization. *N* is the number of species in a group; *H* is the average heterozygosity based on Hardy Weinberg expectations; and SE is the standard error of the mean heterozygosity of that group.

mately 800 mm per year, but it drops to less than 100 mm in the deserts. As the annual rainfall decreases, the year-to-year variability in rainfall increases dramatically. Thus this transect through Israel describes a gradient of rising stress and environmental uncertainty. Nevo and Beiles (1988) summarized allozyme surveys in two plants, two mollusks, two insects, three amphibians, one lizard, and three mammals, testing for correlations with both rainfall and the variation in rainfall across Israel. In the full data set, both heterozygosity and gene diversity increased with the variation in rainfall. Genetic diversity increased toward the deserts in all species studied except the aquatic frog, *Rana ridibunda*. When the data were analyzed by enzyme locus, rather than by species, variation at 10 of the 13 polymorphic loci increased with the variability of rainfall.

Summary

Several population genetic models predict a positive relationship between environmental variability and genetic variability. In laboratory studies with population cages, higher levels of allozyme and additive genetic variation are generally maintained in cages with greater heterogeneity. Among natural populations, single-locus and multilocus patterns suggest that species experiencing higher levels of environmental variation show higher levels of genetic variation. Fault can be found with any one of these summary statements—none of the experiments, standing alone, is sufficient to convince a committed skeptic. But the concordance of these predictions and observations is compelling and leads to the conclusion that environmental variability enhances genetic variation.

4

The Impact of a Single Gene

Human judgment is notoriously fallible and perhaps seldom more so than in facile decisions that a character has no adaptive significance because we do not know the use of it.

G. G. Simpson,
The Major Features of Evolution (1953)

Some theoretical population geneticists have been reluctant to ascribe an important role to balancing selection in the maintenance of protein polymorphisms. Part of this reluctance is due to concern about genetic load. But much of it springs from the difficulty of knowing how a single polymorphism—buried deep in the metabolism of a liver, muscle, or leaf—could substantially contribute to variation in fitness. Many population geneticists have said that the only locus for which we have convincing evidence for balancing selection is the gene for sickle cell hemoglobin in humans, but this statement is no longer true.

Although many studies have explored the adaptive significance of protein polymorphisms, most of them have been indirect and inferential. Koehn conducted the first study that described the geographic variation of enzyme allelic frequencies as a consequence of enzyme kinetics and environmental variation (Koehn 1969; Koehn and Rasmussen 1967). The locus coding for serum esterase in the Gila mountain sucker, *Catostomus clarkii*, segregates two alleles, *a* and *b*, which differ in their rates of migration through a starch gel. Allelic frequencies were estimated from samples taken from natural populations in Arizona, New Mexico, Colorado, Utah, Nevada, Wyoming, and Idaho. Over a distance of 525 miles, the frequency of the *b* allele varied from 0.18 in Preston Springs, Nevada, to 1.00 in the San Pedro River in Arizona, and it decreased linearly with the latitude of the collection locality.

The pattern of geographic variation of serum esterases of the Gila mountain sucker is consistent with the variation in enzyme activity among the three genotypes. Enzyme activities were measured in a spectrophotometer at 0, 8, 25, and 37°C. At 37°C, the *bb* genotype had more than ten times the activity of the *aa* genotype. At 0°C, the pattern reversed; the *aa* genotype had nearly ten times the activity of the *bb* genotype. At both 0 and 37°C, the activity of the heterozygote was intermediate between that of the homozygotes, but in the temperature range of 8 to 25°C, the heterozygote had the highest activity. The *bb* homozygote, which had by far the highest activity at the high temperatures, is the most common genotype in the desert environments of southern Arizona and New Mexico. The *aa* homozygote, which had the highest activity at 0°C, is the most common genotype in Idaho and northern Wyoming.

Circumstantial evidence for selection on the serum esterases was found in a tiny, isolated population at Preston Big Springs, in the headwaters of the Pluvial White River in Nevada. The population currently contains fewer than 50 individuals and has been isolated from the Colorado River for approximately 10,000 years. Because the isolation and small population size enhance the opportunity for genetic drift and fixation of one of the alleles, balancing selection may explain the persistence of genetic variation in this tiny isolate.

Koehn (1978) outlined the evidence needed to demonstrate unequivocally the adaptive significance of protein polymorphism:

1. Functional differences among genotypes must be demonstrated.
2. Functional differences among genotypes must produce physiological differences among genotypes.
3. Ecological variation must influence the enzyme locus via the physiological variation among genotypes.
4. Physiological differences among genotypes must result in variation in some component of fitness.

The most direct way to identify functional differences among allozyme genotypes is to examine the kinetic properties of the genotypes' products (Hall and Koehn 1983). Measurable differences in Michaelis constants, k_{cat}s, maximum velocities, or thermal stabilities have the potential to influence the physiology of whole organisms.

The majority of kinetic studies of protein polymorphisms revealed functional differences among genotypes (Koehn, Zera, and Hall 1983; Zera, Koehn, and Hall 1983), although the functional equivalency of electrophoretically differentiable proteins has also been reported (Dykhuizen, DeFramond, and Hartl 1984; Graves, Rosenblatt, and Somero 1983). The following are several exemplary research programs that meet all the preceding conditions and therefore demonstrate the impact of a single gene on both the physiology of whole animals and relative fitnesses.

LDH in *Fundulus heteroclitus*

Fundulus heteroclitus, known as the mummichog or common killifish, is a euryhaline fish abundant in estuaries, salt marshes, bays, and harbors from the Matanzas River in Florida to Newfoundland. This fish is one of the most sedentary fishes of the Gulf of Maine (Bigelow and Schroeder 1953), for it is virtually never caught more than 100 m from shore, and it has been estimated that individuals typically move less than 100 m in their lifetimes (Lotrich 1975). Individuals become sexually mature at one year of age and may live to be three years old.

Three loci code for lactate dehydrogenase (LDH) in killifish, but only the B locus is expressed in heart, liver, and red blood cells. Two common alleles at this locus, $LDH\text{-}B^a$ and $LDH\text{-}B^b$, segregate at this locus (Mitton and Koehn 1975; Whitt 1969, 1970), and the allelic frequencies are strongly differentiated with latitude. The faster allele, *a*, has a frequency of 1.0 in northern Florida and a frequency of nearly 0.0 in Nova Scotia (Powers and Place 1978).

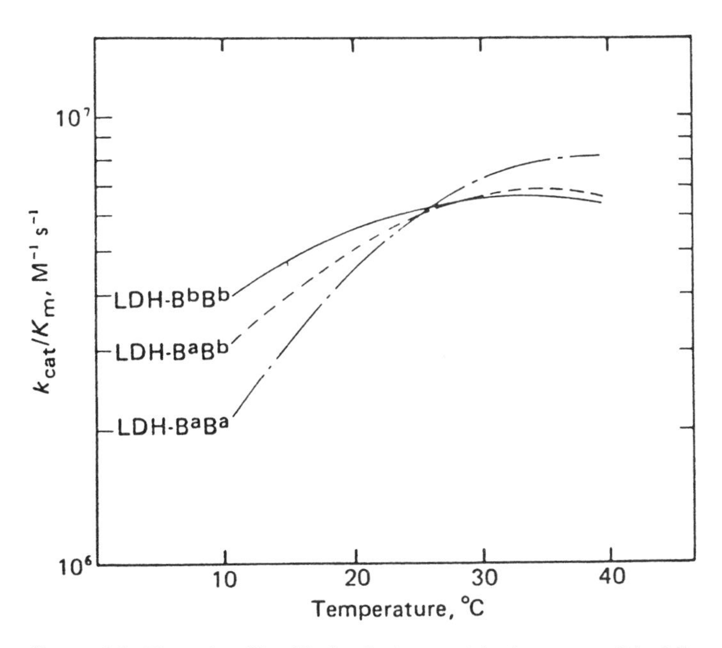

Figure 4.1 The ratio of k_{cat}/K_m for the lactate dehydrogenase of the killifish, *Fundulus heteroclitus*, as a function of temperature. This measure of catalytic efficiency predicts geographic variation along the east coast of North America (figure 4.2), with the *bb* genotype most abundant in the Gulf of Maine, where temperatures are cool, and the *aa* genotype most common along the coasts of South Carolina and Georgia, where temperatures are relatively warm. From Place and Powers 1979.

The catalytic efficiencies of the LDH genotypes were examined to determine whether they were consistent with the patterns of differentiation (Place and Powers 1979). Catalytic efficiency is estimated with the ratio of the rate constant k_{cat} to the Michaelis constant, K_m (Hall and Koehn 1983; Watt 1985). The k_{cat} values were not dependent on pH, and although they did vary with temperature, they were not different among genotypes. Catalytic efficiencies were heterogeneous among genotypes, and when examined at a relative alkalinity of 1.0, they varied with temperature (figure 4.1). At temperatures near 40°C, the catalytic efficiency is highest in the *aa* homozygote, but at temperatures near 10°C, the *bb* homozygote has the highest catalytic efficiency. The efficiencies of the genotypes are essentially equivalent at 25°C. Where there are differences among the homozygotes, the heterozygote is approximately intermediate. The kinetic data are consistent with the latitudinal pattern of differentiation; the genotype with the highest catalytic efficiency at high temperatures is most abundant in warm environments, and the genotype with the highest catalytic efficiency at low temperatures is most abundant in relatively cold environments. There is excellent agreement between the predicted allelic frequencies, based on kinetic analyses and temperatures of collection localities and observed allelic frequencies along a latitudinal transect spanning more than a thousand miles (figure 4.2).

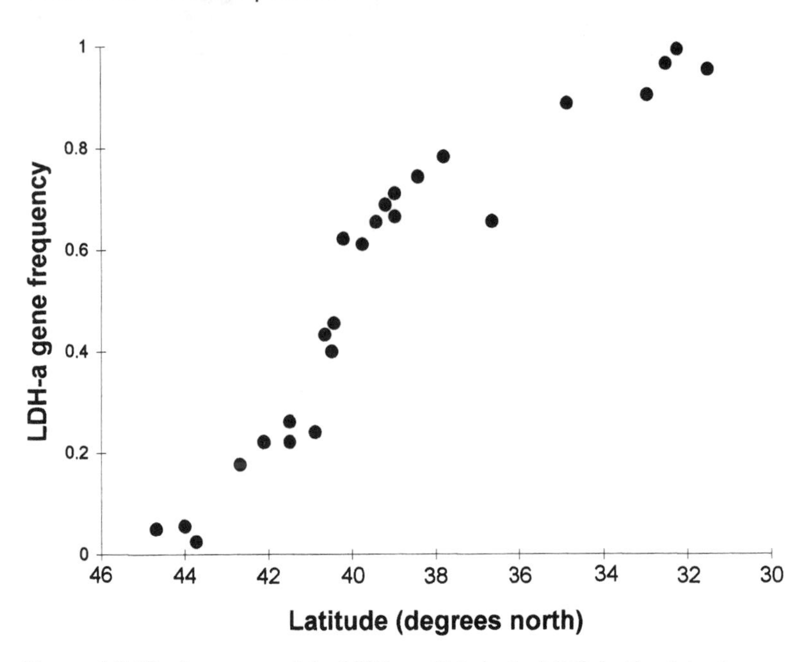

Figure 4.2 The frequency of the LDH-*aa* allele in the killifish, *Fundulus heteroclitus*, as a function of latitude between northern Maine and South Carolina. This pattern of geographic variation is consistent with kinetic variation among the genotypes at this locus (see figure 4.1). From Powers and Place 1978.

The LDH genotype is tightly correlated with the amount of ATP in red blood cells (Powers, Greaney, and Place 1979). ATP is the major allosteric modifier of the hemoglobin of *F. heteroclitus*, decreasing the affinity for hemoglobin (figure 4.3). The relative affinities of hemoglobin for oxygen determine the relative abilities to scavenge oxygen from the water and to deliver oxygen to tissues. The *aa* homozygote, with its lower red blood cell ATP levels and concomitantly higher affinity for oxygen, more efficiently captures oxygen in the gills when the oxygen tension of the water is low. A lower affinity of hemoglobin for oxygen facilitates the release of oxygen, and therefore *bb* homozygotes deliver relatively more oxygen to muscle and other tissues. To determine the impact of variation at the LDH locus, hypotheses based on kinetic data and oxygen equilibrium curves were tested with measures of whole animal physiology.

The eggs of *F. heteroclitus* are often placed in the leaves of the marsh grass *Spartina alterniflora* or in the empty shells of either the blue mussel or the ribbed mussel, and they incubate in the air for much of their development. When sufficient time has passed, hatching is initiated by a combination of immersion in water and low oxygen tension (DiMichele and Powers 1984; DiMichele and Taylor 1980). Those eggs not challenged by low levels of dissolved oxygen will not hatch as soon or may not hatch at all.

The relationship between LDH genotype and hemoglobin–oxygen affinity and the role of low oxygen concentration in hatching led to the hypothesis that LDH genotypes would hatch at different times. The *aa* homozygote, which has the highest affinity for oxygen, was predicted to produce the lowest oxygen tensions and therefore to hatch before the other genotypes did (DiMichele and Powers 1982a). The LDH genotypes and

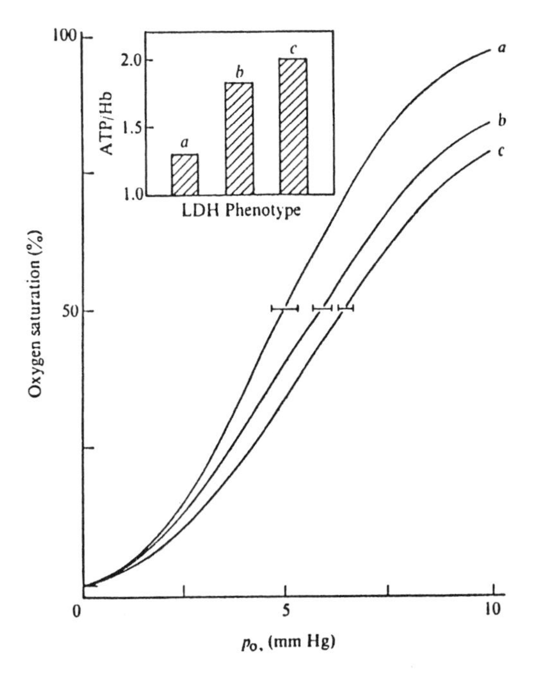

Figure 4.3 The physiological consequences of differences in the amount of ATP per molecule of hemoglobin among the LDH genotypes of the killifish, *Fundulus heteroclitus*. The LDH genotypes *aa*, *ab*, and *bb* are represented as *a*, *b*, and *c*, respectively. ATP modulates the affinity of hemoglobin for oxygen, and the differences in ATP/Hb result in heterogeneous oxygen saturation curves. From Powers, Greaney, and Place 1979.

hatching times of more than 20,000 eggs were examined in a series of mass crosses and pair crosses. The hatching times of the different genotypes were significantly heterogeneous, and as predicted, those eggs bearing the fast homozygote hatched before the other genotypes did. The mean hatching times for the *aa* homozygote, the heterozygote, and the *bb* homozygote were, respectively, 12.2 days, 12.5 days, and 13.6 days.

Further studies of developing eggs revealed the mechanism by which LDH-B genotypes influence the eggs' hatching times. In both air and water, the oxygen consumption of the *aa* homozygotes exceeded that of *bb* homozygotes at 20°C (DiMichele and Powers 1984). These differences in oxygen consumption pointed to differences in rates of development. Within 24 hours of fertilization, the *aa* homozygotes had higher rates of oxygen consumption and levels of lactate than did the *bb* homozygotes (Paynter et. al. 1991). The *aa* homozygotes dramatically increased the metabolic rate between 10 and 15 hours after fertilization, whereas the *bb* homozygotes did not. The metabolic rate of very young eggs appears to depend on lactate accumulation and utilization, both controlled by LDH-B.

The hemoglobin–oxygen affinities of the genotypes were also used to predict heterogeneous swimming endurances in adult animals (DiMichele and Powers 1982b). The higher ATP levels in the red blood cells of *bb* homozygotes depress the affinity of hemoglobin for oxygen, allowing this genotype to deliver more oxygen to tissues active during exercise. This genotype should therefore exhibit greater swimming endurance than the other LDH genotypes should. Animals were swum to exhaustion at 10°C and 25°C, and maximum sustainable swimming speeds of the different genotypes were measured. At 10°C, the maximum sustainable swimming speed of the *bb* homozygote was 20% greater than that of the *aa* homozygote. The amount of oxygen delivered to the muscles of the *bb* homozygote at the maximum swimming speed was estimated to be 18 to 40% greater than that of the *aa* homozygote.

In contrast to these results, no differences in performance or any of the variables measured from blood were detected at 25°C. The differences in physiology and performance at 10°C and the equality of performance at 25°C were consistent with the kinetic data from the LDH genotypes (figure 4.1). The ratios of k_{cat} to K_m differed substantially among genotypes at 10°C, but these values were essentially the same at 25°C.

The LDH polymorphism appears to influence developmental stability, for there is an association between LDH genotype and morphological variance (Mitton 1978). This association was found when population samples of killifish were divided into two groups, one group heterozygous and the other homozygous for a single enzyme polymorphism. In most of the cases, and most clearly for the LDH polymorphism (table 7.3), the heterozygous group exhibited a lower level of morphological variance. This relationship is described in greater detail in chapter 7, which presents data on the relationship between enzyme heterozygosity and developmental stability.

A study of microgeographic variation in LDH frequencies showed genetic differentiation consistent with the latitudinal gradient in LDH frequencies. A population of killifish living in the thermal effluent of a power plant on the north shore of Long Island was found to be genetically (Mitton and Koehn 1975) and morphologically (Mitton and Koehn 1976) differentiated from those in surrounding control localities. In Long Island Sound, the *a* allele, most abundant in warm, southern environments, has a mean frequency of 0.15. However, this allele has a frequency of 0.30 in the population living in the thermal effluent (Mitton and Koehn 1975). The frequency of the *a* allele must have diverged from the frequencies in surrounding control localities since the plant began operation, just a decade earlier. Once again, these results suggest that LDH genotypes influence fitnesses in some thermal environments.

The success of predicting differences in whole animal physiology from kinetic data and oxygen equilibrium curves is stunning. First, the consistent success in predicting physiological differences from kinetic data allow us to attribute directly, with a high degree of confidence, the observed differences in hemoglobin–oxygen affinity, hatching time, developmental rate, and adult swimming performance to the LDH locus. There is no need to ascribe the physiological differences to the effects of linked loci, for the enzyme kinetics of the LDH-B enzymes precisely predict whole animal physiology. Second, the physiological consequences of the LDH polymorphism are great. The homozygotes differ by 35% in red blood cell ATP levels, by 10% in hatching times, by 20% in swimming speeds, by 300% in oxygen consumption at 20 hours in eggs, and by 18 to 40% in the amount of oxygen delivered to muscle during vigorous exercise.

Killifish living in Nova Scotia and Florida often experience temperatures differing by 15°C. Populations at the northern and southern extremes of the distribution are virtually fixed for alternative alleles differing in substrate affinities, reaction rates, heat stabilities, and inhibition constants (Place and Powers 1979, 1984a,b). The dramatic temperature differential between northern and southern extremes of the geographic range causes biochemical reactions to occur more quickly in southern environments, for the rate of chemical reactions increases two to three times with each 10°C increase. To keep reaction velocities roughly equivalent in the different thermal environments, more enzyme would be needed in the colder environment.

Crawford and Powers (1989) tested for further evolutionary adjustments that might complement the kinetic differences. To determine whether evolutionary adjustments had been made in enzyme concentrations, they acclimated fish from Bar Harbor, Maine, and Sapelo Island, Georgia, to 20°C in the laboratory. Fish taken from the colder environment had both higher levels of messenger RNA coding for LDH and higher levels of LDH. Because these differences were observed in fish acclimated to a common environment, they were probably genetically determined. Additional studies of fish from a single locality revealed that acclimatization produced differences that would reinforce the genetic differences identified. As a result of evolutionary adaptations of kinetic parameters and enzyme concentrations, as well as acclimatization to temperature, the northern and southern extremes of this species have similar reaction velocities at the temperatures typical for their localities.

Lap in *Mytilus edulis*

The blue mussel, *Mytilus edulis*, is found typically on hard substrates in intertidal and subtidal regions. In the western North Atlantic, the blue mussel extends from the outer banks of North Carolina to Baffin Island. This dioecious species releases gametes into the water, and fertilized eggs develop into pelagic larvae that drift for a minimum of several weeks. The metamorphosing larvae settle first on algae and then move to a firm substrate, where they anchor themselves with byssal threads.

The blue mussel's extended period of larval dispersal permits high levels of migration among localities, and for this reason, little differentiation is expected among localities in close proximity. Most of the protein polymorphisms investigated in this species conform to this expectation (Levinton and Koehn 1976). The most notable exception is the polymorphism initially referred to as leucine aminopeptidase (*Lap*) but later renamed aminopeptidase-I by Young and colleagues (1979). There are three common (*Lap-94*, *Lap-96*, *Lap-98*) and two relatively rare alleles segregating at this locus. Lap allelic frequencies change dramatically in areas that have abrupt shifts in salinity (Koehn, Milkman, and Mitton 1976; Koehn and Mitton 1972) and temperature (Koehn et al. 1976). The localities that have been studied most intensively are in the regions of changing salinity along the Connecticut shore and the north shore of Long Island at the eastern entrance to Long Island Sound (Hilbish and Koehn 1985; Koehn et al. 1976; Koehn, Newell, and Immerman 1980; Lassen and Turano 1978). Oceanic areas, with average salinities of 33ppt, have an average *Lap-94* frequency of 0.55, and the estuarine sound, which has a salinity of approximately 15ppt, has an average *Lap-94* frequency of 0.15. The areas of sharp clinal transition in allelic frequencies are 10 to 20 km long.

Along the eastern edge of the north shore of Long Island, salinity drops abruptly between the fully salt water of the North Atlantic and the brackish water of Long Island Sound. Mussels from salt and brackish water localities are dramatically differentiated for their allelic frequencies at the Lap locus. Near the boundary between brackish and salt water, salinity, Lap activities, and Lap allelic frequencies are strongly correlated. Among the new immigrants, arriving from both estuarine and oceanic environments, there are no clear correlations between salinity and allelic frequencies. But differential mortality annually reestablishes a cline in allelic frequencies that is stable from year to year (Koehn, Newell, and Immerman 1980). During October, there are differences among genotypes in the ratio of dry weight to shell size, a rough measure of the animal's physiological condition. Genotypes bearing the *Lap-94* allele lose more tissue between summer and October than do genotypes lacking this allele (Koehn, Bayne, Moore, and Siebenaller 1980). This difference is an indication of the stress that produces differential mortality, acting primarily to remove the *Lap-94* allele from localities with relatively low salinity.

Lap activity is found in many tissues, but it is greatest in the brush border of gut epithelial cells and in digestive tubule cells in the hepatopancreas, where Lap is lysosomal (Koehn 1978; Koehn et al. 1980; Moore, Koehn, and Bayne 1980). Lap cleaves neutral and aromatic amino acids from the *N*-terminal end of oligopeptides, releasing free amino acids. The tissue distribution and the function of this enzyme suggest two roles, a function in digestion and a function in osmoregulation. The blue mussel is an osmoconformer, regulating cell volumes in response to changes in salinity by altering the concentration of free amino acids in the cytosol. That is, when a mussel is challenged by an increase in salinity, it regulates its osmotic pressure, and therefore its cell volume, by increasing the free amino acids within the cell. Mussels challenged by a decrease in salinity conform to the change in osmotic pressure by excreting nitrogen.

The activity of Lap changes in response to changes in salinity (Koehn 1978). When mussels are moved from pure sea water (33 ppt) to estuarine water (15 ppt), activity drops over a period of several days. When the mussels are moved from estuarine water to sea water, the activity of Lap increases dramatically within a few hours. Lap has approximately 50% higher activity in salinity of 30ppt than at 15ppt; the increased activity in salt water may be a necessary response to produce and maintain a higher concentration of free amino acids.

The activity of Lap varies among the genotypes (Koehn, Bayne, Moore, and Siebenaller 1980). In sea water, the *Lap-94* homozygote has the highest activity, but in estuarine water, this genotype has the lowest activity. This reversal in ranking is due to a change in the relative concentrations of the enzymes; in sea water, all genotypes produce equal amounts of the enzyme, but in estuarine water, the *Lap-94* homozygote has a concentration 15% lower than do the other genotypes. Nevertheless, the average activity of the genotypes bearing the *Lap-94* allele always exceeds that of the genotypes lacking the *Lap-94* allele, regardless of the salinity (Koehn and Siebenaller 1981). This persistent difference is due the higher catalytic efficiency of genotypes bearing the *Lap-94* allele. The K_{cat}, defined as the maximum velocity divided by enzyme concentration, is 20% higher in genotypes bearing the *Lap-94* allele than in genotypes lacking this allele.

This information can be used to construct a scenario for the maintenance of the Lap polymorphism and the annual reconstruction of abrupt allelic frequency clines. Gametes are released into the water in spring and early summer, and the larvae are carried by currents for three to seven weeks, moving distances of nearly zero to several hundred kilometers. Large areas characterized by oceanic water and estuarine water are differentiated at the Lap locus because of its role in osmoregulation. At full salinity, a relatively high concentration of free amino acids is needed to maintain cell volume, and therefore the Lap enzyme is maintained at the higher concentration, and the frequency of the allele with the highest catalytic efficiency is relatively high. In estuarine water, the concentration of the Lap enzyme is only two-thirds of the level in oceanic water, and the frequency of the allele with the highest catalytic efficiency is relatively low. Genotypes carrying this allele excrete nitrogen at a higher rate than do other genotypes, and when nitrogen is not abundant, the ratio of tissue weight to shell size is lower in genotypes bearing this allele. This difference in the ratio of weight to size indicates a physiological stress that is associated with genotype-dependent mortality. Larvae transported from oceanic nurseries to estuarine settling sites or vice versa bear Lap genotypes that are poorly adapted to the new environments. Immigrants into estuarine environments excrete too much nitrogen, whereas immigrants into oceanic environments are unable to sustain sufficient osmolarities to maintain cell volumes. These physiological consequences of the interaction of Lap genotype and salinity produce a bout of mortality that annually reconstructs the Lap cline where the salinity changes.

Although three common and several rare alleles segregate at the Lap locus, the physiological phenotypes identified to date suggest that this polymorphism behaves as if it were two alleles with dominance (Hilbish and Koehn 1985). The *Lap-94* allele appears to be dominant over *Lap-98* and *Lap-96* for both enzyme activity and physiological traits. When enzyme activity is compared among genotypes, *94/94*, *94/96*, and *94/98* form a homogeneous subset with high activity, and *98/98*, *98/96*, and *96/96* form a homogeneous subset with low activity. Several physiological phenotypes show either the same or similar patterns. The accumulation of free amino acids is homogeneous and high for genotypes bearing at least one *94* allele, and it is lower for other genotypes. Similarly, the excretion of primary amines and ammonia is high for genotypes bearing the *94* allele and low for other genotypes. Alleles *96* and *98* are easily distinguished electrophoretically, but the physiological differences among the genotypes bearing these alleles have not yet been identified.

Parallel patterns of geographic variation among reproductively isolated species may reveal similar evolutionary responses to ecological heterogeneity (Koehn and Mitton 1972). The blue mussel and the ribbed mussel (*Modiolus demissus*) exhibit similar patterns of changes of allelic frequencies along a salinity gradient in a salt marsh in Long Island Sound, suggesting that these polymorphisms respond to ecological heterogeneity in similar ways (Koehn and Mitton 1972). Similarly, there are parallel patterns of biochemical variation between the Lap of the blue mussel and the Lap polymorphism of the quahog, *Mercenaria mercenaria*. Like the Lap of the blue mussel, the Lap locus in the quahog segregates three common alleles. In both species, the activity of the enzyme is substantially higher in pure salt water than in the brackish water of Long Island Sound. In both species, there are differences in enzyme activity among genotypes in full salt water, but few or no detectable differences in brackish water. Today, the Lap polymorphisms in the blue mussel, the ribbed mussel, and the quahog have been differentiated to the point that they do not share any alleles, but each segregates three common alleles, and they exhibit some parallel patterns of response to ecological heterogeneity.

The shape of blue mussel shells varies among individuals within populations, and the variability of mussel shape is associated with the variability of genes coding for enzymes, including the Lap locus (Mitton and Koehn 1985; chapter 7).

PGI in *Colias* Butterflies

A polymorphism for phosphoglucose isomerase (PGI) influences viability, flight time, mating success, and fecundity in *Colias* butterflies. Pgi segregates three common alleles in both *Colias eurytheme* and *C. philodice eriphyle*, and genotypes differ in their heat stabilities, maximal activities (V_{max}), and Michaelis constants (K_m) (Watt 1977).

Watt (1992) reasoned that the best estimate of physiological performance for this enzyme is V_{max} / K_m, and he used this estimate of physiological efficiency to make predictions concerning the rankings of PGI genotypes in viability, flight time, mating success, and fecundity. According to this criterion, some heterozygotes are overdominant, but others are not. Genotypes *44*, *45*, and *55*, with low V_{max} / K_m ratios, are expected to be selected against during most environmental conditions, but exceptionally hot weather is expected to favor these genotypes, for they have unusually high thermal stabilities. Genotypes *22*, *23*, *24*, *34*, and *35* have high V_{max} / K_m ratios and are therefore expected to be favored by selection during periods of normal weather. Genotype *33* is intermediate for both V_{max} / K_m and thermostability. Genotypes *22*, *23*, and *24* have low thermostabilities and are predicted to be selected against during extended periods of hot weather. Consideration of both heat stability and V_{max} / K_m lead to the following predictions for *Colias* Pgi genotypes:

	Favored	**Selected against**
Normal weather	22, 23, 24, 34, 35	44, 45, 55
Hot weather	44, 45, 55	22, 23, 24

At pH 8.75 and at 40°C, the range of V_{max} / K_m (x 10^4) is from 1.70 (for the *44* homozygote) to 3.00 (for the *22* homozygote) (Watt 1983). Under these same conditions, the most common heterozygote, *34*, is overdominant to the most common homozygotes; V_{max} / K_m for the *34*, *33*, and *44* genotypes are 2.37, 1.71, and 1.70, respectively. Note that these predictions concerning performance do not fall out simply as heterozygote versus homozygote and that the predictions differ with the weather conditions. Note also that *33* is never predicted to be either favored or selected against; the *3* allele is the most common in both species, and *33* is the most common homozygous genotype.

A butterfly's age can be estimated from the wear and tear on its wings. Comparisons of animals of different ages within a cohort revealed that the frequency of heterozygotes increased with age during normal weather conditions but that unusually hot weather favored the thermostable genotypes (Watt 1977, 1983; Watt, Cassin, and Swan 1983).

The ability to fly is a critically important component of fitness in *Colias*, for butterflies must fly in order to find food, to mate, and to find oviposition sites. The optimal body temperature for flight is narrow, 35°C to 39°C, but some butterflies are able to fly with body temperatures outside this range, as low as 29°C and up to slightly over 40°C. Examination of V_{max} / K_m revealed that the common *34* heterozygote was more efficient over a broader range of temperatures than the homozygotes were, indicating that heterozygotes would fly over a greater range of temperatures than the other genotypes would. Samples of butterflies collected throughout the day and over a range of temperatures produced data consistent with this hypothesis (Watt 1983). The ability of heterozygotes to fly over a greater range of temperatures than homozygotes can gives heterozygotes advantages in foraging for food, in reaching oviposition sites, and in males' mating success.

Male mating success is highly dependent on the Pgi genotype. Although females may mate as many as three times during their lives, a female uses the sperm from only a single male to fertilize eggs at any one time. This aspect of the biology of *Colias* butterflies allowed careful analyses of the mating system in the field. Males and females were collected in the field and transported to the laboratory. The males were genotyped immediately, and their genotypes represented a sample of the males attempting to mate. The females were placed in cages and allowed to lay eggs. The PGI genotypes of the females and their offspring were then examined to find the genotype of the father (Carter and Watt 1988; Watt, Carter, and Donohue 1986; Watt, Cassin, and Swan 1983). This sampling design obtained estimates of genotypic frequencies for flying females, flying males, and reproductively successful males. The results were consistent over years, over collection localities, and for both *C. eurytheme* and *C. philodice eriphyle*. In the group of males attempting to mate, the frequency of PGI heterozygotes was generally 40 to 56%, but the frequency of heterozygotes in males mating successfully ranged from 67 to 85%. That is, the mating success of PGI heterozygotes was far higher than that for PGI homozygotes. This series of experiments is examined in greater detail in chapter 8.

An experiment with females caged in the field revealed that the fecundity of *Colias* butterflies varies among PGI genotypes (Watt 1992). After depositing each egg on the appropriate host plant, the female must fly to a new site before laying another egg. Thus, fecundity is intimately tied to flight activity time and also to the ability to fly under marginal environmental conditions. Females were captured in the field, and each was placed in an outdoor cage with their host plants for oviposition. Each female was allowed to lay eggs for four or five days while the flight activity time, or the time in which temperature and light permitted flight, was recorded during these days. The female's age was estimated from the condition of her wings. After the effects of both flight activity and age were accounted for, fecundity was tightly associated with the Pgi genotype ($P < .001$; Watt 1992). Once again, differences in fecundity were consistent with predictions based on V_{max} / K_m. The number of eggs laid per day for the *34*, *33*, and *44* genotypes was approximately *16*, *12*, and *6*, respectively. Note that fecundity provides more than a twofold advantage to a favored over a disfavored genotype.

A short digression is needed here. Many biologists familiar with the literature on enzyme polymorphisms may remember that the genotypic distributions for most outbreeding species conform to Hardy–Weinberg expectations. In addition, many writers, noting that the Hardy–Weinberg expectations are based on the assumption of no selection, concluded that equilibrium genotypic distributions in their species meant that their genetic markers were effectively neutral. I used a simplified model of the *Colias* Pgi polymorphism to demonstrate that equilibrium genotypic distributions can be found in a population experiencing strong selection.

To simplify the model of the dynamics on the PGI polymorphism, I considered only male mating success and female fecundity and only the most common alleles, *3* and *4* (table 4.1). I estimated the male mating success of the three genotypes by noting that although approximately 45% of the males flying were heterozygous, about 65% of the males siring broods were heterozygous. Very roughly, then, the fitness of the heterozygotes is 0.65/0.45, and the fitness of homozygotes is 0.35/0.55, yielding estimates of relative fitness of 0.45, 1.0, and 0.45 for the *33*, *34*, and *44* genotypes, respectively. The fecundity of these genotypes estimated in field cages was 12, 16, and 6, yielding estimates of relative fitness of 0.75, 1.0, and 0.37. To estimate total fitness, I multiplied the relative fitnesses, obtaining total fitnesses of 0.34, 1.0, and 0.17 for the *33*, *34*, and *44* genotypes, respectively. Note that the fitness of the *34* genotype is six times the fitness of the *44* homozygote. Although theoretical population geneticists are not inclined to

Table 4.1 A model of mating in *Colias* butterflies, with overdominance for both male mating success and female fecundity

	heterozygotes	homozygotes
Male mating success (W_m) =	$\frac{0.65 \text{ of copulations}}{0.45 \text{ of population}}$	$\frac{0.35 \text{ of copulations}}{0.55 \text{ of population}}$
W_m =	1.00	0.45

	Pgi genotypes 33	34	44
Male mating success (W_m)	.45	1.0	.45
Female fecundity (eggs/day)	12	16	6
Female fecundity (W_f)	.75	1.0	.37
$W_m * W_f$	.34	1.0	.17
Selection coefficients	$s = .66$	—	$t = .83$

Allelic frequencies at equilibrium $f(3) = p = \frac{t}{s+t} = .56$ $f(4) = q = \frac{s}{s+t} = .44$

Genotypic frequencies at equilibrium $P^2 = .31$ $2pq = .50$ $q^2 = .19$

♀♀

Genotype	33	34	44	
Frequency	.31	.50	.19	
W_f	.75	1.0	.37	
Product	.23	.50	.07	$\Sigma = .80$
Relative contribution	.29	.62	.09	$\Sigma = 1.0$

	Genotype	Frequency	W_m	Product	Relative contribution	♀ 33	♀ 34	♀ 44
♂♂	33	.31	.45	.14	.19	.06(33)	.06(33) .06(34)	.02(34)
♂♂	34	.50	1.0	.50	.70	.10(33) .10(34)	.11(33) .22(34) .11(44)	.03(34) .03(44)
♂♂	44	.19	.45	.08	.11	.03(34)	.03(34) .03(44)	.01(44)
				$\Sigma = .72$	$\Sigma = 1.0$			

	Pgi genotype 33	34	44	p	q
Observed frequencies of larvae	.33	.49	.18	.575	.425

consider fitness differentials of this magnitude, this is a reasonable approximation for the *Colias* of Gothic, Colorado.

If the fitnesses of the *33*, *34*, and *44* homozygotes are represented as $1 - s$, 1, and $1 - t$, then the equilibrium frequency of the *3* allele is $p = t / (s + t) = 0.55$, and, by subtraction, the frequency of the *4* allele is $q = 0.45$. At Hardy–Weinberg equilibrium, the expected frequencies of the *33*, *34*, and *44* genotypes are p^2, $2pq$, and q^2, or 0.30, 0,50, and 0.20. These are the frequencies expected in a large population practicing random mating and no selection.

The frequencies of larvae produced in a population of *Colias* with overdominance for both male mating success and female fecundity are shown in table 4.1. The frequencies of the genotypes were multiplied by the relative fitnesses for fecundity to represent the contribution of females bearing alternative genotypes. The frequencies of the genotypes were multiplied by the relative fitnesses for male mating success to represent the contribution by males bearing alternative genotypes. The production of young by the various genotypes is illustrated as the product of the contributions of the males and females. The genotypic frequencies accumulated from this table are the same as those of the Hardy–Weinberg equilibrium. Strong overdominance for both male mating success and female fecundity produced frequencies that might be misinterpreted as revealing selective neutrality.

Additional Examples

Although the number of examples of kinetic and physiological studies of enzyme variation and their demographic and biogeographic consequences is small, the list is not limited to the three just presented (Koehn et al. 1983; McDonald 1983; Zera et al. 1983).

Perhaps the most extensively studied polymorphism is alcohol dehydrogenase (ADH) in *Drosophila melanogaster* (Kreitman, Shorrocks, and Dytham 1992; McDonald 1983; Van Delden 1982). The two common alleles at this polymorphism (Grell, Jacobson, and Murphy 1965) differ at only 1 of 255 amino acids; the *F* allele contains threonine at amino acid number 192, and the *S* allele contains lysine (Fletcher et al. 1978). Like the LDH in *Fundulus*, the ADH of *D. melanogaster* has both variation in the regulation of the alleles and kinetic variation among the genotypes. The production of mRNA is higher in *FF* homozygotes than in *SS* homozygotes, producing a higher concentration of ADH in *FF* homozygotes (Anderson and McDonald 1983; Lewis and Gibson 1978) and a two- to threefold difference in enzyme activity. The genotypes differ in thermostability (Anderson, Santos, and McDonald 1980), and at very high temperatures, these differences may have a direct effect on fitness (Sampsell and Simms 1982). In addition, the enzymes differ in values of K_m for ethanol, propanol, and butanol (Cavener and Clegg 1981; Day, Hillier, and Clarke 1974; McDonald, Anderson, and Santos 1980).

The influence of Adh activity on the synthesis of lipid from ethanol was measured in third-instar larvae of *D. melanogaster* (Freriksen et al. 1991). The larvae encounter ethanol as a natural product in rotting fruit, and they tolerate high levels of ethanol and use it as a food source. Nuclear magnetic resonance spectroscopy measured the accumulation of monounsaturated fatty acids from ethanol. ADH, which catalyzes the first reaction in the pathway from ethanol to lipid, exerts a high degree of control on the flux, or the flow, of carbon through this metabolic pathway. The flux control coefficient of the ADH enzyme was estimated to be approximately 1.0, the maximum possible. Flux increases regularly with ADH activity. The tight control by the ADH activity of flux through an important pathway led Freriksen and colleagues (1991) to conclude that ADH could be a major target of natural selection. ADH is thought to help *D. melanogaster* exploit environments with high concentrations of alcohols, and for this reason the majority of studies of ADH concern environments supplemented with alcohols. The ADH locus appears to respond to selection, for the *F* allele generally increases in frequency in environments with high alcohol or with food supplemented with alcohol (Cavener and Clegg 1978; Van Delden 1982). In natural populations, this polymorphism has repeated latitudinal clines in allelic frequencies in Australia, North America, and Asia (Kreitman et al. 1992; Oakeshott et al. 1982).

Environments that contain high levels of ethanol may evoke fitness differentials among ADH genotypes in *D. melanogaster*, but differences in fecundity also appear in environments that contain little or no alcohol (Bijlsma-Meeles and Bijlsma 1988; Serradilla and Ayala 1983b). In the first of these studies, reproductively mature females were exposed to males and allowed to lay eggs for several days. Heterozygotes deposited 40 to 50% more eggs than homozygotes did (Serradilla and Ayala 1983a). Similar results were obtained in a study of components of fitness associated with the alcohol dehydrogenase polymorphism (Bijlsma-Meeles and Bijlsma 1988). Relative fecundities estimated for slow homozygotes (*SS*), heterozygotes (*SF*), and fast homozygotes (*FF*) were 0.88, 1.00, and 0.61, respectively.

The Adh polymorphism of the tiger salamander, *Ambystoma tigrinum*, is associated with oxygen consumption, growth rate, and success at metamorphosis (Carter 1992; Carter et al. 1997). In the ephemeral ponds of western Colorado, salamanders that do not metamorphose risk dying of dessication when the ponds dry up. Before a salamander can successfully metamorphose, it must grow beyond a minimal size threshold, absorb its tail fin and external gills, and modify its skin. In Colorado, the Adh locus segregates two common alleles, *F* and *S*. Enzyme products were purified from gilled larvae, and the K_{cat} / K_m ratio was determined for each genotype, in both directions. In the direction of acetaldehyde to ethanol, the K_{cat} / K_m ratio was higher in the heterozygote than in either homozygote, and the ratio was higher in Adh-*SS* than in Adh-*FF*. In the opposite direction, the heterozygote and the Adh-*SS* had similar ratios, and both of these were higher than the ratio in the Adh-*FF*.

Levels of adenosine triphosphate (ATP) in the red blood cells, which modulate the affinity of hemoglobin for oxygen, are associated with Adh genotypes. Adh-*FF* homozygotes have lower levels of ATP/red blood cells and higher rates of oxygen consumption than do the Adh-*SS* homozygotes, and heterozygotes are intermediate for both variables. The frequencies of the homozygous genotypes in ponds near Gothic, Colorado, are correlated with the amount of oxygen in the water. Studies in the field and in the laboratory indicate that Adh-*FF* homozygotes metamorphose more slowly in supersaturated water than in normoxic water and that Adh-*SS* homozygotes metamorphose more slowly in hypoxic water than in normoxic water. Heterozygotes metamorphose at the same rate in all oxygen environments. These rates of metamorphosis explain the heterogeneity in Adh frequencies among ponds. The frequency of Adh-*SS* homozygotes increases and the frequency of Adh-*FF* homozygotes decreases with the level of dissolved oxygen in the ponds. The frequency of heterozygotes does not vary with the level of dissolved oxygen.

Allelic frequencies at the phosphoglucose isomerase locus in the sea anemone, *Medtridium senile*, are correlated with latitude and temperature (Hoffmann 1981a,b, 1983). Gpi^{f} reaches high frequencies or approaches fixation in localities south of Cape Cod, where summer temperatures are relatively high. North of Cape Cod, the Pgi^{s} allele increases in frequency with decreasing temperatures. Kinetic properties vary among genotypes (Hoffmann 1981b, 1983), with differences in both K_m and V_{max} / K_m,. Relative to the Gpi^{ff} homozygote, the Gpi^{ss} homozygote is inhibited more by the pentose intermediate 6-phosphogluconate, and differences among genotypes are greater at low temperatures. The differential inhibition by metabolic intermediates led to the prediction that Gpi^{ss} homozygotes would have relatively higher fluxes through the pentose shunt than other genotypes would. Measurements of metabolic flux through glycolysis and the pentose shunt were consistent with this prediction; relative to Gpi^{ff} homozygotes, the flux of Gpi^{ss} homozygotes through glycolysis was 90%, and the flux through the pentose shunt was 114% (Zamer and Hoffmann 1989). The fractional contribution of the pentose shunt to carbohydrate metabolism in Gpi^{ff} homozygotes was only 78% of that in Gpi^{ss} homozygotes. The differences in flux were highest at low temperatures and may explain the increase in frequency in the Gpi^{s} allele in regions with low temperatures.

The glutamate-pyruvate transaminase (Gpt) of the copepod, *Tigriopus californicus*, plays a role maintaining cell volume and responding to hyperosmotic stress (Burton and Feldman 1983). This copepod lives in isolated pools in the intertidal zone, where salinity can be decreased by heavy rains or increased by the splash of waves and the subsequent evaporation of water. Like the blue mussel, *Tigriopus* releases free amino acids to balance osmotic pressure in response to increasing salinity. Tolerance of salinity change has a high cost in energy; the costs of generating typical levels of alanine and proline after hyperosmotic stress are approximately 12% of the daily energy budget (Goolish and Burton 1989). Alanine is one of the amino acids that appears quickly in response to hyperosmotic stress, and Gpt catalyzes the final step of alanine biosynthesis. Alternative genotypes of Gpt accumulate alanine at different rates and consequently exhibit different survival after hyperosmotic shock (Burton and Feldman 1982, 1983). The allele that provides the faster accumulation rate of alanine appears in higher frequencies in sites that have highly variable salinities and reach extremely high levels of salinity (Burton 1986). Laboratory populations maintained at various constant levels of salinity do not exhibit changes in Gpt frequencies, suggesting that the physiological differences among genotypes are important only when salinity changes abruptly. Studies of both survival after hyperosmotic shock (Burton 1986) and developmental rate (Burton 1987) revealed adaptation to local environments. Isolated pools in the high intertidal are heterogeneous environments, and therefore the allelic frequencies of Gpt and other enzymes form a complex, highly differentiated mosaic (Burton 1986; Burton and Feldman 1981).

The esterase-6 polymorphism of *D. melanogaster* appears to play a role in mating behavior (Richmond et al. 1980). Values of k_{cat} / K_m are heterogeneous for the substrate β-napthylpropionate. The enzyme is concentrated in the anterior ejaculatory duct of males, and it is capable of cleaving a substrate to produce a pheromone in the male that is transferred to the female during copulation (Mane, Tompkin, and Richmond 1983). The hormone influences the time to remating in females; the latency of a female is a function of the EST-6 genotype of her mate.

Polymorphisms at both 6-phosphogluconate dehydrogenase and glucose-6-phosphate dehydrogenase influence the relative fluxes of glucose metabolism through glycolysis and the pentose shunt in *D. melanogaster*. Dilocus genotypes differ in thermostability, activity, and the relative utilization of the pentose shunt (Bijlsma 1978; Cavener and Clegg 1981). These results seemed to be largely attributable to the 6-phosphogluconate dehydrogenase polymorphism, but when the pentose shunt was partially blocked by a genotype with low 6-phosphogluconate dehydrogenase activity, differences in viability attributable to the glucose-6-phosphate dehydrogenase locus were detected (Eanes 1984).

The variation at superoxide dismutase (SOD) in *Drosophila melanogaster* may be balanced by different levels of ionizing radiation. SOD segregates two common allelomorphs in *D. melanogaster*, SOD^S and SOD^F, differing by one, two, or three amino acids, depending on the geographic locality (Lee, Misra, and Ayala 1981). Studies of purified enzymes revealed that SOD^S has a specific activity two to three times higher than that of SOD^F and that SOD^S scavenges O_2 2.7 times faster than SOD^F. The thermostability of SOD^F exceeds that of SOD^S. To test whether the SOD polymorphism was adaptively relevant in the context of ionizing radiation, Peng, Moya, and Ayala (1986) measured the viability of SOD genotypes as a function of ionizing radiation and monitored allelic frequencies in populations exposed to X-irradiation. Larval viabilities differed significantly, with the highest viability in SOD^{SS} homozygotes, which had the highest activities. Similarly, the frequency of the SOD^S allele was maintained at higher frequencies in an irradiated population than in a control population. Comparisons of the SOD^S and SOD^F alleles with a null allele further supported the hypothesis that SOD activity was correlated with fitness in irradiated environments. Under normal laboratory conditions, heterozygotes had higher fecundities than did homozygotes at all temperatures and densities (table 4.2; Peng, Moya, and Ayala 1991). Comparative studies of the amino acid sequence of SOD found that the rate of amino acid replacement per 100 residues per 100 million years varied from 5.8 to 30.9 in different comparisons, indicating that natural selection might influence this locus at some times or in some lineages (Lee, Friedman, and Ayala 1985).

A complex hemoglobin polymorphism directly influences whole animal physiology and contributes to adaptation to elevation in the deer mouse, *Peromyscus maniculatus* (Snyder 1981). Deer mice have an extraordinary range in elevation, from below sea level in Death Valley to above 4300 m in the White Mountains of California and the Rocky Mountains of Colorado and Wyoming. This elevational range presents a formidable challenge, for at 4300 m the partial pressure of oxygen is only 55% of its value at sea level. The potential for hypoxia at high elevations is exacerbated by the need for sustained activity. Foraging, courtship, territorial defense, and escape from predators all require substantial, prolonged activity for *Peromyscus*. High-altitude hypoxia is a source of stress that might interact with the complex hemoglobin polymorphism in *Peromyscus*. Two loci coding for a- hemoglobin, *HbA* and *HbC*, are each polymorphic for two alleles, Hba^0, Hba^1, Hbc^0, and Hbc^1 (Snyder 1978a,b, 1979). These loci are tightly linked and are in strong linkage disequilibrium, so that in most populations, the frequencies of the haplotypes a^0c^0 and a^1c^1 sum to approximately 99% (Snyder 1979). In

Table 4.2 Number of offspring surviving from day 1 of eggs' deposition for Sod genotypes in *Drosophila melanogaster*

		Low Density		High Density	
Genetic Background	Sod Genotype	20°	28°	20°	28°
Low variability	*SS*	6.09±.10	5.42±.11	3.22±.04	2.65±.03
	SF	6.74±.11	6.13±.11	3.62±.04	3.20±.05
	FF	6.25±.12	5.79±.11	3.52±.04	2.95±.04
High variability	*SS*	7.12±.10	6.44±.10	3.31±.03	3.12±.04
	SF	7.74±.10	7.36±.12	3.85±.04	3.84±.05
	FF	7.34±.10	6.81±.11	3.83±.04	3.45±.04

Note: Data from Peng, Moya, and Ayala 1991.

western North America, the frequency of the a^1c^1 haplotype is negatively correlated with elevation ($r = -.77$, $P < .001$) (Chappell and Snyder 1984) so that below 1750 m, populations are generally fixed for a^1c^1, whereas above 2750 m, populations are nearly fixed for a^0c^0. Physiological studies were conducted on strains created identical by descent for hemoglobin haplotype in an effort to test for differences among hemoglobin genotypes on similar genetic backgrounds. These studies revealed that hemoglobin genotypes influence P_{50} (the partial pressure of oxygen needed for 50% saturation of hemoglobin) and V_{O_2max} (maximal oxygen consumption during exercise), but not P_{CO_2}, the CO_2 Bohr effect, blood buffering capacity, hematocrit, or hemoglobin concentration (Chappell and Snyder 1984). Values of P_{50} were heterogeneous ($P < .001$) among genotypes, with mean values for the genotypes a^0c^0/a^0c^0, a^0c^0/a^1c^1, and a^1c^1/a^1c^1 of 31.5, 32.7, and 34.0, respectively. The values of P_{50} did not vary with elevation. V_{O_2max} was tested at a low (340 m) and a high (4300 m) elevation. The values differed among genotypes, and the ranking of the genotypes reversed between the high and low elevations. At the low elevation, a^1c^1/a^1c^1 homozygotes had higher values of V_{O2max} than other genotypes ($P < .001$) did, but at high elevations, the a^0c^0/a^0c^0 homozygotes had the highest V_{O_2max}. Thus, the ranking of genotypes for V_{O_2max} varied with elevation, and high and low elevational sites were essentially fixed for the genotypes that maximized V_{O_2max} in those environments. This program of research, like the studies of sickle cell hemoglobin, demonstrated selection on hemoglobin loci. The balance of selective forces on these loci is intuitive—hemoglobin delivers oxygen to tissues in a scurrying mouse, and mice at high elevations have about half the partial pressure of oxygen as do mice at sea level. High-elevation hypoxia demands that the delivery system be adjusted, and the adjustments are produced by the changes in hemoglobin genotype.

The hemoglobin locus in the house mouse, *Mus musculus*, is associated with both variation in blood chemistry and viability differentials during periods of cold stress. The locus coding for the *b* subunit of hemoglobin segregates two alleles, Hbb^s and Hbb^d. The *ss*, *sd*, and *dd* genotypes have hemoglobin concentrations of 16.2, 16.4, and 17.2 g/100 ml and values P_{50} of 47.0, 46.7, and 45.1, respectively (Berry, Jakobson, and Peters 1987; Newton and Peters 1983). Comparisons among strains of mice polymorphic for the *d* and *s* alleles revealed that the *s* allele generally conveyed a lower oxygen affinity and so a higher P_{50} than the *d* allele did (Newton and Peters 1983). Multiple samples from the mice on Skokholm Island, Wales, revealed a general pattern of change among seasons (Berry et al. 1987), with the heterozygotes increasing during the breeding season but decreasing during the summer months. Calculations based on changes in genotypic frequencies over time yielded estimates for overall selection coefficients of 0.49 and 0.24 for the *ss* and *dd* homozygotes, respectively (Petras and Topping 1983). Selection clearly acts on the hemoglobins in deer mice, as it does in humans (Templeton 1982). Even though the links between genetic variation and fitness may be less intuitive for enzymes, they may be just as strong.

Demographic studies (Sassaman 1978) and physiological studies (Mitton, Carter, and DiGiacomo 1997) suggest that LDH variation has a major influence on physiological variation in the sow bug, *Porcellio scaber*. *P. scaber* has a worldwide distribution, and in all population samples studied in North America, two alleles segregate at the LDH locus, and the same allele predominates in all populations studied. The LDH genotypes appear to be active at different times during the mating season, and consequently their mating system is partially assortative at the LDH locus (Sassaman 1978).

Table 4.3 Relative viabilities at the *ldh* locus in *Porcellio scaber* from six natural populations in California

	Genotype		
Sample	*ff*	*fs*	*ss*
Stanford 1	0.89	1.00	0.37
Stanford 2	0.86	1.00	0.63
Stanford 2	1.48	1.00	1.83
Stanford 3	0.64	1.00	0.63
Carmel Beach	0.52	1.00	0.09
Little Sur River	0.87	1.00	0.30
Average	0.87	1.00	0.64

Note: Data summarized from Sassaman 1978.

In addition, Sassaman detected a superior postpartum viability of LDH heterozygotes (table 4.3). Samples from six populations showed a consistent pattern of strong viability differentials among the LDH genotypes, with the heterozygote usually superior to both homozygotes and the common homozygote usually superior to the rarer homozygote. Viability selection at this locus is intense, for the average viabilities of the two homozygotes are 0.86 and 0.64, relative to 1.00 for the heterozygotes (table 4.3). Rates of respiration, measured as milliliters of CO_2 produced per gram per hour, were measured for the common Ldh homozygote and the heterozygote in population samples from Burlington, North Carolina, and Pacific Grove, California. In the sample from Burlington, rates of respiration for homozygotes and heterozygotes were 9.62 and 5.68, respectively ($P < .001$). In the sample from Pacific Grove, the respiration rates for homozygotes and heterozygotes were 10.58 and 6.84, respectively ($P < .05$). Thus, in both experiments, CO_2 production was approximately 35% lower in heterozygotes than in homozygotes. Data on rates of respiration were consistent with the viability differentials (Sassaman 1978) measured two decades earlier (table 4.3).

The PGM polymorphism in the meadow brown butterfly, *Maniola jurtina*, may influence flight capability (Goulson 1993), just as the Pgi polymorphism in *Colias* butterflies is related to flight capacity. Although kinetic studies have not been performed on the PGM locus, allelic frequencies change with elevation in southeastern England, Italy, and Switzerland, suggesting a regular response to temperature at this locus. The homozygous genotype most common at low elevations has by far the greatest endurance for hovering at 29°C (10.5 seconds versus 2.2 seconds for all other genotypes).

Although most studies of the physiological and demographic consequences of protein polymorphisms have used animals, there is no reason to suspect that these relationships are limited to animals or are more frequent in animals. None of the studies employing plants is as extensive as the studies of LDH in *Fundulus*, Lap in *Mytilus*, or Pgi in *Colias*, but there is sufficient evidence to indicate that variation at 6-phosphogluconate dehydrogenase influences dark respiration rate and other components of fitness in perennial ryegrass, *Lolium perenne* (Rainey, Mitton, and Monson 1987; Rainey et al. 1990). Rates of dark respiration are highly heritable and are inversely correlated with growth rate in annual ryegrass (Day et al. 1985; Wilson 1981; Wilson and Jones

1982). Studies of the regulation of the variation in dark respiration suggest that control is not in mitochondria, but in glycolysis (Day et al. 1985). The 6PGD locus segregates two common alleles, and the K_ms are heterogeneous among the three common genotypes (Rainey-Foreman and Mitton 1995). Rates of dark respiration differ among 6PGD genotypes, with the dark respiration rate of *6PGD-11* exceeding the rate of *6PGD-22* by more than 50% ($P < .01$; Rainey et al. 1990). The Q_{10} of dark respiration, or the increase in dark respiration with a 10°C increase in temperature, is highest in *6PGD-11*, intermediate in *6PGD-12*, and lowest in *6PGD-22*. The different values of Q_{10} may provide an advantage of *6PGD-12* over *6PGD-11* during heat stress (Rainey, Mitton, and Monson 1987). Studies with double-labeled carbon revealed that the flux through glycolysis is heterogeneous among the genotypes (Rainey-Foreman and Mitton 1995). Although 6PGD is not in glycolysis, the reaction catalyzed by 6PGD produces ribulose 5-phosphate, which inhibits phosphoglucose isomerase and consequently reduces flux through glycolysis (Noltman 1972).

A locus with strong overdominance for viability in rapid cycling brassica, *Arabidopsis thalliana*, was revealed in an interval-mapping study using molecular markers (Mitchell-Olds 1995). The locus was mapped to a short segment on chromosome I, but this type of study does not indicate anything about the function or product of the gene. Homozygotes at this locus have viabilities 50% lower than those of heterozygotes.

Summary

Kinetic studies of both hemoglobin and enzyme polymorphisms typically reveal biochemical differences among the gene products of alternative genotypes at a locus. Biochemical differences among genotypes can have measurable effects on the physiology of whole individuals. Although the time at which an egg hatches, the swimming endurance, the ability to adjust cell volume when salinity changes, or the ability to take flight on a chilly morning may at times be trivial variations among individuals, at other times they may play a direct and major role in the determination of fitness. The kinetic properties and physiological effects of LDH in killifish, Lap in blue mussels, and PGI in *Colias* butterflies clearly contribute to fitness and, in combination with environmental variation, contribute to the formation and maintenance of geographic patterns of allelic frequencies.

5

Patterns of Variation Among Loci

We conclude that those characters that exhibit greater variation in local populations show greater divergence of extremes between populations, and that such a relationship is not an artifact. . . . In providing a possible explanation for how within-population variation limits divergence, we postulate that there is an inverse relationship between the variability of a character in a local population and the effect it is likely to have on individual fitness.

A. G. Kluge and W. C. Kerfoot
"The Predictability and Regularity of Character Divergence," (1973)

Population geneticists often assume that proteins constitute a random sample of the genome and that an electrophoretic survey of proteins can be used to estimate the level of genetic variation within a strain, a population, or a species. For some biologists, this assumption implies that genetic variation is either random or homogeneous across loci. However, the consistent patterns in the genetic variation help us understand the forces that act on proteins. This chapter summarizes consistent patterns of genetic variability observed at polymorphic enzyme loci and among groups of enzyme loci.

Unimodal Electrophoretic Profiles

At most of the enzyme polymorphisms segregating more than two alleles, the electrophoretic mobility of the most common allele is intermediate, and the mobilities of rare alleles tend to be extreme. Bulmer (1971) first reported rare alleles to have extreme mobilities, and this observation has been extended in a series of subsequent studies that explored the charge-state model for protein polymorphism (Brown, Marshall, and Weir 1981; Haldorson and King 1976; Richardson, Richardson, and Smouse 1975; Weir, Brown, and Marshall 1976). An extreme example of this relationship between mobility and allelic frequency is seen in the muscle esterases of the sand launce, *Ammodytes dubius* (figure 5.1). Population samples taken in successive years from the Emerald Bank, off Nova Scotia, yielded homogeneous allelic frequencies (Mitton and Odense 1985);

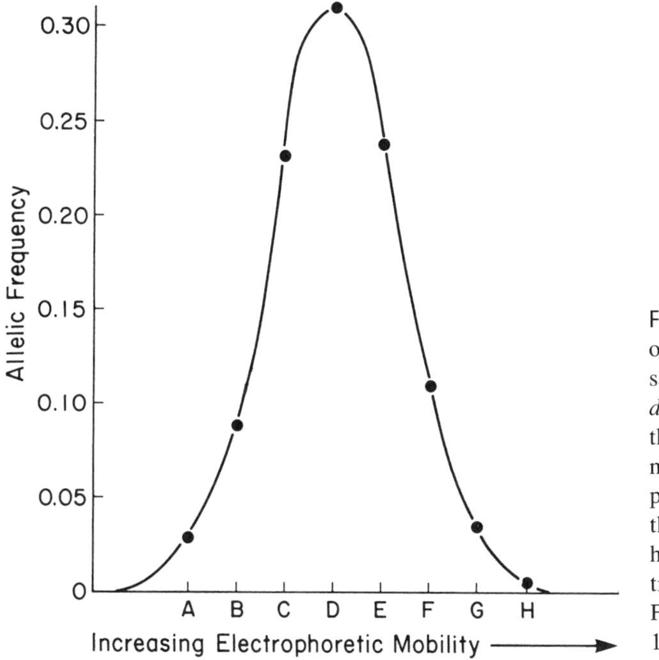

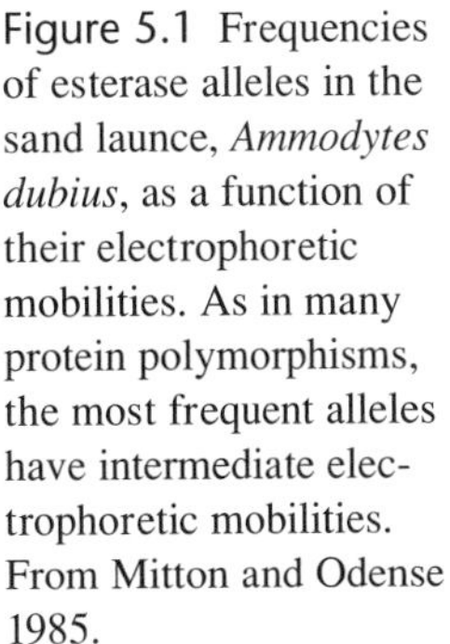
Figure 5.1 Frequencies of esterase alleles in the sand launce, *Ammodytes dubius*, as a function of their electrophoretic mobilities. As in many protein polymorphisms, the most frequent alleles have intermediate electrophoretic mobilities. From Mitton and Odense 1985.

the pooled data are presented in the figure. The profile of the frequencies of these alleles, when ordered by their electrophoretic mobility, is unimodal and symmetric.

A symmetrical, unimodal profile of allelic frequencies could be produced by either natural selection or neutral mutations. In either case, alleles may differ from adjacent alleles by single amino acid replacements. Natural selection could produce this distribution by favoring alleles with a particular set of biochemical characteristics (at the center of the charge distribution) and selecting against, in progressive degrees, those alleles that deviate more and more from the favored set of characteristics. Or all the alleles might be adequately and equivalently functional, but the series of alleles may be growing by successive neutral mutations. Consider a population with a single allele susceptible to adaptively inconsequential unit-charge mutations. The single allele can be transformed into one faster-migrating and one slower-migrating allele. Mutation of these new alleles is a rare but recurrent event, for the alleles themselves are rare, but the products may have electrophoretic mobilities outside the range of existing alleles. This scenario describes the evolution of a symmetric, unimodal electrophoretic profile. Discrimination between these hypotheses is just the sort of problem that will never be settled by statistical analyses of shapes of allelic profiles. Demographic comparisons among genotypes are needed to determine whether alternative genotypes produce any significant consequences. In the example of sand launce esterases (figure 5.1), fish with the most common homozygous genotype were larger than fish with other homozygous genotypes (Mitton and Odense 1985), suggesting that these genotypes are associated with differences in either growth rate or survival.

Classes of Proteins

Enzymes may be somewhat arbitrarily assigned to functional groups with heterogeneous levels of genetic variation. Gillespie and Kojima (1968) first defined and contrasted Group I (metabolic) and Group II (nonmetabolic) enzymes in *Drosophila ananassae* and found higher levels of genetic variation in the Group II enzymes. Similarly, the heterozygosities of Group II enzymes were higher than those of Group I enzymes in 44 species of decapod crustacea (Nelson and Hedgecock 1980). In addition, the patterns of differentiation among decapod species appears to vary between these two groups: cluster analysis based on correlations of heterozygosities among loci perfectly separates Group I and Group II enzymes on a minimum spanning tree. Johnson (1974) contrasted regulatory and nonregulatory enzymes and found higher levels of variation within the regulatory group of enzymes.

Although this pattern of variation appears to be general (Ayala and Powell 1972; Gillespie and Langley 1974; Johnson 1973a, 1976; Kojima, Gillespie, and Tobari 1970), it is not observed in all studies (Selander 1976). There are problems with defining these groups (Selander 1976), for biochemists and enzymologists may not have enough information to place an enzyme definitively into one group (see, for example, Burton and Place 1986). In the future, enzymes may be moved between the regulatory and metabolic groups as more is learned about how, when, and where they act. Nevertheless, the current classification carries some useful information, for genetic variation tends to be lowest in nonregulatory metabolic enzymes, intermediate in nonmetabolic enzymes, and highest in regulatory enzymes.

The explanations for heterogeneous levels of genetic variation among groups of proteins rely on natural selection that acts on the proteins. An enzyme in Group II may act on many substrates, some of them originating in the external environment, whereas an enzyme in Group I usually works on a single metabolic product. Balancing selection may maintain greater genetic variation at loci coding for enzymes that use a greater diversity of substrates. Regulatory enzymes influence the flow of metabolites through pathways, and the control that they impose may be an important aspect of the metabolic phenotype. The regulation of heterozygous genotypes may be superior to the regulation of homozygous genotypes (Berger 1976; Johnson 1974; Koehn and Bayne 1988).

Reliably Polymorphic Loci

Surveys of genetic variation lead us to expect some loci to be polymorphic and others to be monomorphic. This assertion may best be illustrated with the results obtained by O'Brien, Gail, and Levin (1980), who examined the genetic variation of a group of 34 homologous loci in domestic cats, house mice, and humans. They tested the hypothesis that the pattern of polymorphism versus monomorphism is independent across these species, but they found significant excesses of enzymes monomorphic in all three species and polymorphic in all three species. In general, polymorphism in one of the three species predicts polymorphisms in the other species. These species diverged a minimum of 75 million years ago, so it does not seem likely that this concordance in the pattern of polymorphism could be explained by the retention of the pattern carried by the common ancestor. The forces generating and maintaining variation at these loci must be similar in these species.

Quaternary structure

The rate of amino acid replacement varies among proteins by more than a thousandfold and also varies among the different parts of a protein. In general, the rate of amino acid replacement is high in relatively unimportant, exterior, hydrophilic portions of molecules and is low in interior, hydrophobic areas and important areas such as folding and binding sites. Furthermore, the majority of amino acid substitutions are exchanges of amino acids with similar net charges (Clarke 1979). Thus, restrictions on the sorts of mutations are recorded in molecular lineages.

Monomers are proteins that function with a single subunit, or a single polypeptide. Dimers, trimers, and tetramers need two, three, and four subunits, respectively, to form functional proteins. As the number of subunits increases, steric restrictions on the shape of the subunits increase as well—they have to fit together to work together. These observations have led some biologists to propose that proteins with different quaternary structures would exhibit different levels of genetic variation. The steric restriction of oligomers might limit the number of acceptable amino acid replacements, effectively reducing the neutral mutation rate in those proteins. Thus, the level of genetic variation should be highest in monomers, lower in dimers, lower yet in trimers, and so on. This hypothesis has been tested with data from *Drosophila* (Zouros 1975), vertebrates (Ward 1978), and humans (Harris, Hopkinson, and Edwards 1977). In each study, the data were consistent with this expectation. The most recent and most comprehensive analysis (Ward, Skibinski, and Woodwark 1992) showed strong decreases in genetic variation with increasing quaternary structure in both vertebrates and invertebrates. For monomers, dimers, tetramers and hexamers, the average heterozygosities were 0.10, 0.06, 0.04, and 0.03, respectively, in vertebrates and were 0.12, 0.10, 0.06, and 0.02 in invertebrates. These results are certainly consistent with the hypothesis that steric restrictions limit genetic variability and the proteins' rate of evolution.

Subunit size

Larger proteins, all else being equal, commit a lower proportion of their structure to critical areas such as binding sites and thus should be able to accommodate a greater number of mutations that do not disrupt enzymatic function. Large proteins should therefore exhibit greater numbers of alleles and higher levels of heterozygosity than small proteins do. These expectations are met in virtually all the tests that use empirical data (Eanes and Koehn 1978; Koehn and Eanes 1977, 1978; Nei, Fuerst, and Chakraborty 1978; Ward 1978; Ward et al. 1992). For example, among 11 dimeric enzymes in *Drosophila*, approximately 20% of the variance in heterozygosity among loci is explained by subunit size, and if the heterozygosities are averaged across species for each enzyme, the proportion of the variance explained by subunit size jumps to 60% (figure 5.2). In humans, average heterozygosity is not related to subunit size (Harris et al. 1977), but the number of alleles at a locus increases with the protein's subunit size (Eanes and Koehn 1978; Koehn and Eanes 1978), so that 36% of the variance in the number of alleles of monomers and 72% of the variance in the number of alleles of dimers are explained by subunit size.

In combination, quaternary structure and subunit size explain much of the variation among enzymes in heterozygosity and the number of alleles. Genetic variation at a

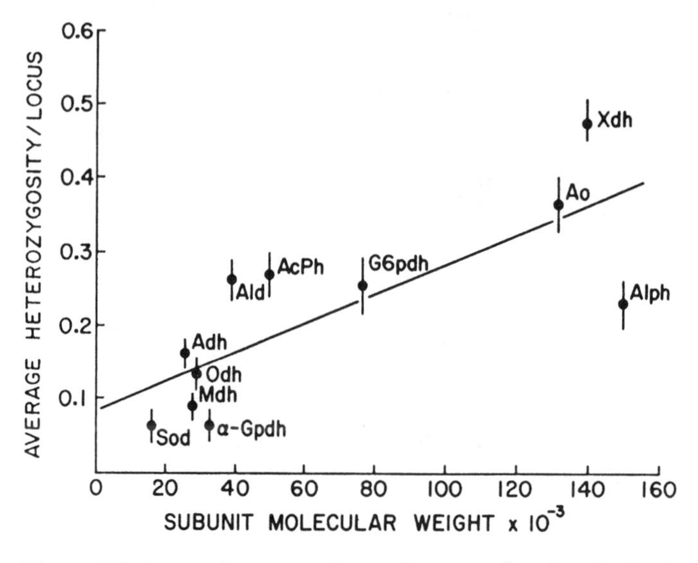

Figure 5.2 Average heterozygosity per locus as a function of subunit size for 11 enzymes in 31 species of *Drosophila*. From Koehn and Eanes 1978.

locus decreases with increasing quaternary structure and decreasing subunit size. These major patterns are consistent with predictions based on the assumption of neutral mutation (Nei, Fuerst, and Chakraborty 1978).

The Kluge–Kerfoot Phenomenon

A strong and general relationship links traditional measures of variation within populations with the degree of divergence among populations. This relationship was first established for morphological measures taken from seven species of vertebrates (two fishes, one snake, three lizards, and one bird) (Kluge and Kerfoot 1973). The degree of differentiation of characters among populations rises with the average level of variation within populations. An example of the Kluge–Kerfoot phenomenon is presented in figure 5.3, which shows the pattern of morphological variation within and among populations of *Fundulus heteroclitus* (Pierce and Mitton 1979). The characters used in this example are meristic characters such as the number of fin rays in the dorsal fin and the number of scales in the lateral line. Very clearly, those characters that are more variable within populations are more differentiated among populations.

This phenomenon has remarkable predictive power. Imagine discovering the first population of an undescribed species. By measuring morphological variability within that population, you could predict which characters would vary most throughout the undescribed range of your new species. And although it was first described for morphological variation, it should be applicable to all forms of variation, such as physiological and life-history variation.

Morphological characters are influenced by both genetic and environmental variation, and hypotheses relying on environmental variation, adaptive genetic variation,

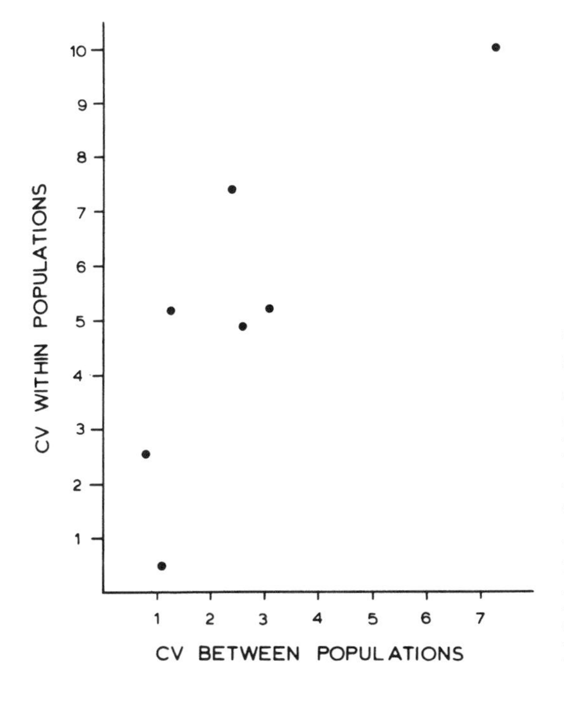

Figure 5.3 Coefficients of variation of seven meristic characters within and among four populations of *Fundulus heteroclitus*. Population samples from Northport, NY; Mystic Isle, NJ; Ladies Island, SC; and St Augustine, FL, were measured for lateral line scales, scales above the lateral line, circumpeduncular scales, and fin rays of the dorsal, caudal, anal, and pectoral fins. The Kendall rank-order correlation coefficient is .52. From Pierce and Mitton 1979.

and genetic drift all were consistent with the Kluge–Kerfoot phenomenon. Variation within populations, however, also is positively correlated with divergence among populations for strictly genetic characters. Among blood group loci and allozyme loci, divergence among populations increases with variability within populations in three species of *Drosophila*, a killifish, a lizard, a pocket gopher, and humans (Pierce and Mitton 1979). Pierce and Mitton's study demonstrated that the Kluge–Kerfoot phenomenon is not dependent on environmental control of development. An example of the Kluge–Kerfoot phenomenon for strictly genetic characters is illustrated in figure 5.4, which shows the increase in genetic distance with the heterozygosity of 12 protein polymorphisms in *Fundulus heteroclitus* sampled at 11 localities in Long Island Sound (Pierce and Mitton 1979).

The Kluge–Kerfoot phenomenon is apparent in other studies as well, even though it may not have been identified by this specific appellation. For example, in analyses of the influence of molecular weight on the genetic variation of enzymes, Koehn and Eanes (1978) found relationships among molecular weights of proteins, their average levels of heterozygosity, and their rates of divergence. The genetic variation at a locus increased with the molecular weight of the protein that it produced. Within a *Drosophila* species, the larger, more heterozygous loci exhibited the greatest extent of differentiation among populations.

The Kluge–Kerfoot phenomenon also was apparent in genetic comparisons of pheromone strains of the European corn borer, *Ostrinia nubilalis*. The two forms of this species, identified by different sex pheromones, were compared at 30 enzymatic loci (Ciandri, Maini, and Bullini 1980). Fourteen loci produced variable substrate and regulatory enzymes, and 16 coded for nonregulatory enzymes. The variable substrate and

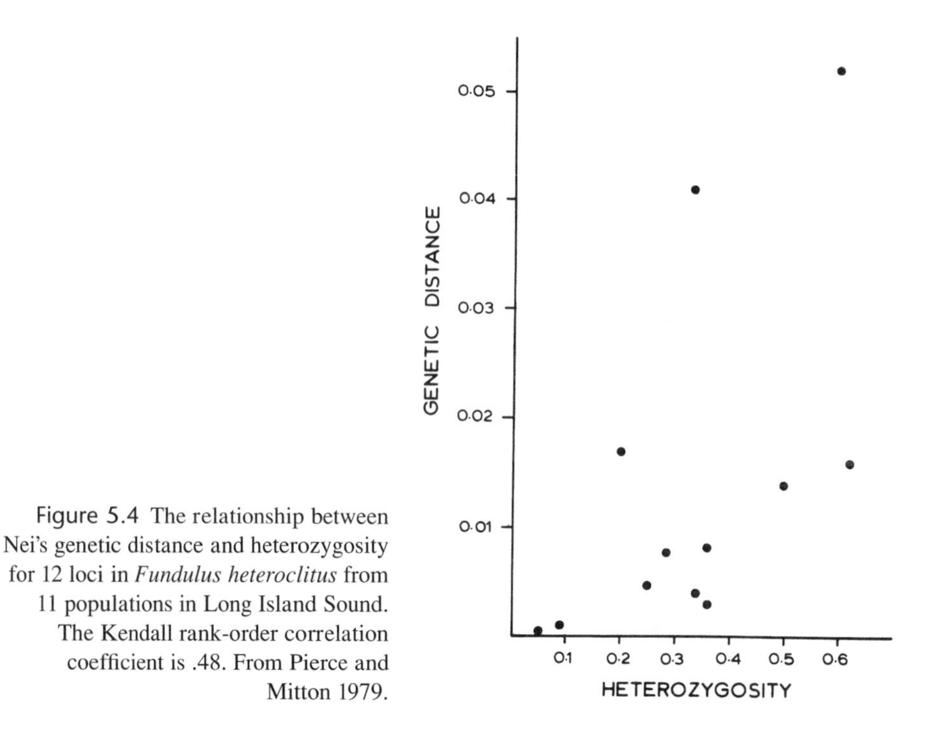

Figure 5.4 The relationship between Nei's genetic distance and heterozygosity for 12 loci in *Fundulus heteroclitus* from 11 populations in Long Island Sound. The Kendall rank-order correlation coefficient is .48. From Pierce and Mitton 1979.

regulatory group was more variable than the nonregulatory group, and the more variable group was also differentiated to a greater degree between strains (a genetic distance of 0.056 for the variable substrate and regulatory group and 0.015 for the nonregulatory group).

Does the Kluge–Kerfoot phenomenon describe an evolutionary relationship between variation at a locus and its degree of differentiation in space and time, or is this phenomenon simply a statistical artifact resulting from relationships among allelic frequencies, variances of allelic frequencies, heterozygosity, and genetic identities? The phenomenon is at least partly dependent on traditional measures of within- and among-population variation (Riska 1979; Rohlf, Gilmartin, and Hart 1983; Sokal 1976), for these measures are not independent. For single loci, this dependence is produced by two relationships. First, the standard error of an allelic frequency, p, is

$$SE = \sqrt{p(1 - p)/2N}$$

where N is the sample size of individuals. The standard error of intermediate allelic frequencies exceeds that of extreme allelic frequencies, and so we expect genetic drift to increase with the heterozygosity of a locus. Second, there is a necessary relationship (Skibinski and Ward 1981) between the heterozygosity at a locus and the values of genetic identity between pairs of loci (figure 5.5). Genetic identities of 0 and 1 can result from a locus's being monomorphic in the two populations, but intermediate identities necessarily reflect intermediate or high average heterozygosities at that locus in the two populations. Genetic identities of zero are less common than values between 1 and 0.5, and for this reason, empirical data often contain a correlation between heterozygosity and genetic identity.

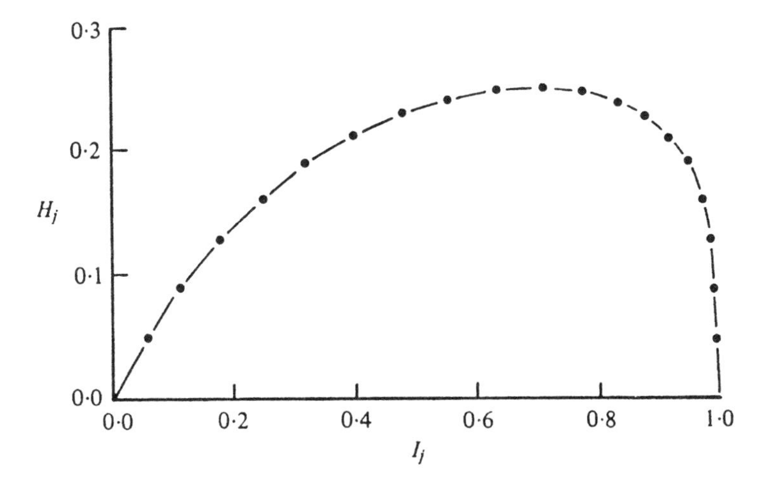

Figure 5.5 The relationship between heterozygosity (H_j) and genetic identity (I_j) for a pair of populations diverging in time. Initially, both populations are fixed for the *A* allele, and thus both *H* and *I* are zero. Then one population remains fixed for the *A* allele, and the frequency of the *a* allele increases to fixation in the other population. From Skibinski and Ward 1981.

The possible causes of the Kluge–Kerfoot phenomenon have been considered in several analyses of geographic variation. To minimize the dependency of genetic drift on allelic frequency, Pierce and Mitton (1979) measured the differentiation of populations with the standardized variance of Cavalli–Sforza (1969), but the Kluge–Kerfoot phenomenon persisted undiminished in all but one of their six data sets. They concluded that, by itself, genetic drift was insufficient to explain the phenomenon. A more complex analysis was conducted by Skibinski and Ward (1981) to examine the second nonbiological mechanism for generating the Kluge–Kerfoot phenomenon. They organized data from *Drosophila* and vertebrate allozyme surveys into three-way contingency tables, looking at the overall genetic identity between populations, the per locus identities, and the per locus heterozygosities. By comparing groups of populations with different degrees of overall genetic identity and then examining the distributions of per locus identity and heterozygosity in those groups, they concluded that the more variable loci diverged more rapidly.

Even after reproductive isolation evolves, the variation at a locus in a population predicts the relative magnitude of divergence at that locus among species. This was first demonstrated in a comparative study of morphological and genetic variation in five species of *Menidia* (Johnson and Mickevich 1977). Twenty-four polymorphic enzyme loci were used to estimate genetic variation within and among the five species. Variation within a population was described with the information statistic

$$H_j = -\sum_{i=1}^{n} p_i \log p_i,$$

where p_i is the frequency of the *i*th allele at the *j*th locus, and the total amount of evolution separating species was estimated with the patristic distance measure from the

Wagner tree describing the phylogenetic relationships among the five species of *Menidia* (Mickevich and Johnson 1976). The correlation between variability within populations and distance among species was $r = .77$, similar to the value obtained with morphological characters.

Further observations of the relationship between the heterozygosity of a locus and its rate of evolution were obtained with multiple regression analyses of data from allozyme surveys of amphibians, reptiles, birds, and mammals (Skibinski and Ward 1982). For neutral genetic variation, the genetic distance between populations or species is dependent on both the heterozygosity of the loci and the time since the groups diverged. In accordance with neutral theory, we expect D to be a function of u, the neutral mutation rate, and t, the time elapsed since the species became reproductively isolated; $D = 2ut$ in the infinite allele model (Nei 1972). By substituting for u, we can show that $D^* = H_j t / 2 N_e (1 - H_j)$ (Skibinski and Ward 1982). The divergence among species or populations is thus expected to be a function of both the heterozygosity of a locus and the length of time that the populations have been separated.

The compilation of surveys examined by Skibinski and Ward (1982) contained data from $j = 31$ polymorphic loci. For each pair of populations for which comparable data were available, H_j and I_j, the heterozygosity and genetic identity for locus j, were calculated. The mean values, $\overline{H_j}$ and $\overline{I_j}$, were calculated from all the available estimates of H_j and I_j. The average genetic distance, $\overline{D_j}$, was calculated as $\overline{D_j} = -\log e\overline{I_j}$ for each locus. The correlation of $\overline{D_j}$ with $\overline{H_j}$ was $r = .759$ ($P < .001$), indicating that the degree of genetic distance between species increased with the average heterozygosity of that locus. Not all loci were available for all combinations of species, however, and there was some concern that heterozygosities or distances would not be homogeneous among all 31 loci. To rule out this potential bias, a standardized genetic distance, D^*_j, was calculated, taking into account the average values of D in all the studies and the values of D in the comparisons used to calculate $\overline{D_j}$. The correlation between D^*_j and $\overline{H_j}$ (figure 5.6) was $r = .828$ ($P <.001$).

The fact that divergence at a locus between populations and between species is related to the heterozygosity at that locus in contemporary populations suggests that the relative levels of variation at individual loci must be stable over evolutionary spans of time.

One of the implications of the association between the heterozygosity of a locus and its evolutionary rate is that species maintaining high levels of genetic variation would be expected to exhibit high rates of molecular evolution. This expectation comes simply from the statistical summation of many highly heterozygous loci, each with its own relatively high evolutionary rate. Skibinski and Ward (1982) reported a clear relationship between H_j and D_j, but their loci ranged in heterozygosity over an order of magnitude. It is doubtful that a comparable range of average heterozygosity would be found among the species in a group of vertebrates such as mammals or birds, and thus the relationship between heterozygosity and genetic distance might be less apparent among species than it is among loci.

Avise and Aquadro (1982) surveyed the literature to summarize levels of allozyme heterozygosity and genetic distances in a genus for fishes, amphibians, reptiles, birds, and mammals. They tabulated values of genetic distance in a genus as simple means and a mean weighted by the number of comparisons per genus (table 5.1). They also summarized mean allozyme heterozygosity for each of the five groups of vertebrates

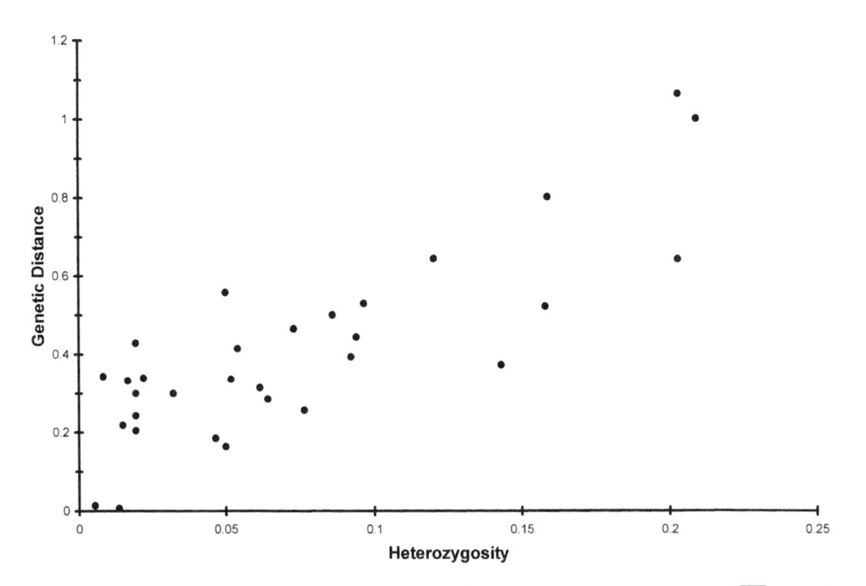

Figure 5.6 Genetic distance (D^*) plotted against average heterozygosity ($\overline{H_j}$) for 31 enzyme loci. Data are from allozyme surveys of fishes, amphibians, reptiles, birds, and mammals. Loci maintaining high levels of variation within populations tend to exhibit greater degrees of genetic distance among species. Modified from Skibinski and Ward 1982.

and, for comparison, listed the corresponding estimates from surveys by Powell (1975) and Nevo (1978). Amphibians are by far the most variable group, and they have the greatest genetic distances among species in a genus. Both birds and mammals have relatively low levels of genetic variation and relatively low levels of genetic differentiation among species in a genus. Correlations between the two estimates of genetic dis-

Table 5.1 Genetic distances within genera and estimates of allozyme heterozygosity in five groups of vertebrates

	Mean *D* per Genus		Average Heterozygosity		
Group	Unweighted	Weighted[a]	H^a	H^b	H^c
Fishes	0.36	0.60	0.058	0.051	0.054
Amphibians	1.12	1.75	0.105	0.078	0.058
Reptiles	0.51	0.67	0.043	0.047	0.041
Birds	0.10	0.08	0.031	0.047	0.037
Mammals	0.30	0.41	0.039	0.036	0.039

Note: Data from Avise and Aquadro 1982. [a]Weighted by the number of pairwise comparisons per genus; H_a from Powell 1975; H_b from Nevo 1978; H_c from Avise and Aquadro 1982. The Pearson product moment correlation coefficient between H_c and unweighted D is nonsignificant, but H_c and both measures of D are significantly correlated by both the Kendall (r = .80) and the Spearman (r = .90) rank-order correlation coefficients. Product moment correlation coefficients between estimates of heterozygosity and unweighed D fall between .76 and .94, and two of the three correlation coefficients are statistically significant.

tance and the three estimates of genetic variation fall in the range of $r = .76$ to $r = .94$, and most are statistically significant.

A more direct test for an association between heterozygosity and genetic distance would be a test within, rather than among, groups. Does genetic distance between species increase with genetic distance in groups of fishes, amphibians, reptiles, birds, and mammals? One problem with this analysis is the confounding effects of time and heterozygosity on genetic distance. If the time of divergence among species could be statistically removed, a relationship between the average heterozygosity of a small group of species and the genetic distance among those species might become apparent. But the lack of sufficient data on the ages of genera prevents this correction. However, the influence of time can be minimized by (1) conducting the analyses within groups of vertebrates and (2) using only the measures of genetic distance among the species in a genus. If the genera in a group of vertebrates differ substantially in age, the problem has not been resolved.

This test arranged data from the literature (appendix 1) in the following way: It used published studies of vertebrate groups that reported both average heterozygosities of species and genetic distances among species in a genus. For example, the mean average heterozygosities and genetic distances among nine species of miniature salamanders of the genus *Thorius* (Hanken 1983) constitute one data point. Within a group of vertebrates such as reptiles and amphibians, a nonparametric correlation coefficient tested the null hypothesis of no relationship between genetic diversity and genetic divergence. The Spearman rank-correlation coefficient was positive for birds ($r_S = .28$, $N = 18$), reptiles and amphibians ($r_S = .22$, $N = 25$), and fishes ($r_S = .15$, $N = 12$) and was negative in mammals ($r_S = -.25$, $N = 19$), but none of these relationships was statistically significant. Furthermore, although the relationship in the data set combining all the vertebrates was positive ($r_S = .15$, $N = 74$), it did not reach statistical significance.

The positive association between genetic variation within a population and differentiation among populations or species is most apparent at the level of the individual locus (Skibinski and Ward 1982) and can be detected in contrasts of the major vertebrate groups (Avise and Aquadro 1982) but is not statistically significant among genera in the major groups of vertebrates. The difficulty at this last level is the restricted variability in both the genetic variation and differentiation within genera. Appendix 1, which contains the data and the citations for this analysis, shows the variation among vertebrate genera in both heterozygosity and genetic distance in a genus. The heterozygosities vary from 0.007 in sunfish of the genus *Elassoma* to 0.264 in toads of the genus *Bufo*. Genetic distances vary from zero in sturgeon, genus *Scaphirynchus*, to 1.238 in rats of the genus *Rattus*. Although variation among populations increases with variation within populations at individual loci, this pattern may be obscured in some analyses. Other factors, such as the mating system and the potential for gene flow among populations, certainly influence the differentiation of populations.

Discordant Patterns of Geographic Variation

An unexpected discovery in molecular studies of geographic variation is that different sets of markers indicate dramatically different levels and patterns of geographic variation. Protein polymorphisms and DNA markers reveal discordant patterns of geographic variation in deer mice, cod, horseshoe crabs, oysters, lodgepole pine, the closed-cone pines of California, and limber pine (Mitton 1994).

In the deer mouse, *Peromyscus maniculatus*, allozyme polymorphisms reveal little genetic differentiation, whereas mtDNA reveals five well-differentiated geographic areas (Avise, Lansman, and Shade 1979; Lansman et al. 1983). In populations stretching from central Mexico to central Canada, the value of F_{st} at six allozyme polymorphisms varies from 0.04 to 0.38, suggesting moderate amounts of gene flow among populations. However, mtDNA haplotypes are diagnostic for large geographic areas. For example, the mtDNA haplotypes in the eastern states and northern Michigan differ from the haplotypes in the central states by approximately 20 mutational steps.

Allozyme polymorphisms in the horseshoe crab, *Limulus polyphemus*, exhibit only slight differentiation of populations between Cape Cod and the Florida panhandle. The value of F_{st} at these loci was used to estimate gene flow (Mitton 1994) by the relationship of Wright (1931): $F_{st} = 1 / (1 + 4\,Nm)$. Gene flow was estimated to be $Nm = 5$ individuals exchanged between populations during each generation. This is a relatively high amount of gene flow, much more than the critical value of $Nm \approx 1$ needed to prevent the differentiation of neutral loci by drift. Studies of mtDNA (Avise 1992, 1994; Saunders, Kessler, and Avise 1986), however, showed a striking discontinuity in northeastern Florida, marked by diagnostic markers differing by a 2% sequence divergence. This variation tells us that migration is not moving mtDNA across this recognized biogeographic boundary between warm-temperate and tropical marine faunas. The allozyme data suggest extensive gene flow, and the mtDNA data uncover a barrier to gene flow.

Different degrees of population structure are apparent in the Atlantic cod, *Gadus morhua*, in allozyme polymorphisms (Mork et al. 1985) and nuclear RFLPs (Pogson, Mesa, and Boutilier 1995). Both surveys studied populations throughout the range of cod in the north Atlantic. Only 2 of 10 allozyme loci revealed significant heterogeneity among localities, but 10 of 11 RFLP loci were significantly heterogeneous. Genetic distance increased with geographic distance at both sets of markers, but at a distance of 8000 km, Rogers's genetic distance was 0.03 for allozymes but 0.17 for RFLPs. Estimates of gene flow from values of F_{st} were fivefold higher for proteins than for RFLPs. Highly significant spatial patterns were revealed by RFLPs that were not detected by the allozymes. Pogson and colleagues attributed the discordant patterns of geographic variation to be a consequence of balancing natural selection acting on allozyme polymorphisms.

Patterns of geographic variation and inferences about the magnitude of gene flow differ between protein polymorphisms and DNA markers in the American oyster, *Crassostrea virginica*. Oysters have pelagic larvae and therefore can have high levels of gene flow. Allozyme frequencies in oysters exhibit little variation from Maine to Texas, a geographic pattern initially interpreted as evidence for extensive gene flow among populations (Buroker 1983). A survey of mtDNA variation discovered a strikingly different pattern of geographic variation, however. Eighty-two mtDNA haplotypes in 212 oysters clustered neatly into two highly differentiated geographic groups. From the original allozyme data, *Nm* was estimated to be approximately 6 (Reeb and Avise 1990), but the geographic pattern of mtDNA revealed a barrier to gene flow. Clearly, the estimates of extensive gene flow from allozymes and very limited gene flow from mtDNA cannot both be correct. To determine which set of data yielded the more accurate inference of gene flow, data from four anonymous single-copy nuclear RFLPs were collected (Karl and Avise 1992). These markers described the same pattern of geographic variation as did the mtDNA. Thus the contrasting patterns of geographic variation seen in allozymes and mtDNA cannot be attributed to the different modes of inheritance, for the biparentally inherited nuclear RFLPs followed the same pattern of geographic variation as did the mtDNA. After considering alternative hypotheses for the discordant patterns of geographic variation in oysters, Karl and Avise (1992) tentatively concluded that balancing selection acting on enzyme loci caused this set of loci to diverge from the pattern of geographic variation exhibited by the mtDNA and nuclear RFLPs. This series of empirical studies demonstrates that different compartments of the genome can exhibit strikingly different patterns of geographic variation and suggests that evolutionary forces may affect various compartments of the genome in different ways.

A novel opportunity for comparing patterns of geographic variation is afforded by conifers' nuclear, chloroplast, and mitochondrial genomes. The conifers are pollinated by the wind, and pollen may disperse hundreds of meters to many kilometers, whereas the distance of seed dispersal is generally a distance similar to the height of the maternal parent. CpDNA is inherited paternally and mtDNA is inherited maternally in the species examined next. Because pollen moves farther than seeds do, cpDNA has a much greater potential for dispersal than mtDNA does. The paternal contribution of nuclear, biparental markers is moved by both pollen and seeds, and the maternal contribution is moved only by seeds. Consequently, we expect the relative degrees of differentiation among localities to be least in cpDNA, intermediate in nuclear markers, and greatest in mtDNA.

Provenance studies and allozyme surveys show discordant patterns of variation in conifers. Striking patterns of phenology, often associated with environmental variation, necessitate seed zones for the major timber species, such as ponderosa pine, *Pinus ponderosa*, and Douglas-fir, *Pseudotsuga menziesii* (Namkoong, Kang, and Brouard 1988). Common garden studies of ponderosa pine (Rehfeldt 1993), Douglas-fir (Rehfeldt 1989), lodgepole pine, *Pinus contorta* (Rehfeldt 1988), and western larch, *Larix occidentalis* (Rehfeldt 1982) revealed that survival and growth suffered dramatically if seeds were planted more than 250 to 300 meters higher or lower than the site from which they were collected (Rehfeldt 1993). In contrast, the differentiation of enzymes, at least measured with single loci, is typically slight (El-Kassaby 1991; Guries and Ledig 1982; Hiebert and Hamrick 1983; Loveless and Hamrick 1984; O'Malley, Allendorf, and Blake 1979; Yeh 1988; Yeh and El-Kassaby 1980; Yeh and Layton 1979; Yeh and O'Malley 1980; Yeh et al. 1986). We know that important patterns of phenological variation are genetically determined, but the intensity and patterning of phenological variation do not appear to match the pattern of enzyme variation (but see Millar and Westfall 1992; Westfall and Conkle 1992).

The degrees of differentiation of populations of Bishop pine, *Pinus muricata*, differ among allozymes, cpDNA, and mtDNA. Allozyme surveys of genetic variation in *P. muricata* disclosed the typical apportionment of genetic variation for pines—high amounts of variation within populations and relatively little differentiation, approximately 22%, among populations (Millar et al. 1988). Bishop pine has little cpDNA variation within populations; rather, the majority of variation, 87%, is among populations (Hong, Hipkins, and Strauss 1993). Similarly, 96% of the variation revealed by mtDNA was among populations (Strauss, Hong, and Hipkins 1993). Thus, the degrees of differentiation of populations are heterogeneous among the allozymes, cpDNA, and mtDNA, and the relative degrees of differentiation are not consistent with the potential gene flow of the sets of markers.

The degrees of allozyme, cpDNA, and mtDNA differentiation appear to fit expectations based on gene flow in lodgepole pine (Dong and Wagner 1993, 1994; Wheeler and Guries 1982). Allozyme differentiation in lodgepole pine is typical of conifers; values of F_{st} estimated from allozymes rarely exceeded 6% (Wheeler and Guries 1982) and were generally less than 10% in conifers (Hamrick and Godt 1990). Large proportions of the mtDNA variation within lodgepole pine were among populations within subspecies (F_{st} = 66%) and among subspecies (F_{st} = 31%). CpDNA variation was high within populations, with little differentiation—less than 5%—among populations. Thus, the differentiation of populations was strong for mtDNA, but slight for both allozymes and cpDNA.

Allozymes and DNA markers demonstrate contrasting patterns of genetic variation in limber pine, *Pinus flexilis* (Latta and Mitton 1997). Allozyme variation was used to describe variation within and among populations from the lower tree line (1650 m) to the upper tree line (3350 m) in Colorado (Schuster, Alles, and Mitton 1989). The average for these loci was 0.02, indicating little differentiation among sites and gene flow on the order of ten migrants between populations per generation. But this estimate of gene flow did not seem to be possible, for the pollination phenology of limber pine varied dramatically with elevation. Most sites along the elevational transect that differed by 400 m or more did not have overlapping pollination periods and therefore could not exchange genes during a single generation. A survey of RAPD markers yielded two patterns of variation. Approximately two-thirds of the markers had a pattern similar to that of the allozyme markers, with values of F_{st} very close to zero. But approximately one-third of the markers had values of F_{st} near 0.50, far higher than values for the other RAPD markers and the allozymes.

Summary

An allele's electrophoretic mobility at a protein polymorphism is related to its frequency, with rare alleles tending to migrate either more quickly or more slowly than the common alleles do.

The loci coding for proteins are often assumed to be a random sample of the genome, but it is not clear that this assumption is valid. Certainly, some groups of proteins are more genetically variable than other groups. Regulatory enzymes appear to be more variable than nonregulatory enzymes, and enzymes that work on many substrates appear to be more variable than enzymes that utilize a single substrate. Genetic variation tends to decrease from monomer to dimer to tetramer as the steric restrictions on the molecule increase. The number of alleles at a locus and the heterozygosity at a locus tend to increase with the subunit size of the protein.

For both morphological characters and single-gene polymorphisms, the divergence among populations increases with the average variability within populations. This generalization appears to describe both patterns of geographic variation and divergence among species.

Protein polymorphisms and DNA markers show discordant patterns of geographic variation in several plant and animal species. When estimates of gene flow are estimated from values of F_{st}, allozyme markers usually return higher estimates of gene flow than do DNA markers.

6

The Axis of Individual Heterozygosity: Theory

Whether nature ranks and truncates, or approximates this behavior, is an empirical question, yet to be answered.

J. F. Crow and M. Kimura,
"Efficiency of Truncation Selection" (1979)

For several decades before population geneticists used electrophoresis to estimate the levels of genetic variation in natural populations, empirical geneticists believed that natural populations contained an abundance of genetic variation (Dobzhansky 1970). Nevertheless, the first quantitative estimates from electrophoretic surveys seemed to surprise many geneticists. One of their first reactions to the empirical estimates was that if the abundant genetic variation were maintained by balancing selection, it would generate an intolerable genetic load (Lewontin and Hubby 1966). The new data thus forced a sweeping reanalysis of the population genetic models of fitness determination.

Segregational genetic load is defined as the decrement in average fitness within a population in comparison with a population composed solely of the most fit genotype. When population geneticists first discovered that one-third to one-half the proteins in population samples of *Drosophila* and humans were polymorphic, consideration of genetic load led some biologists to propose that the genetic variation was functionally and adaptively neutral. They proposed neutrality because the estimated level of segregational genetic load was so high that it could be borne only by the most fecund species. Kimura and Crow (1964) presented the problem, but when Lewontin and Hubby (1966) restated it after estimating the level of genetic variation in *Drosophila pseudoobscura*, the message hit home. If a representative protein polymorphism were maintained by balancing selection favoring the heterozygote,

genotype	*AA*	*Aa*	*aa*
relative fitness	$1-s_1$	1	$1-s_2$

then the average fitness would be $(1-s)^L$, where s is the average selection coefficient over all genotypes and L is the number of loci experiencing this form of balancing selection. There are more than 1,000 enzymes in a eukaryote but probably fewer than 2,000. If we estimate that the average selection coefficient is .05 and the number of enzymes is 1,500 and that one-third, or 500, are maintained by this form of balancing selection, the average fitness in the population will be $(.95)^{500} = 7.27 \times 10^{-12}$. This value of fitness assaults the intuition of population biologists and natural historians. Something is clearly wrong here.

This dilemma disappears if we assume that the protein genetic variation is predominantly produced by neutral mutations, so that alleles segregating at a locus code for biochemically and adaptively equivalent proteins. We can make several predictions concerning the rate of evolution and the amount of genetic variation present in populations, based on estimates of the neutral mutation rate, effective population size, and time (Kimura 1983; Nei 1975). Many of these predictions appear to be consistent with patterns in the empirical data (Chakraborty, Fuerst, and Nei 1978; Fuerst, Chakraborty, and Nei 1977). But there are other solutions to the dilemma as well.

Estimates of segregational load depend heavily on the model of fitness determination (Crow and Kimura 1979; Kimura and Crow 1978; King 1967; Milkman 1967, 1978, 1982; Sved, Reed, and Bodmer 1967; Wills 1978, 1981). The use of multiplicative fitness determination, as in this example, leads us to conclude that the number of loci that can be maintained by overdominance or marginal overdominance is severely limited. This line of reasoning led many biologists to decide, on the basis of no direct empirical information, that protein polymorphisms did not play a role in adaptation. But other models of fitness determination return much lower estimates of segregational load. For example, if fitness is determined by truncation selection (Wills 1978), many polymorphisms can be maintained by balancing selection. Wills considered a diploid, panmictic population in which all loci had two alleles and identical equilibrium frequencies. When the population size is 10^5, equilibrium frequencies are 0.50, and selective mortality truncates the 5% of the population that is most homozygous, 10^7 polymorphisms can be maintained in the population. This number of loci is two orders of magnitude greater than the number of loci in the human genome. Thus, segregational load is no longer a sufficient justification for denying an adaptive role for protein polymorphisms or any other sort of polymorphism. Still, we need to determine which models of fitness determination most closely approximate the determination of fitness in natural populations.

Empirical results consistent with the simulation by Wills (1978) were found in studies using balancer chromosomes in population cages of either *Drosophilia melanogaster* or *D. pseudoobscura* (Seager and Ayala 1982; Seager, Ayala, and Marks 1982; Sved 1971; Sved and Ayala 1970; Tracey and Ayala 1974). Tracey and Ayala (1974) tested for balancing selection on both the second and third chromosomes. Wild-type chromosomes were captured in males (male *Drosophilia* have no recombination) from natural populations and kept intact, or identical by descent, with the aid of balancer chromosomes. They selected 23 second chromosomes and 14 third chromosomes from those chromosomes with quasi-normal fitnesses as homozygotes. They then established 92 population cages with equal frequencies of a balancer chromosome and a wild-type chromosome and maintained the population cages until the frequencies came to equilibrium. Because the balancer chromosomes are lethal when homozygous, the trajectories of chromosome frequencies and equilibrium frequencies are dependent on the relative fitnesses of heterozygous and homozygous genotypes for the wild-type chromosomes. In comparison to relative fitnesses of 1.0 in chromosomal heterozygotes, the mean fitnesses of wild-type homozygotes was 0.12 for the second chromosome and 0.13 for the third chromosome. This magnitude of chromosomal overdominance is consistent with the hypothesis that many loci are maintained by heterotic selection in natural populations of *D. melanogaster* (Tracey and Ayala 1974).

Multilocus Models with Multiplicative Interactions

Population genetics theory typically begins with models that treat drift or selection at a single locus. Organisms are not single loci, however, but complex functioning associations of thousands of loci. Moreover, many biologists, including geneticists, have had serious reservations concerning the relevance of single-locus models to the real world. The construction of multiplicative multilocus models correctly revealed the potential for unforeseen complexity but may have misled us with respect to the impact of genetic load and the magnitude of genetic organization in natural populations.

The earliest models of multilocus systems revealed that multilocus systems did indeed contain dynamic interactions among loci, which could not be anticipated with single-locus models (Kojima and Lewontin 1970; Lewontin 1964). For example, models incorporating many overdominant, linked loci showed that linkage disequilibrium, a measure of the correlation between alleles at separate loci, could accumulate along a chromosome. Consequently, variation at a locus might reflect not only selection at that locus but also selection at adjacent loci and, to a smaller degree, more distantly linked loci. The most startling results came from a model of 32 overdominant, linked loci (Franklin and Lewontin 1970), which demonstrated that within a narrow range of recombination among loci, selection would "crystallize" chromosomes into highly organized, complementary units that nearly maximized the amount of genetic variation in the population. The prediction from this model prompted biologists to measure linkage disequilibrium in natural populations.

With specific exceptions, studies of genetic organization in natural populations found that most loci were distributed independently of one another, that linkage disequilibrium was not significantly different from zero (Mukai, Mettler, and Chigusa 1971). In sexually reproducing organisms, two common classes of exception to this generalization have emerged. Inbreeding species of grasses and cereal crops (Allard 1975) exhibit linkage disequilibrium due primarily to the effective restriction of recombination by the mating system. Another case in which linkage disequilibrium is commonly seen is between protein polymorphisms and adjacent inversions in *Drosophila* (reviewed in Hedrick, Jain, and Holden 1978). In contrast to this last observation, studies of syntenic polymorphisms usually reveal little or no disequilibrium among loci (Charlesworth and Charlesworth 1976).

Studies of the dynamics of linked genetic systems found a systematic departure from theoretical expectations for linkage disequilibrium (Clegg, Kidwell, and Horch 1980). In cages of *Drosophila melanogaster* established with known levels of linkage disequilibrium, the associations between loci decayed faster than theoretical expectations, which assumed no selection acting on the system. Effective recombination between a pair of loci can occur only in those individuals heterozygous for both loci. In the population cages, the superior viability and fecundity of highly heterozygous individuals consistently enhanced the level of recombination, hastening the decay of linkage disequilibrium. Linkage will forever be an important component of multilocus models, but the assumption of multilocus equilibrium is a reasonable first approximation for many natural populations.

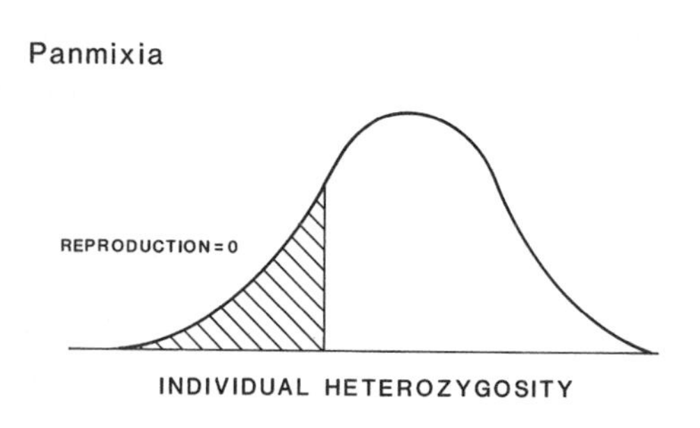

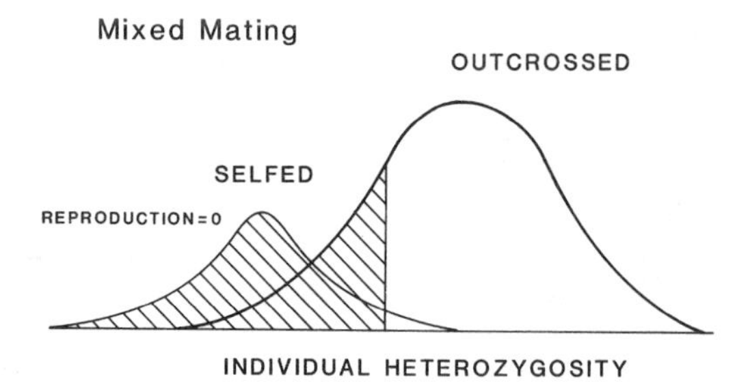

Figure 6.1 Diagrammatic representations of truncation selection of individual heterozygosity in a panmictic population and in a population with a mixed mating system, producing offspring by selfing and by outcrossing. All genotypes falling below some threshold of individual heterozygosity have been removed from the population.

Multilocus Models Incorporating Truncation Selection

A model of fitness determination that appeals to the intuition of naturalists and experimental population biologists is rank-order selection (Wills 1978, 1981) or, more traditionally, truncation selection (Crow and Kimura 1979; Kimura and Crow 1978; Milkman 1978, 1982), as presented diagrammatically in figure 6.1. We can easily envision this form of selection imposed on a domesticated species by a plant or animal breeder. A dairy farmer, for example, may cull all the cows that fall below an arbitrary level of milk production. But it is difficult to imagine that selection in natural populations can precisely separate the fit and the unfit. Modified forms of truncation selection, as in figure 6.2, retain most of the characteristics of pure truncation selection, with only a slight reduction in the efficiency of maintaining genetic variation (Crow and Kimura 1979; Milkman 1978).

Truncation selection is a form of soft selection (Wallace 1975), for fitness is not a constant but is determined by an interaction between the genotype and the environment in which it performs. Milkman (1978) described truncation selection in the following way: Consider first a character, called *fitness potential*, determined solely by the genotype. Rank the individuals in a population so that they form a single axis spanning the lowest to the highest fitness potential. Impose on this ranking a fitness function that does not decrease with fitness potential, as in figure 6.2. The shape of the fitness func-

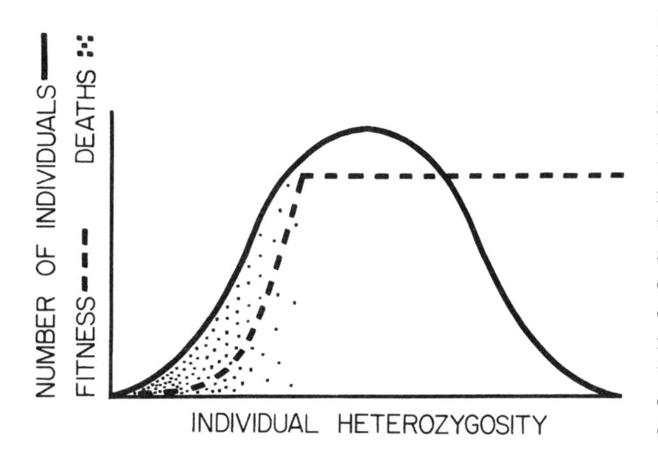

Figure 6.2 A modified model of truncation selection in which the threshold is not a distinct line but a range of individual heterozygosity in which fitness is changing dramatically. In this representation, fitness reaches a plateau and does not change with individual heterozygosity. In other representations, fitness might continue to increase with heterozygosity, but less dramatically than in the area of the threshold.

tion is determined principally by the environment and perhaps also by the size of the population in relation to what the environment can support. Weather conditions are variable, but in some years the weather may be so severe that few individuals survive to reproduce, and in other years it may be so salubrious that the population becomes dense. Species with low fecundity—particularly those that have few offspring in any breeding season—occasionally experience environments that allow virtually all their progeny to survive. Thus, in species with low fecundity, the threshold might be either low or high on the ranking of fitness potentials, depending on environmental conditions and the local population density. The situation is dramatically different in highly fecund species, for the breeding females of blue mussels, American eels, and ponderosa pines produce several million eggs or seeds, and it is inconceivable that all offspring could be accommodated by any environment; the threshold would typically be high on the ranking of fitness potentials in highly fecund species. An excess of young individuals is reduced to the local environment's carrying capacity, by removing individuals from the lower portion of the ranking of the fitness potentials. The intensity of selection is determined by the placement of the threshold delimiting those that do not have the opportunity to reproduce. If the threshold falls at the bottom of the ranking, natural selection is absent, but if the threshold falls near the top of the distribution, natural selection is intense.

It is difficult to measure components of fitness in natural populations, and although it is heuristic to imagine a ranking of fitness potentials, I am not aware that this has been done in either natural or laboratory populations.

Fitness and Heterozygosity

Plant and animal breeders and quantitative geneticists are familiar with the consequences for fitness of gross changes of heterozygosity (Mitton 1993a; Wright 1978). Prolonged inbreeding reduces the amount of genetic variation within populations, and when inbreeding is imposed on species adapted to random outcrossing, severe reductions in fitness, collectively called *inbreeding depression*, are commonly observed. Heterosis is antithetical to inbreeding depression. Crosses between inbred strains typically produce highly heterozygous progeny, and these highly heterozygous individuals exhibit levels of vigor, production, and luxuriance superior to both the parental inbred strains and individuals in large, panmictic populations. This extensive corpus of theory and empirical observations of inbreeding depression and heterosis have had surprisingly little impact on population biology or evolutionary genetics.

The range of individual heterozygosity produced by sexual reproduction in panmictic populations is surprisingly narrow. For example, if a population has 2,000 polymorphic loci, all with a heterozygosity of 50%, 95% of the population will have between 955 and 1,044 heterozygous loci. This range is small in comparison with what professional breeders can produce by modifying the breeding system. For example, one generation of selfing imposed on this population drops the mean individual heterozygosity from 1,000 to 500. Perhaps the narrow range of individual heterozygosity in outcrossed species was the main reason that population biologists have paid little attention to the extensively documented effects of heterozygosity on fitness in domesticated species. But both theory and empirical studies (chapter 7) now indicate that the axis of individual heterozygosity, which underlies the continuum between inbreeding depression and heterosis, has a major impact on the fitnesses of individuals in natural populations.

An incisive observation was reported by Ginzburg (1979, 1983) and Turelli and Ginzburg (1983) on the relationship between fitness and heterozygosity. This development began with the triangle inequality (Ginzburg 1983), which limits the patterns of fitnesses that allow balancing selection to maintain alleles segregating at a locus. The triangle inequality states that the fitness of a heterozygote bearing alleles i and j (W_{ij}) must not exceed the sum of fitnesses of heterozygotes bearing either of these alleles and a third allele, k, so that $W_{ij} < W_{ik} + W_{jk}$. Within the bounds of the triangle equality, computer simulations revealed a general result relating fitness and heterozygosity. For loci whose variation was maintained by balancing selection, average fitness increased with the number of heterozygous loci.

Truncation of the Axis of Individual Heterozygosity

Empirical evidence, presented in the next chapter, suggests that some form of truncation selection acts on the axis of individual heterozygosity. Individual heterozygosity is the number of loci at which an individual is heterozygous (Mitton and Pierce 1980). Consider the example in table 6.1, which shows the distribution of individual heterozygosity in a population of the killifish, *Fundulus heteroclitus*. The genotypes of approximately 1,000 fish were obtained at each of 12 polymorphic protein loci at Flax Pond, a salt marsh on the north shore of Long Island, New York. Thirteen individuals were homozygous at all these loci, and 237 individuals were heterozygous at 3 of 12 enzyme loci. Analyses of these data, summarized in chapter 7, showed that overwintering mortality fell more heavily on the individuals with low heterozygosity, thereby enhancing the heterozygosity of the surviving population.

Linkage Disequilibrium and Zygotic Disequilibrium

Little or no linkage disequilibrium lingers in a large, randomly mating population exposed to truncation selection. This assertion does not deny that linkage disequilibrium

Table 6.1 The distribution of individual heterozygosity in the common killifish, *Fundulus heteroclitus*

	No. of Heterozygous Loci												
	0	1	2	3	4	5	6	7	8	9	10	11	12
No. of individuals	13	62	160	237	235	166	78	35	8	2	1	0	0

Note: mean = 3.68, variance = 2.56. Data are from 12 polymorphic allozyme loci scored from fish collected from five localities in Long Island Sound, NY.

could ever be generated, for selection or drift in finite populations can inadvertently generate linkage disequilibrium. But this form of selection does not necessarily generate disequilibrium, and by favoring highly heterozygous individuals, it breaks down disequilibrium faster than neutral theory would predict. Linkage disequilibrium between two genes is broken down during gametogenesis in individuals heterozygous for both loci. Selection assigning higher viability or fecundity to highly heterozygous genotypes accelerates the decay of linkage disequilibrium resulting from stochastic events. This result has been observed in laboratory populations initiated with known levels of linkage disequilibrium (Clegg et al. 1980).

If natural selection works simply on the axis of individual heterozygosity, counting only the number of heterozygous loci and judging all homozygous genotypes to be of equivalent fitness, then viability selection will build zygotic disequilibrium while it actively erodes linkage disequilibrium. A simple example illustrates this assertion, but first we will compare and contrast linkage disequilibrium and zygotic distribution.

Imagine a pair of loci, each with two alleles with frequencies of 0.5, at both single-locus and dilocus equilibrium (figure 6.3). The gametic frequencies can be estimated from the zygotic matrix summarizing the distribution of dilocus genotypes (figure 6.3). Here the estimate of linkage disequilibrium, D, is calculated as the determinant of the gametic matrix $D = g_{AB}g_{ab} - g_{Ab}g_{aB}$ (Lewontin 1988; Weir 1990). Depending on the form of natural selection, equilibrium may be characterized by $D = 0.0$, or D may be significantly different from zero. Directional selection and stochastic events in finite populations can generate substantial levels of linkage disequilibrium. D will range from -0.25 to 0.25 if all allelic frequencies are equal to 0.5, but when allelic frequencies are not equal to 0.5, D is further limited within this range. To standardize this measure, D can be divided by the maximum value of D for the given allelic frequencies (Spiess 1977), yielding D', which ranges from -1.0 to 1.0.

Zygotic disequilibrium is a measure of the departure of heterozygosity classes from equilibrium expectations. Zygotic disequilibrium is estimated with a procedure similar to the estimation of linkage disequilibrium, differing only in the formation of a reduced matrix from the original matrix of dilocus genotypes. Starting from the dilocus zygotic matrix in figure 6.3, the data are transferred to a 2×2 zygotic matrix by recoding the dilocus genotypes. Genotypes at the first locus are condensed into the rows, with the first row labeled 0 for homozygotes and the second row labeled 1 for heterozygotes. Similarly, the first and second columns are labeled 0 and 1 for homozygosity and heterozygosity at the second locus. All the double homozygotes are placed into the 0,0 cell; the double heterozygotes are placed into the 1,1 cell; and the

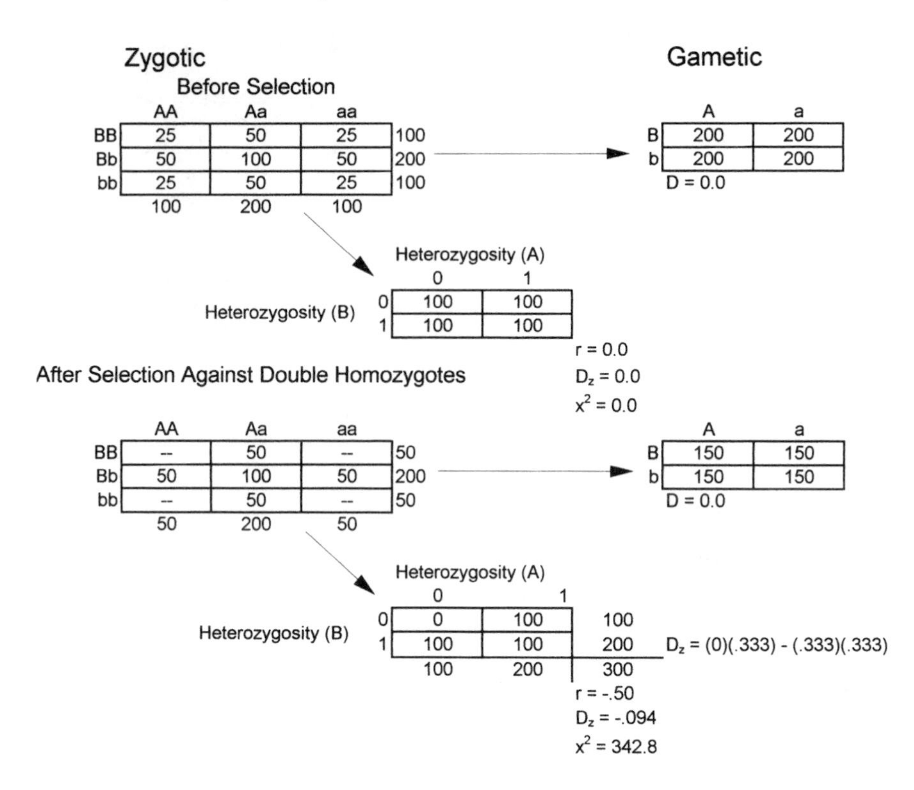

Figure 6.3 Truncation selection of individual heterozygosity may leave gametic disequilibrium undisturbed but generate zygotic disequilibrium. Here gametic disequilibrium is measured with D, and zygotic disequilibrium is measured with the correlation of heterozygosity between loci, r, and D_z.

single heterozygotes are placed into either the 0,1 cell or the 1,0 cell, depending on which locus is heterozygous. D_z is then calculated as $D_z = f_{0,0}f_{1,1} - f_{1,0}f_{0,1}$, and the calculation of D_{zmax} and D'_z are the same as the analogous calculations for linkage disequilibrium. Statistical significance of both D and D_z is tested with a row-by-columns test of independence.

If the distribution of individual heterozygosity is truncated so that all double homozygotes are removed from the zygotic matrix (figure 6.3), linkage disequilibrium will still be zero, even though selective mortality, acting on dilocus genotypes, has removed 25% of the population. However, zygotic disequilibrium, D_z, is changed from its value of zero before selection to $D_z = -0.11$ and $D'_z = 1.0$ ($X^2 = 342.8$, $P < .001$), indicating that the value of zygotic disequilibrium is maximal for the allelic frequencies in the population sample.

A simple demonstration of the accumulation of zygotic disequilibrium with increasing intensities of truncation selection on the axis of individual heterozygosity is shown in table 6.2. Two examples are provided, the first illustrating selection culling only dou-

Table 6.2 The accumulation of negative values of zygotic disequilibrium, D_z and D'_z, when natural selection acts as truncation selection on the axis of individual heterozygosity

Bout of Selection	Proportion Remaining: Double Homozygotes	Proportion Remaining: Single Homozygotes	D_z	D'_z
A. Selection against double homozygotes				
0	1.0	1.0	0.0	0.0
1	0.9	1.0	–.006	.028
2	0.8	1.0	–.013	.062
3	0.7	1.0	–.022	.104
4	0.6	1.0	–.031	.156
5	0.5	1.0	–.041	.222
6	0.4	1.0	–.052	.306
7	0.3	1.0	–.064	.414
8	0.2	1.0	–.078	.555
9	0.1	1.0	–.094	.744
10	0	1.0	–.111	1.0
B. Selection against single and double homozygotes				
0	1.0	1.0	0.0	0.0
1	0.9	0.95	–.0001	.001
2	0.8	0.9	–.001	.003
3	0.7	0.85	–.002	.009
4	0.6	0.8	–.004	.020
5	0.5	0.75	–.007	.040
6	0.4	0.7	–.011	.074
7	0.3	0.65	–.018	.136
8	0.2	0.6	–.028	.250
9	0.1	0.55	–.042	.479
10	0	0.5	–.065	1.0

Note: The zygotic matrix (see figure 6.3) was initialized with two polymorphic loci, each locus segregating two alleles with frequencies of 0.5, with genotypes at multilocus outcrossing equilibrium. Selection was simulated by progressively taking only double homozygotes (A) or by taking double homozygotes and a smaller proportion of single homozygotes (B). In both simulations, double heterozygotes were untouched by selection.

ble homozygotes and the second illustrating culling both double homozygotes and genotypes homozygous for a single locus. In both cases, selection creates negative values of $D_{z'}$, and D'_z rises to a value of 1.0 when the double homozygotes disappear.

A Search for Zygotic Disequilibrium

If natural selection truncates the axis of individual heterozygosity, not distinguishing among the alternative homozygous genotypes at a locus, then we might expect to find dilocus genotypic distributions characterized by negligible linkage disequilibrium and substantial zygotic disequilibrium. This search would be most fruitful in outcrossing, randomly mating species with large populations, to avoid the linkage disequilibrium associated with inbreeding and recovery from population bottlenecks. The data should

Table 6.3 Linkage disequilibrium and zygotic disequilibrium among five polymorphic loci in 332 *Fundulus heteroclitus* from a thermal effluent in Northport, NY

Loci	D	D'	D_z	D'_z
SERE PGM	–.010	.109	–.010	.080
LDH PGM	–.009	.062	.003	.014
MDH PGM	.009	.061	.006	.025
PGI PGM	–.004	.052	.015	.105
SERE PGI	–.016	.139	.005	.030
LDH PGI	.007	.063	–.013	.159
MDH PGI	–.024***	.228	.031**	.232
SERE MDH	–.001	.002	.013	.067
LDH MDH	.006	.096	–.002	.011
SERE LDH	.003	.050	.002	.008

Note: D and D_z measure linkage of disequilibrium and zygotic disequilibrium, respectively, and D' and D'_z are the ratios of those estimates to the maximum values that they can obtain given the distribution of genotypes. ** = P<.01, *** = P<.001.

also be restricted to a sample for which a hypothesis of selection is plausible, such as very old trees or very big fish. I have examined several data sets from killifish, blue mussels, ponderosa pine, and Engelmann spruce but have not found data demonstrating zygotic disequilibrium and the absence of linkage disequilibrium. Data from killifish sampled from a thermal effluent at Northport, Long Island, New York, are representative of the data that I examined (table 6.3). This population sample was restricted to adult fish. Data from both allozyme genotypes and morphology suggest that the population made adaptive responses to the novel thermal environment (Mitton and Koehn 1975, 1976) and that overwintering mortality during the first year favored heterozygous genotypes (Mitton and Koehn 1975). These data do not reveal either linkage disequilibrium or zygotic disequilibrium to be common; each has one statistically significant value. Most of the other data sets examined had no statistical significance. Furthermore, there is no general tendency for D_z to be negative.

I interpret these results to indicate not that selection does not favor predominantly heterozygous genotypes but that the simplistic representation of all genotypes on the axis of individual heterozygosity is not an accurate model of the ranking of fitness potentials in natural populations. In particular, the next section suggests that selection discriminates among different homozygotes, favoring common homozygotes and selecting against rare homozygotes.

The Adaptive Distance Model

A growing number of empirical studies have reported positive correlations between individual heterozygosity and some component of fitness (chapter 8), but it is never clear whether those associations are due to heterozygosity (1) at the enzyme loci, (2) at linked loci in linkage disequilibrium with the enzyme loci, or (3) at loci sprinkled throughout the genome. For a few loci, such as Ldh in *Fundulus heteroclitus*, Lap in *Mytilus edulis*, and PGI in *Colias* butterflies, the accumulated data are so thorough that we can be confident that selection acts differentially on the genotypes at the locus coding for the enzyme. But we have these extensive data for only a handful of cases, and the number of loci examined so thoroughly will probably not grow very much—the data are simply too hard to get. These studies have established that single loci can have a major impact on whole animal (and presumably plant) physiology, but they do not reveal what proportion of polymorphic protein loci is balanced by selection.

The adaptive distance model (Smouse 1986) provides a test to make inferences concerning the presence or absence of selection at polymorphic loci. This test is particularly useful when correlations between heterozygosity and components of fitness cause an investigator to wonder whether selection is acting directly at one locus or a small group of loci. The test determines whether the pattern of fitnesses estimated for the genotypes is consistent with the allelic frequencies.

Three assumptions lead to a clear relationship between allelic frequencies and relative fitnesses. These assumptions are as follows:

1. Selection is acting directly on the polymorphism.
2. There is overdominance or heterozygote superiority at the locus.
3. Allelic frequencies are at equilibrium.

Along with these explicit assumptions comes the implied assumption that fitness variation at the component of fitness used in the study is similar to the variation in total fitness.

Consider a locus with two alleles, A and a, whose frequencies are p and q.

	AA	**Aa**	**aa**
Frequency	p^2	$2pq$	q^2
Fitness	$1-s$	1	$1-t$
Adaptive distance (X)	$1/p$	0	$1/q$

If we consider the classic case of overdominance, in which the fitness of the AA genotype is depressed by s and the fitness of the aa genotype is depressed by t relative to the heterozygote, then the equilibrium allelic frequencies are $p = t/(s+t)$ and $q = s/(s+t)$, and the segregational load is $L = st/(s+t)$.

Under this model of selection, the fitness of the more common homozygote exceeds the fitness of the rarer homozygote. Smouse (1986) defined the adaptive distance X of the heterozygote to be zero and of a homozygous genotype to be the inverse of the frequency of its allele. He went on to define a linear relationship relating the log of fitness to the product of segregational load and adaptive distance:

$$\ln(\mathit{fitness}\ AA) = -LX = -[(st/(s+t)][1/p]$$
$$\ln(\mathit{fitness}\ Aa) = -LX = -[st/(s+t)][0]$$
$$\ln(\mathit{fitness}\ aa) = -LX = -[st/(s+t)][1/q].$$

Furthermore, if fitnesses are assumed to be multiplicative across loci, the fitnesses of multilocus genotypes for loci 1 and 2 can be represented as

$$\ln(\mathit{fitness}) = -L_1X_1 - L_2X_2.$$

The adaptive distance model is consistent with the prediction that fitness generally increases with heterozygosity at selected loci (Ginzburg 1979, 1983; Turelli and Ginzburg 1983). The predictions of the adaptive distance model differ, however, from a ranking of individuals by individual heterozygosity, for homozygotes are not judged to have equal fitnesses. In addition to the highest fitness falling to the multilocus heterozygote, the common multilocus homozygote should have a relatively high fitness, and the rarest multilocus homozygote should have the lowest fitness. If the assumptions of the adaptive distance model are appropriate, the adaptive distance plot reveals points on a line with a negative slope equal to the segregation load.

If a polymorphism is not influenced by natural selection, there should be no significant differences among the observed fitnesses and no relationship between observed fitnesses and adaptive distances. If the data deviate substantially from the model's prediction, then either the estimates of the component of fitness are poorly correlated with total fitnesses, or one or more of the assumptions is inappropriate. Although the marker locus may be neutral, in linkage disequilibrium with a locus directly influenced by selection, linkage disequilibrium appears to be a rare phenomenon in outbreeding species (Charlesworth and Charlesworth 1976; Clegg et al. 1980; Mukai et al. 1971), and there should be no predictable relationship between fitnesses at a selected locus and allelic frequencies at a linked marker locus. If selection were acting at a locus in linkage disequilibrium with the marker locus, variation at the marker locus would not necessarily define a straight line with a negative slope. Therefore, a good fit to the adaptive distance model may be used to make a weak inference concerning the action of selection on the locus. Adaptive distance analyses need to be interpreted with caution, however, for inbreeding and assortative mating can produce patterns that mimic overdominance (Houle 1994).

The adaptive distance plot is illustrated in figure 6.4, which uses data from the PGI polymorphism of the butterfly *Colias philodice* (chapter 4), the Ldh polymorphism of the sow bug *Porcellio scaber* (chapter 4), and the serum esterase locus of the killifish *Fundulus heteroclitus* (chapter 7).

The PGI genotypes of *Colias eriphyle* differ significantly for enzyme kinetics, viability, fecundity, and mating success (chapter 4). The summary in table 4.1 estimates the relative fitnesses of heterozygotes, common homozygotes, and rarer homozygotes to be 1.00, 0.34, and 0.17, respectively. The vertical axis of the adaptive distance plot is the natural logarithm of fitness. To estimate the frequencies of the alleles in natural populations, I combined data from several studies (Watt 1977, 1983) and pooled alleles of similar electrophoretic mobilities to reduce the data to two alleles, as in table 4.1. The frequencies of the synthetic alleles 3 and 4 were estimated to be 0.69 and 0.31, respectively. The inverses of these frequencies are the adaptive distances, which are plotted along the horizontal axis.

The Ldh genotypes of sow bugs differ significantly for their times of activity, resting metabolic rates, and viabilities (chapter 4). From the experiments conducted by Sassaman (table 4.3), relative viabilities of the heterozygotes, common homozygotes, and rare homozygotes were 1.00, 0.87, and 0.64, respectively. If we assume that these populations are at equilibrium, we can use the average selection coefficients in table 4.3 in the equation

$$\hat{p} = t \,/\, (s+t)$$

to estimate that the equilibrium frequency of the fast allele to be 0.73. The average frequency of the *f* allele at the sites that Sassaman sampled was 0.80—a reasonably good fit. The natural logarithms of the relative fitnesses and adaptive distances from the allelic frequencies produced the adaptive distance plot in figure 6.4.

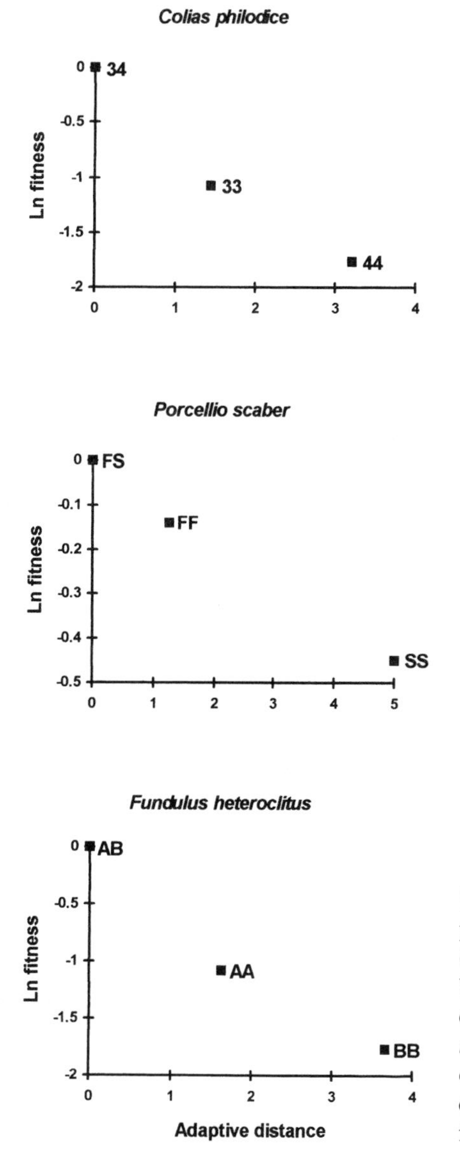

Figure 6.4 Adaptive distance plots for the PGI polymorphism of the butterfly *Colias philodice* (chapter 4), the Ldh polymorphism of the sow bug *Porcellio scaber* (chapter 4), and the serum esterase locus of the killifish *Fundulus heteroclitus* (chapter 7). in each case, the highest estimate of fitness is for the heterozygote, the second highest for the common homozygote, and the lowest for the rare homozygote.

The viabilities of serum EST genotypes in *F. heteroclitus* differ significantly. Comparisons of genotypic frequencies in young and mature fish yield estimates of 1.0, 0.94, and 0.46, respectively, for the relative viabilities of the heterozygote, common homozygote, and rare homozygote. Frequencies of the alleles were estimated from the same population samples (table 7.8). The dilocus fitnesses of EST and PGM and their fit to the adaptive distance model are examined in the next chapter.

The plots in figure 6.4 reveal that fitnesses of homozygotes decline with the frequency of the allele carried by the homozygote: common homozygotes have high fitnesses, and rare homozygotes have low fitnesses. Because the data points in these plots approximate straight lines, the allelic frequencies are close to the equilibria described by the fitness differentials. These plots are consistent with selection at these loci.

The adaptive distance plots for these polymorphisms reveal both the utility and the shortcomings of condensing data into an axis of individual heterozygosity. On the positive side, pooling homozygous genotypes yields a class of individuals with fitnesses lower than those of heterzygotes, and this has been helpful for detecting fitness differentials in multilocus data (chapter 7). These plots also reveal how much information is lost by pooling the common and rare homozygotes. Consequently, the variation in fitness explained by an adaptive distance plot can substantially exceed the variation explained when genetic data are reduced to individual heterozygosity (Bush and Smouse 1992; Bush, Smouse, and Ledig 1987; Mitton 1989, 1993a; Smouse 1986). For example, applying the adaptive distance model to data on heterozygosity and growth in pitch pine (Ledig, Guries, and Bonefield 1983) more than doubled the proportion of variance explained (Bush et al. 1987).

Summary

Evolutionary biologists are still groping with the problem of fitness determination. Genetic variation is abundant in natural populations, but there is no consensus concerning whether fitness variation is determined by a few or many loci.

It was once thought that segregational load would severely limit the number of loci contributing to variation in fitness, but segregational load may have been more of a burden to evolutionary biologists than it has been in natural populations. Although some models of fitness determination generate sufficient genetic load to severely limit the number of loci contributing to variation in fitness, other models, such as truncation selection, allow all the polymorphic loci in a population to be balanced by selection.

Theoretical studies suggest that for loci whose variation is balanced by selection, fitness generally increases with heterozygosity. This expectation is consistent with many empirical studies, summarized in the next chapter. But even though fitness may generally increase with heterozygosity, a ranking of genotypes by their numbers of heterozygous loci lumps common and rare homozygous genotypes into the same class and may not adequately describe fitness differentials in natural populations.

7

The Axis of Individual Heterozygosity: Empirical Data

Indeed the evidence for greater fitness of heterozygotes is stronger than that for most other kinds of natural selection.

J. B. S. Haldane,
The Biochemistry of Genetics (1954)

Ever since biologists began to use protein polymorphisms to study variation in populations, they have accumulated data indicating that highly heterozygous individuals enjoy fitness advantages. At first, biologists viewed these data as curiosities, suggestive of strong selection within a life cycle but inconsistent with the convenient assumption of neutrality. Such observations are numerous, have a broad systematic base (Allendorf and Leary 1986; Mitton 1989, 1993a,b; Mitton and Grant 1984; Zouros and Foltz 1987), and appear to be consistent with predictions from recent models in population genetics (Gillespie 1977; Gillespie and Langley 1974; Ginzburg 1979, 1983; Milkman 1978, 1982; Turelli and Ginzburg 1983; Wills 1978, 1981). A sampling of these data, biased by personal experience, follows.

Heterozygosity and Viability

The early mortality characteristic of highly fecund species is striking. Consider a single female blue mussel, depositing several million eggs in the water. If she succeeds in leaving behind two offspring that live through reproductive age, she will have attained the mean fitness in a population that is neither growing nor shrinking. Much of the mortality in this sort of life cycle must be independent of genotype; currents carry many larvae to areas unsuitable for settlement and growth, and planktivores harvest them with little taste for their genotype. But within the life cycles of highly fecund species, even if 99% of the mortality is independent of genotype, selection has ample opportunity to modify genotypic proportions.

Mortality in the blue mussel, *Mytilus edulis*, modifies the genotypic proportions of the leucine aminopeptidase locus (Koehn, Milkman, and Mitton 1976). The sizes and Lap genotypes of 1,350 mussels were examined at a site near the Bourne Bridge, in the Cape Cod Canal, to determine whether genotype varied with size. Strong tidal currents

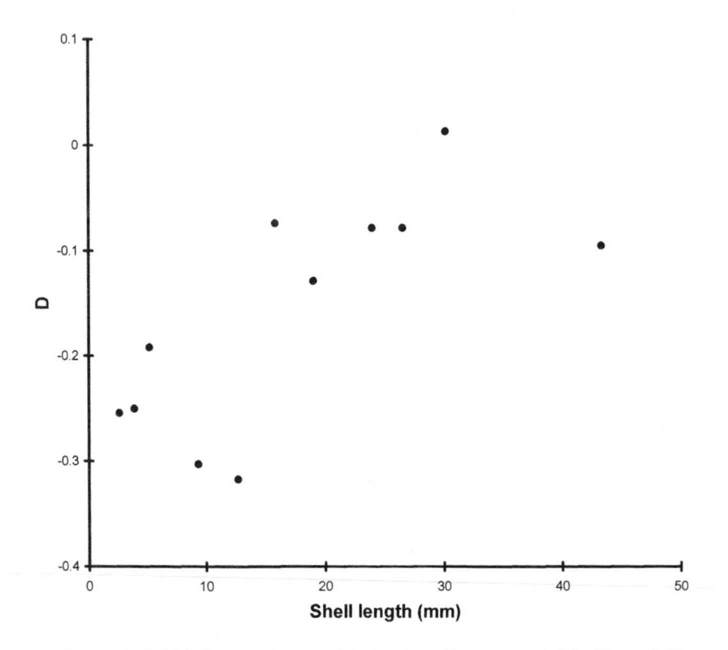

Figure 7.1 Differential mortality in the blue mussel, *Mytilus edulis*, favors heterozygotes at the leucine aminopeptidase locus. This analysis treats 1,350 mussels collected near the Bourne Bridge, in the Cape Cod Canal, grouped by shell length. *D* is a relative measure of heterozygosity, calculated as (observed-expected)/(expected), where observed and expected refer to proportions of heterozygotes. Modified from Koehn, Milkman, and Mitton 1976.

bring a heterogeneous set of larvae to this site each year. Postsettlement selection then modifies allelic frequencies and, concomitantly, increases the relative frequency of heterozygotes (figure 7.1).

Selective mortality favoring heterozygous genotypes may be common in marine pelecypods. The superior viability of heterozygotes has been reported for the ribbed mussel (figure 7.2), *Modiolus demissus* (Koehn, Turano, and Mitton 1973), the American oyster, *Crassostrea virginica* (Singh 1982; Zouros et al. 1983), *Macoma balthica* (Green et al. 1983), *Mytilus californianus* (Tracey, Bellet and Gravem 1975), the Pacific oyster, *Crassostrea gigas* (Fujio, Nakamura, and Sugita 1979), and the palourde, *Ruditapes decussatus* (Borsa, Jousselin, and Delay 1992). Shell size is positively correlated with allozyme heterozygosity in the zebra mussels, *Dreissena polymorpha*, invading the Great Lakes, but we do not yet know whether this correlation is produced by differential growth rate, differential viability, or a combination of the two (Garton and Haag 1991).

Selective mortality favors the highly heterozygous individuals of the killifish, *Fundulus heteroclitus* (Mitton and Koehn 1975). Unlike many other fishes that inhabit the bays and marshes of the western North Atlantic, killifish do not escape from the extreme cold of winter by migrating to deeper water but instead burrow into the mud of marshes, reappearing in the spring. This behavior and the killifish's general sedentary

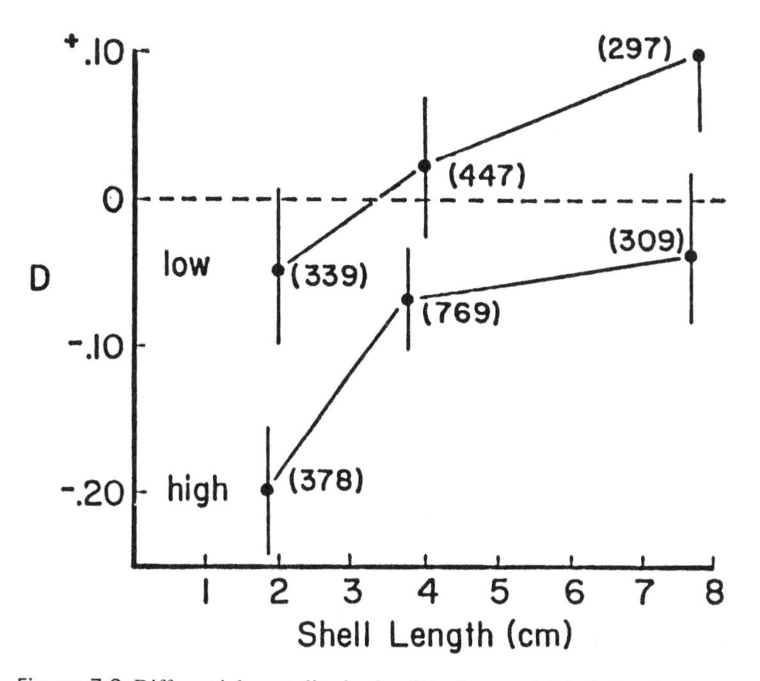

Figure 7.2 Differential mortality in the ribbed mussel, *Modiolus demissus*, favors heterozygotes at the tetrazolium oxidase locus. Population samples were taken low and high in the intertidal zone in the estuary of the Nissequoque River, Long Island, NY. Sample sizes are in parentheses. *D* is a relative measure of heterozygosity, calculated as (observed-expected)/(expected), where observed and expected refer to proportions of heterozygotes. At both positions in the intertidal zone, heterozygotes exhibited superior viability. From Koehn, Turano, and Mitton 1973.

nature (Lotrich 1975) permit studies of single-age classes through time. Twelve polymorphic proteins were used to study two populations, one in a natural environment and the other in an industrially heated environment, the effluent of a power plant that uses water from Long Island Sound to cool condensing tubes. In both these environments, overwintering mortality during the first winter significantly increased individual heterozygosity (figure 7.3).

Differential mortality during acute exposure to anoxia, heat, and cold revealed balancing selection in the topminnow, *Poeciliopsis monacha* (Vrijenhoek, Pfeiler, and Wetherington 1992). *P. monacha* lives in the springs and small streams of northwestern Mexico. These pools and streams are extremely heterogeneous environments, whose temperature and dissolved oxygen dramatically fluctuate both seasonally and daily. Fluctuations in weather cause local extinctions and permit recolonization, but despite the opportunity for genetic drift, allozyme frequencies have been relatively stable for 15 years. Researchers studying the factors that stabilize these polymorphisms exposed fish from natural populations to hypoxia, extreme heat, and unusual cold. Four enzyme polymorphisms revealed differential survival in response to the hypoxic and heat stresses. The responses of genotypes to the heat and hypoxia were very similar

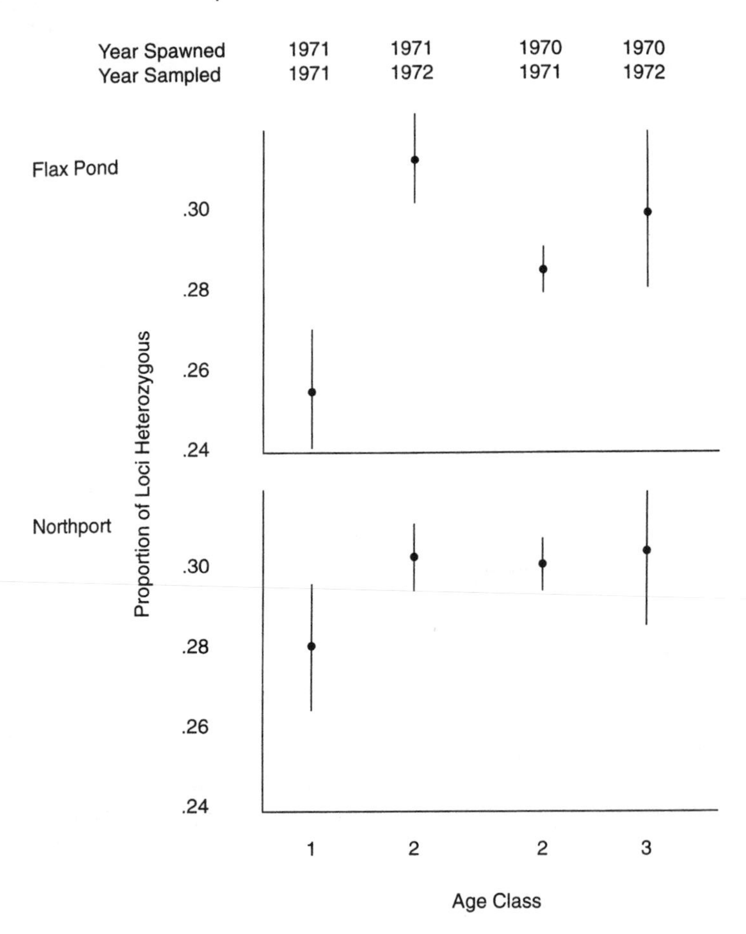

Figure 7.3 Overwintering mortality favors allozyme heterozygosity in the killifish, *Fundulus heteroclitus*. Samples were taken from a natural population at Flax Pond and the thermal effluent of a power plant at Northport, both on Long Island, NY. Regular differences in size revealed the age class of each fish. Each fish was genotyped at 12 polymorphic allozyme loci. At both localities, heterozygosity increased during the first winter but not during the second winter. Modified from Mitton and Koehn 1975.

(Kendall's Tau = 0.73, $P < .001$). This is a satisfying result, for as water temperature increases, the amount of available dissolved oxygen typically declines, so these stresses are usually linked. The responses of genotypes to heat stress and cold stress were negatively correlated (Kendall's Tau = –0.79, $P < .001$). Heat and hypoxic stress generally favored homozygotes for the most common allele, and the cold stress favored homozygotes for a less common, endemic allele. Ck-A, which is in very high concentrations in muscle, was strikingly overdominant during hypoxia. Overall, heterozygotes exhibited less variation in survival than did homozygotes, and this could provide the balancing selection to stabilize allelic frequencies.

Selective mortality favoring heterozygous genotypes can be found in plants as well as animals, and some of the best examples have been reported for species of grasses. Although the fecundity of annual grasses does not rival that of the marine invertebrates just discussed, grasses can produce hundreds or thousands of seeds, and the density of seedlings in early spring is typically far higher than the density of mature plants at the end of summer. For example, the density of seedlings of annual ryegrass may reach 600 to 800 seedlings per square meter in early spring, but the same space may contain only one or a few mature plants in June and July when seeds are being shed.

I used five protein polymorphisms to study the mating system and viability of genotypes of annual ryegrass, *Lolium multiflorum*, along two soil-moisture gradients on the shore of Lake Berryessa in the Coast Range of California (Mitton 1989, 1993a). Panicle samples, or seed from a single maternal parent, were collected in the field and sown in a greenhouse, and the seedlings were examined electrophoretically. The genotypes of more than 6,000 seedlings from 696 open-pollinated families were determined at five polymorphic loci. The distribution of genotypes in a sample of progeny arrays was examined with a maximum likelihood estimation procedure (Brown and Allard 1970; Clegg 1980) to infer the genotypes of the maternal parents and to estimate both the frequencies in the effective pollen pool and the outcrossing rate. If the population is assumed to be at equilibrium, then the genotypes in seedlings and the inferred maternal genotypes can be compared to find changes in genotypic frequencies within the life cycle. Differences in the genotypic frequencies of seedlings and maternal genotypes were used to estimate viability differentials (table 7.1).

Annual ryegrass is partially self-incompatible, and approximately 80% of the seeds at Lake Berryessa were produced by outcrossing, 20% by selfing. Consequently, the distribution of genotypes in the seedlings differed from that expected in a randomly outcrossing population. The values of F in seedlings were typically positive and statistically different from zero (table 7.1), indicating a deficiency of heterozygotes. In contrast, the values of F in parents never differed from zero, indicating a satisfactory fit to Hardy–Weinberg expectations. Thus, selective mortality preferentially culled homozygous genotypes, increasing the frequency of heterozygotes at each of the polymorphic loci. In five of the eight cases, and in each case in which a striking deficiency of heterozygotes was observed in seedlings, the heterozygotes exhibited the highest fitnesses (Mitton 1989).

In each case in which the heterozygote had the highest fitness, the pattern of selection coefficients was consistent with the predictions of the adaptive distance model (Smouse 1986; chapter 6). That is, the fitness of the most common homozygote exceeded that of the rarer homozygote, and the fitness of the homozygote was proportional to the inverse of the frequency of its allele. The estimated fitnesses of the homozygotes listed in table 7.1 declined with the frequencies of their alleles—homozygotes with common alleles have high fitnesses, but homozygotes with rare alleles have low fitnesses (figure 7.4). Although selection against selfed genotypes might produce this general pattern, the elimination of selfed homozygotes with 20% selfing can produce only a small proportion of the fitness differentials observed here. These observations suggest that selection is acting at these loci or at loci in strong linkage disequilibrium with these loci.

Comparisons of genotypic frequencies between seedlings germinated in the greenhouse and mature individuals in the field indicate that viability favors heterozygotes at the Mdh locus in the endangered Hawaiian endemic *Brighamia rockii* (Chrissen Gemmill, pers. comm.). This species is restricted to steep sea cliffs on Kauai and Molokai,

Table 7.1 Genotypic frequencies in seedlings and adults, and estimates of relative fitness for annual ryegrass, *Lolium multiflorum*, from two sites at Lake Berryessa, CA

			Frequencies, Genotype					Relative Fitnesses, Genotype		
	Enzyme	Sample	11	12	22	F	X^2	11	12	22
Site A	PER	seedlings	257	1355	2954	0.088	***	0.59	1	0.87
		adults	19	171	323	−0.027				
	AP	seedlings	4247	411	46	0.137	***	0.64	1	0
		adults	455	68	0	−0.069				
	PGM	seedlings	4425	251	35	0.19	***	0.66	1	0.19
		adults	481	41	1	0.006				
	PGI	seedlings	583	1997	1958	0.031	*	0.94	0.98	1
		adults	62	222	221	0.024				
	GOT	seedlings	47	737	3872	0.026		0.78	0.86	1
		adults	4	72	441	0.024				
Site B	PER	seedlings	278	626	758	0.178	***	0.57	1	0.72
		adults	22	87	76	−0.028				
	AP	seedlings	1529	123	14	0.147	***	0.85	1	0.54
		adults	168	16	1	0.066				
	PGI	seedlings	132	644	890	0.025		0.83	0.86	1
		adults	13	66	106	0.045				

Note: From Mitton 1989. There was not sufficient genetic variation for either GOT or PGM at site B to infer maternal genotypes. F is the inbreeding coefficient, estimated as $F = 1 - O/E$, where O and E are observed and expected proportions of heterozygotes. X^2 tests homogeneity of genotypic frequencies in seedlings and adults with a rows by column test of independence. * = $P < .05$, ** = $P < .01$, and *** = $P < .001$.

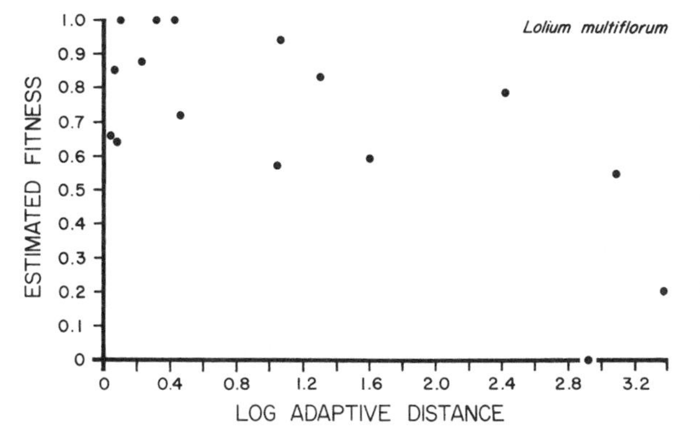

Figure 7.4 A modified adaptive distance plot for the data on annual ryegrass, *Lolium multiflorum* (table 7.1). Only homozygous genotypes are plotted. Fitness is estimated from viability differentials, and adaptive distance is the reciprocal of the frequency of the allele carried by that homozygote. The fitness of the homozygotes decreases with the frequencies of their alleles.

where the number of individuals has declined to approximately 200. Genotypic frequencies are in outbreeding equilibrium in plants germinated in the greenhouse, but mature individuals on the sea cliffs of Molokai have excesses of heterozygotes at the MDH locus. Heterozygosity at this locus is only 15% in the greenhouse, but it is 90% on the sea cliffs.

Mortality associated with moisture stress in the outcrossing annual herb *Echium plantagineum* favors highly heterozygous genotypes (Burdon, Marshall, and Brown 1983). The number of plants in a study plot on a riverbank ranged from a high of 1,491 to a low of 95 near the end of the growing season. Genetic changes were minimal during the early periods of natural thinning, but during the last three censuses, heterozygosity increased, and values of *F* decreased at three of the four polymorphic loci. This selection favored genotypes with two or more heterozygous loci and selected against genotypes with three or four homozygous loci.

Viability differentials favoring heterozygous seedlings have been widely reported in the plant literature. Viability selection favors heterozygous genotypes in slender wild oat, *Avena barbata* (Clegg and Allard 1973), barley, *Hordeum vulgare* (Clegg, Allard, and Kahler 1972; Clegg, Kahler, and Allard 1978; table 3.2), soft bromegrass, *Bromus mollis* (Brown, Marshall, and Albrecht 1974), and ponderosa pine, *Pinus ponderosa* (Farris and Mitton 1984). The frequency of heterozygous individuals increases with age in the perennial herb *Liatris cylindracea* (Schaal and Levin 1976).

When randomly sampled, the genotypic proportions of coniferous forest trees generally fit Hardy–Weinberg expectations. But when only mature, only the largest, or only the oldest trees are sampled, analyses often reveal excesses of heterozygotes (summarized in Mitton and Jeffers 1989). Significant excesses of heterozygotes have been reported in mature stands of ponderosa pine, *Pinus ponderosa* (Linhart et al. 1981), jack pine, *Pinus banksiana* (Cheliak et al. 1985), black spruce, *Picea mariana* (Boyle and Morgenstern 1986; Yeh et al. 1986), Monterey pine, *Pinus radiata* (Plessas and Strauss 1986), Douglas-fir, *Pseudotsuga menziesii* (Shaw and Allard 1982) balsam fir, *Abies balsamea* (Neale and Adams 1985), Polish larch, *Larix decidua* (Lewandowski, Burczyk, and Meinartowicz 1991), *Pinus leucodermis* (Morgante et al. 1993), quaking aspen, *Populus tremuloides* (Cheliak and Dancik 1982; Jelinski and Cheliak 1992; Mitton and Grant 1995), and the stone pines *Pinus sibirica* and *P. cembra* (Politov and Krutovskii 1994). Whereas the decline in values of *F* from initial positive values to zero is consistent with selection against inbred individuals, the production of excesses of heterozygotes must be a different process in which heterozygotes are favored (Mitton and Jeffers 1989; Shaw and Allard 1982). If selection acted only to eliminate selfed genotypes, then all outcrossed genotypes would have equal fitnesses, and F would come to rest at zero. But when selection produces excesses of heterozygotes, then fitness must increase with heterozygosity in outcrossed genotypes.

It appears that foresters use the advantageous traits associated with highly heterozygous genotypes when they select trees for breeding. When foresters established orchards of Engelmann spruce, *Picea engelmannii* (Mitton and Jeffers 1989), Norway spruce, *Picea abies*, (Bergmann and Ruetz 1991), Sitka spruce, *Picea sitchensis* (Chaisurisri and El-Kassaby 1994), and Douglas-fir, *Pseudotsuga menziesii* (El-Kassaby and Ritland 1996), they selected trees according to the normal criteria used by foresters, but they unknowingly chose trees with higher individual heterozygosities than would be found in a random sample of trees.

A survey of studies of plant population biology showed that fitness differentials in natural populations of plants generally favor enzyme heterozygotes (Lesica and Allendorf 1992). In a summary of eight population studies of plants in which selection coefficients were calculated for allozyme loci, 12 of 38 estimates of selection coefficients were significantly different from zero, and all revealed higher fitnesses in heterozygotes. When the relative fitness of heterozygotes was set at 1.0, the average fitness of homozygotes at the 38 loci was 0.77, or conversely, the average selection coefficient for homozygotes was $s = .23$. Although most population geneticists may judge this intensity of selection as high, it is in the middle of the range for studies of natural populations (figure 1.1). This result certainly has implications for evolutionary and conservation biology, and the physiological mechanisms underlying this generality are receiving more attention (Clark and Koehn 1992; Feder and Watt 1992; Mitton 1989, 1993b, 1995b).

Heterozygosity and Developmental Stability

Developmental homeostasis is not a constant—or a fixed attribute of a population or strain—but a variable that reflects both the genotype of the individual and the environment in which that genotype develops. Developmental homeostasis declines with virtually any environmental change that saps energy and thus imposes a stress on development. Insufficient food, minerals, or vitamins, high or low temperature, and disease all can challenge a developing embryo and disturb normal development (Soulé 1982 and references therein). These stresses may disrupt cellular structures, such as the endoplasmic reticulum, but in this book, the currency of stress is considered to be the net energy balance with which an individual must work (Mitton and Koehn 1985). The organism's capability of meeting the energetic demands of development and environmental stress is related to its energy reserves or net energy balance. Developing individuals with critically low energy reserves are vulnerable to any stresses that require yet more energy to maintain homeostasis. If the available energy is not sufficient to buffer against the insult, the event will be recorded as the development of an unusual character or an asymmetry.

Highly heterozygous individuals or races exhibit greater buffering of developmental processes and hence lower morphological variability (summarized in Lerner 1954).

Table 7.2 The squared coefficient of variation of wing length in *Drosophila melanogaster* as a function of chromosomal heterozygosity

	No. of Heterozygous Chromosomes			
Sex	0	1	2	3
Female	2.54	2.23	1.44	1.20
Male	1.64	1.31	1.19	

Data: from Robertson and Reeve 1952.

This generalization was made on the basis of work with inbred strains and the highly heterozygous progeny of crosses between inbred strains. Thus the difference between the "homozygous" and "heterozygous" individuals was relatively great, leading biologists to suspect that developmental stability was a function of the entire genome's heterozygosity. But the same pattern of developmental stability was observed in *Drosophila* when the categories of "heterozygous" and "homozygous" were based on a single chromosome (Dobzhansky and Spassky 1963; Dobzhansky and Wallace 1953; Reeve 1960; Robertson and Reeve 1952). For example, when the variability of wing shape in *D. melanogaster* was tabulated for individuals heterozygous for zero, one, two, or three chromosomes, morphological variability clearly declined with chromosomal heterozygosity (table 7.2). More recently, studies distinguishing groups on the basis of heterozygosity at a single protein polymorphism disclosed the same patterns of differences in developmental stability.

Protein heterozygosity was first related to developmental stability in the killifish, *F. heteroclitus* (Mitton 1978). Variation was examined at five polymorphic enzymes, and developmental stability was estimated with the variance at seven meristic characters. For each locus, the morphological variation of heterozygotes and homozygotes was compared. The variation in the two groups was then compared with the average coefficient of variation ($\overline{CV}$) and the generalized variance (Σ, estimated with the determinant of the variance–covariance matrix). Σ has the advantage of using both the variances of the characters and the covariances among them, and it is proportional to the volume occupied by a set of data points in morphological hyperspace (Sokal 1965). Males and females were treated separately, and the study was replicated with samples from three localities. In most of the comparisons, the heterozygotes exhibited lower levels of phenotypic variation than did the homozygotes. This pattern was most evident at the *LDH-B* locus (table 7.3), which has a major impact on the physiology of killifish (chapter 4). When morphological variance was estimated with Σ, the heterozygotes consistently had lower levels of morphological variation than did the homozygotes. For example, for males at Flax Pond, the value of Σ in heterozygotes was only 58% (e.g., 0.032/0.055) of the value for homozygotes. For the six samples in table 7.3, the average value of Σ in the heterozygotes was 43% of the value in the homozygotes. Thus, when these populations were divided into two groups according to their genotypes at a single locus, the heterozygous group had consistently lower phenotypic variation.

The analysis of morphological variation as a function of LDH genotypes (Mitton 1978) was analyzed further with additional samples (Mitton 1993b). A sample of 219

Table 7.3 Estimates of morphological variance from seven meristic characters in the killifish, *Fundulus heteroclitus*

Locality	Sex	Variable	*N*	Het	*N*	Hom
Flax Pond	♂♂	*CV*	31	5.53	51	5.82
	♂♂	Σ	31	0.032	51	0.055
	♀♀	*CV*	54	5.43	123	5.65
	♀♀	Σ	54	0.024	123	0.064
Northport	♂♂	*CV*	42	5.62	69	5.82
	♂♂	Σ	42	0.036	69	0.098
	♀♀	*CV*	71	5.44	120	5.56
	♀♀	Σ	71	0.033	120	0.042
Centerport	♂♂	*CV*	14	5.11	31	5.37
	♂♂	Σ	14	0.0002	31	0.012
	♀♀	*CV*	20	5.75	34	5.31
	♀♀	Σ	20	0.006	34	0.013

Note: Data from Mitton 1978. *CV* is the average coefficient of variation, Σ is the determinant of the variance–covariance matrix, *N* is the sample size, and Het and Hom are groups heterozygous and homozygous for the lactate dehydrogenase-B locus.

killifish from Flax Pond, Long Island, New York, was divided into two groups, those heterozygous and those homozygous for LDH (table 7.4). Although the differences were not statistically significant for any of the characters, the variances of the seven morphological characters tended to be smaller in heterozygotes than in the homozygotes. This trend was also apparent for the generalized variance, which was 0.041 for heterozygotes and 0.072 for homozygotes. The generalized variances suggest that the volume occupied in morphological hyperspace by LDH heterozygotes was only 57% of the volume occupied by LDH homozygotes.

The same pattern of morphological variation in LDH genotypes was found in a population sample from West Meadow Creek, Long Island, New York. The generalized variances for heterozygotes and homozygotes were 0.006 and 0.024, respectively. At West Meadow Creek, heterozygotes occupied 25% of the volume of hyperspace occupied by homozygotes. Morphological shape differed among the LDH genotypes as well, for the correlation between dorsal fin rays and scales below the lateral line differed between LDH heterozygotes and homozygotes. At Flax Pond, the number of dorsal fin rays and scales below the lateral line was negatively correlated in heterozygotes ($r = -.34$, $P < .01$) but positively correlated in homozygotes ($r = .21$, $P < .01$). These correlation coefficients were significantly different ($P < .001$). The same pattern was evident at West Meadow Creek. The correlation between dorsal fin rays and scales below the lateral line was negative in heterozygotes ($r = -.41$, $P < .10$) and positive in homozygotes ($r = .40$, $P < .05$), and once again, the correlation coefficients were different ($P < .02$) for LDH heterozygotes and homozygotes.

These analyses uncovered several consistent patterns. When populations are divided into groups heterozygous or homozygous for a single locus, univariate variances of morphological characters tend to be lower in heterozygotes than in homozygotes. These comparisons are only occasionally statistically significant with a test of the ratio of

Table 7.4 Measures of morphological variance and the correlation between scales below the lateral line and dorsal fin rays for *Fundulus heteroclitus* grouped by *Ldh* genotype

	Flax Pond		West Meadow Creek	
	Het	Hom	Het	Hom
Σ	0.041	0.072	0.006	0.024
N	69	150	18	34
det	0.590	0.702	0.098	0.349
r	−0.34**	0.21**	−0.41†	0.40*

Note: Σ is the generalized variance (determinant of the variance–covariance matrix), *N* is the sample size, *det* is the determinant of the correlation matrix, and *r* is the correlation between the number of scales below the lateral line and the number of dorsal fin rays. Het and hom refer to heterozygous and homozygous at the *Ldh* locus. The correlations are different between heterozygotes and homozygotes at both localities.

variances, but the general trend is apparent when a sign test is used to summarize many comparisons (Eanes 1978; Fleischer, Johnston, and Klitz 1983; Mitton 1978; Yezerinac, Lougheed, and Handford 1992). Consistent with this pattern is the decrease in the generalized variance (Σ) with individual heterozygosity. Although this pattern can be partially attributed to the pattern of univariate variances, correlations among morphological characters also vary among genotypes.

The most consistent evidence that a single enzyme polymorphism can influence development is found at the LDH locus in *Fundulus heteroclitus*, and additional observations suggest that the LDH locus modulates metabolism at the appropriate time in development to influence morphological variability and symmetry. At hour 20 of development, the respiration rate of the *LDH-aa* genotype is four times as high as the respiration rate of the *LDH-bb* homozygote (Paynter et al. 1991). There are substantial differences in lactate accumulation, lactate utilization, glucose utilization, and heat dissipation as well. Injections of purified enzymes into *Fundulus* eggs (DiMichele, Paynter, and Powers 1991) clearly demonstrated that these differences were due to the lactate dehydrogenase enzymes—not to genomic heterozygosity, not to the heterozygosity of particular chromosomes, and not to some mysterious gene in linkage disequilibrium with the LDH locus. The physiological differences produced by the LDH enzymes are profound and are probably responsible for the differences in phenotypes measured in adult fish.

The variability of shell shape declines with enzyme heterozygosity in the blue mussel (Mitton and Koehn 1985). To determine whether variation in shape was related to enzyme genotype, a population sample of mussels from Cape Cod was measured for 14 shell characters and examined with electrophoresis to identify their genotypes at loci coding for three aminopeptidases and glucose phosphate isomerase. Genetic variation was summarized with the number of loci heterozygous for each individual. The variation at the 14 shell characters was summarized with four principal axes from a principal-components analysis. The first principal axis summarized the variation in size, and the remaining axes described some component of variation in shape. Heterozygosity was not related to size in this sample, nor was heterozygosity related to the mean of the projections of individuals on the principal axes summarizing shape variation. But the heterozygosity classes were heterogeneous for the variance of the projections on the shape axes; the coefficients of variation declined with increasing heterozygosity (table

Table 7.5 Individual heterozygosity and coefficients of variation in shell shapes of the blue mussel, *Mytilus edulis*

	Principal Axis			
Heterozygosity	2	3	4	Mean
0,1	110.0	108.6	188.6	135.7
2	103.5	98.3	79.1	93.6
3,4	91.2	91.2	79.4	87.2

Note: Data from Mitton and Koehn 1985. Shell shape was measured with 14 morphological characters, and summarized with principal components analysis. This first principal axis was a size axis, and the remainder of the axes summarized shape. The means of the heterozygosity classes did not differ on these axes, but the variances were significant heterogeneous. The average coefficient of variation for each heterozygosity class is tabulated under "mean." The coefficients of variation decrease with heterozygosity.

7.5). Heterozygous individuals reliably developed the modal shape of the blue mussel, whereas more homozygous individuals deviated to a greater degree from the modal shape. Imagine the mussels measured in this study as points in a multidimensional morphological space. Highly homozygous individuals were dispersed around the periphery, and highly heterozygous individuals were concentrated at the nucleus.

Morphological variation decreases with allozyme heterozygosity in the monarch butterfly, *Danaus plexippus* (Eanes 1978). Morphological variation was described with two characters—the length of the forewing and the diameter of the major forewing spot divided by the length of the adjacent wing vein. When the variances were compared in the heterozygotes and homozygotes at a locus, the heterozygotes had lower variances in 19 of 24 tests, and 5 of these reached statistical significance.

Several researchers (Chakraborty 1987; Chakraborty and Ryman 1983; Leary, Allendorf, and Knudsen 1984b, 1985) do not consider associations between protein heterozygosity and phenotypic variability to be relevant to the phenomenon of developmental stability. They assert that allozyme heterozygotes are expected to have intermediate phenotypic values and lower phenotypic variance than homozygotes do. That is, they argue that the differences in phenotypic variance between heterozygotes and homozygotes is attributable to additive genetic variation at the genes controlling these characters. These objections, however, apply only to those genes that directly control the morphological characters and to those genes in strong linkage disequilibrium with the genes controlling the characters' development. Direct involvement is not possible in this particular case, for the gene products of the allozyme loci are proteins, not meristic and quantitative characters. Furthermore, if the increased variance of homozygotes were attributable to pooling genotypes with heterogeneous phenotypes, then an analysis of variance would reveal different mean values of phenotypic traits for the alternative homozygotes. Several studies have reported that heterozygotes had lower phenotypic variance than homozygotes did and that the means of alternative homozygous genotypes were statistically homogeneous (Eanes 1978; Fleischer et al. 1983; Mitton and Koehn 1985). Thus there is no evidence that the allozyme loci contribute directly to the additive genetic variation of morphological characters.

Fluctuating asymmetry is yet another measure of developmental stability that can be derived from morphological variation. Fluctuating asymmetry, or the nondirectional departure from bilateral symmetry, is measured as small, random deviations from symmetry in otherwise bilaterally symmetrical characters (Palmer and Strobeck 1986). For example, in a sample of sunfish, the left pectoral fin may sometimes have more fin rays than does the right fin, and sometimes fewer, but in general the numbers of fin rays on the left and the right are not statistically different (Van Valen 1962). Fluctuations from bilateral symmetry are interpreted as measurements of the ability to buffer development. A study of protein heterozygosity and both the variance of morphological characters and fluctuating asymmetry concluded that both measured developmental stability and that the latter of the two measures was better (Leary et al. 1985).

The fluctuating asymmetry of scales, fin rays, gill rakers, and mandibular pores decreases with protein heterozygosity in trout (Leary, Allendorf, and Knudsen 1983, 1984b, 1985). The first of these observations used 13 polymorphic loci and five measures of asymmetry in a population of rainbow trout, *Salmo gairdneri* (now *Oncorhynchus mykiss*). Individual heterozygosity was negatively correlated with both the number of asymmetric characters and the magnitude of this asymmetry. The correlation between heterozygosity at a single locus and asymmetry was statistically significant at 2 of the 13 loci. The initial observation of a negative correlation between protein heterozygosity and fluctuating asymmetry was replicated in rainbow trout and extended to cutthroat trout and brook trout (Leary et al. 1984b). Protein heterozygosity is also negatively correlated with fluctuating asymmetry in the side-blotched lizard (Soulé 1979).

An enzyme locus segregating a null allele provided new insight into the relationship between allozyme genotypes and developmental stability in trout (Leary, Allendorf, and Knudsen 1994). Levels of asymmetry in individuals homozygous for active alleles and heterozygous for the null allele were compared. Unlike heterozygotes for active alleles, heterozygotes for the null allele had significantly more, not significantly less, asymmetry than homozygotes did. The results obtained with the null allele suggest that the association between developmental stability and allozyme heterozygosity is directly attributable to the enzyme loci rather than to some indirect measure of inbreeding depression or a contribution of additive genetic variation.

The genetic variation that influences developmental homeostasis may also be associated with other components of fitness. For example, a study of geographic variation in the Sonoran topminnow in Arizona revealed that allozyme variation was associated with both survival and developmental homeostasis (Quattro and Vrijenhoek 1989). Three populations of the endangered Sonoran topminnow, *Poeciliopsis occidentalis occidentalis*, were compared for allozyme heterozygosity and components of fitness. Populations from Monkey Springs, Tule Springs, and Sharp Springs had average heterozygosities of 0.0%, 1.5%, and 3.7%, respectively. Ten gravid females were captured from these remnant populations, and their offspring were raised in the laboratory. The viability of fish from Monkey Springs was significantly lower than the viabilities of the other populations, which did not differ from each other. Growth rates differed among localities and increased regularly with heterozygosity. Fecundities also differed among populations and again increased with heterozygosity. Fluctuating asymmetry at seven morphological characters was used to estimate developmental stability: fluctuating asymmetry differed among populations and decreased with heterozygosity.

Messier and Mitton (1996) chose a single enzyme polymorphism to study the relationship between enzyme polymorphism and developmental stability in the honeybee, *Apis mellifera*. They chose the MDH polymorphism for two reasons. First, although the haplodiploidy of honeybees reduces their genetic variability, the MDH polymorphism is unusual in that it is polymorphic throughout the geographic range of honeybees. Second, the genotypes of MDH differ in respiration rates during hovering (Coelho and Mitton 1988). A subsequent study of metabolic rates has also revealed heterogeneous rates among MDH genotypes (Harrison, Nielsen, and Page 1996), and clinal patterns of MDH frequencies on three continents suggest that MDH variation responds to environmental variation (Nielsen, Page, and Crosland 1994). In diploid workers, oxygen consumption is highest in FF, intermediate in FS, and lowest in SS. In haploid drones, which are larger and therefore hover with a higher internal temperature for any given ambient temperature, the oxygen consumption of the F haplotype is significantly lower than that of the S haplotype. Selection favoring one homozygous genotype in workers and the alternative haplotype in males could balance this genetic variation, explaining the occurrence of a common, widespread polymorphism in a species with relatively little genetic variation. Heterogeneous rates of respiration among MDH genotypes suggest that the MDH polymorphism (or a locus in linkage disequilibrium with it) has an important impact on metabolism. To test whether MDH influences developmental homeostasis, measurements of five wing veins were taken from both wings of workers from a single hive, and their MDH genotypes were identified with starch gel electrophoresis. The experiment was replicated in two hives. In both replicates, the MDH heterozygotes had lower levels of fluctuating asymmetry than did the homozygotes. Analysis of variance was used to compare the means of the characters measured from the right wings. In both replicates wing shape differed between heterozygotes and homozygotes.

Because most of the observations linking developmental stability with allozyme variation have been reported in poikilotherms, the importance of this phenomenon in homeotherms has been suspect. Development occurs in a wider range of temperatures in poikilotherms than in homeotherms, conceivably reducing the associations between heterozygosity and developmental homeostasis in homeotherms. Handford (1980) was the first to test this association in a homeotherm, and he found no relationship between protein heterozygosity and morphological variation in the rufous-collared sparrow, *Zonotricchia capensis*. His study was criticized, however, for its small sample sizes and lack of statistical power (Eanes 1981).

A later study using the same species but with large samples ($N = 479$) yielded very different results (Yezerinac et al. 1992). Approximately 20 males were captured at each of 25 collection localities. Heterozygosity was monitored at 12 allozyme polymorphisms, and morphological variation was measured with 8 external and 12 skeletal measurements. The dimensionality of the morphological data was then reduced and summarized with principal components. Among populations, morphological variation increased slightly with heterozygosity, indicating that the range of morphological variation increases with the level of genetic variation. Within populations, in 95 of 100 single-locus tests, the variability of heterozygotes was less than the variation in homozygotes (sign test, $P < .001$). The results were heterogeneous among loci, for comparisons with 6-phosphogluconate dehydrogenase did not reveal the differences seen in the other loci. The low level of genetic variation, which is typical for birds, restricted multilocus comparisons to groups with 0, 1, and >1 heterozygous loci. Sixteen of 20 tests of homogeneity of variance were heterogeneous among the heterozygosity classes. Morphological variation did not decrease linearly with heterozygosity, but the most heterozygous group consistently had the lowest level of morphological variability.

Enzyme heterozygosity is related to both size and the variability of shape in the house sparrow, *Passer domesticus* (Fleischer et al. 1983). Size did not differ among the genotypes at any single locus, but it did increase with individual heterozygosity in females and in two of four analyses of males. Variances generally decreased with individual heterozygosity, so the most heterozygous birds tended to be both the largest and the least morphologically variable.

Enzyme heterozygosity is related to developmental stability in the brown hare, *Lepus europaeus*, in natural populations in Austria (Hartl, Willing, and Nadlinger 1994; Hartl et al. 1995). Fluctuating asymmetry was estimated for 27 meristic traits and 9 metric traits from 742 hares taken from 30 populations. The genotype of each hare was identified at 39 enzyme loci. Enzyme heterozygosity was not related to the amount of morphological variation in populations, but it was related to fluctuating asymmetry. Within populations, no consistent relationship was found between heterozygosity and fluctuating asymmetry. Among populations, in juveniles, heterozygosity was positively correlated with the fluctuating asymmetry of both meristic characters ($r_s = .48$, $P < .05$) and metric characters ($r_s = .51$, $P < .05$). In adults, heterozygosity was negatively correlated with the fluctuating asymmetry of meristic characters ($r = -.41$, $P < .05$) but was not correlated with the fluctuating asymmetry of metric characters.

The changes in the associations across the age classes show that the relationship between heterozygosity and fluctuating asymmetry is dynamic. When genetic variability was estimated with the Shannon–Weaver information index (Hartl et al. 1995), the correlations between genetic variability and fluctuating asymmetry of meristic characters were stronger (juveniles: $r_s = .53$, $P < .01$; adults: $r_s = .67$, $P < .01$). This study should serve as a model for future studies of fluctuating asymmetry, indicating that distinctions should be made between meristic and metric characters and that sampling should be stratified with respect to age class. Changes in fluctuating asymmetry with age have also been reported for the antlers of white-tailed deer (Smith et al. 1982) and the skulls of cotton rats (Novak et al. 1993).

In a sample of 200 Jewish men of similar age and ancestry, heterozygosity at blood group loci was clearly associated with morphological variation (Livshits and Kobyliansky 1984). The men were placed in five heterozygosity classes on the basis of their genotypes at the MN, Ss, and Duffy systems and the C locus of the Rhesus system. Morphological variation was measured with ten morphological characters. For each of the characters, men were assigned to the lower extreme (= 25%), the upper extreme (= 25%), or the modal group (±0.67 standard deviations of the mean, = 50%). For nine of the ten characters, the frequency of individuals in the modal group increased with heterozygosity, and seven of these correlations were statistically significant (mean of nine r values = .85). Similarly, the coefficients of variation were negatively correlated with heterozygosity at nine of the ten characters, and four of these were statistically significant (mean of nine correlations = .82). Only one character deviated from this general pattern: the modal frequency at this character decreased with heterozygosity ($r = -.975$, $P < .01$), and the coefficient of variation increased with heterozygosity ($r = .959$, $P = .01$). In a multivariate analysis of morphological and genetic variation, researchers found that heterozygotes were closer than homozygotes to the multivariate mean in multidimensional morphological space (Kobyliansky and Livshits 1985).

In a multivariate study of fluctuating asymmetry in humans, no relationships between heterozygosity and fluctuating asymmetry or size or extremes of shape were found (Livshits and Smouse 1994). The data consisted of 14 genetic loci and 26 anthropometric traits measured on approximately 200 elderly people. Ten of the anthropometric traits measured size or mass, and eight pairs of bilateral measurements estimated fluctuating asymmetry. For each individual, the distance to the multivariate centroid of the population sample was calculated for two measures of size and two measures of shape distance. Fluctuating asymmetry was not related to either heterozygosity or any of the four multivariate measures of morphological distance. It is not clear whether the discordant results among the studies of humans can be primarily attributed to the differences in statistical analyses or to the differences among population samples.

Heterozygosity and Growth Rate

An association between enzyme heterozygosity and growth rate was first reported for the American oyster, *Crassostrea virginica* (Singh and Zouros 1978; Zouros, Singh, and Miles 1980). The designs of the first two studies were quite similar. Oyster larvae, called *spat*, were placed in a tray that was anchored in a bay and left undisturbed for about a year. This gave all animals, which were essentially the same age, the same opportunity to feed and grow in a relatively natural environment. By the time the oysters were harvested, their differential growth rates had produced a substantial range of sizes. The animals were weighed and then subjected to electrophoresis to determine their genotypes at five loci in the first experiment and at seven loci in the second experiment. With the exception of glutamate-oxaloacetate transaminase polymorphism, the heterozygotes grew more quickly than did the homozygotes at each locus, and, consequently, growth rate increased with individual heterozygosity (figure 7.5). In these studies, the variance of growth rate declined with increasing heterozygosity, so that both the mean growth rate and the reliability of the growth rate increased with heterozygos-

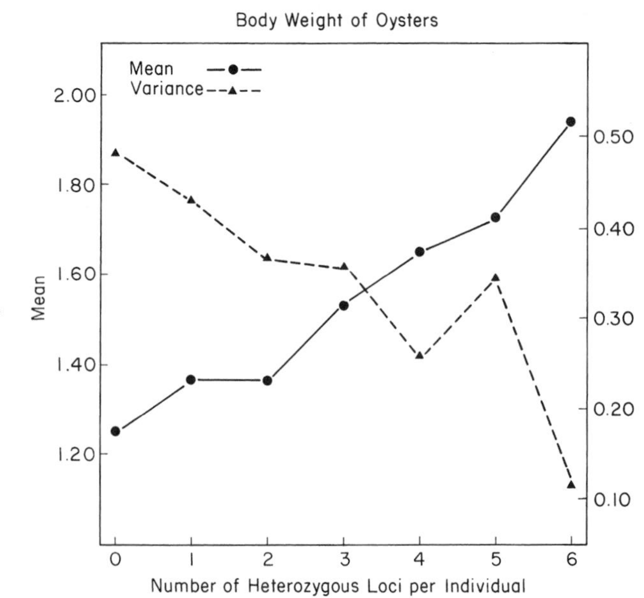

Figure 7.5 Weight and the variance of weight of American oysters, *Crassostrea virginica*, grouped by their heterozygosity at allozyme loci. Because all individuals are the same age, differences in weight are caused by differences in growth rate. From Singh and Zouros 1978.

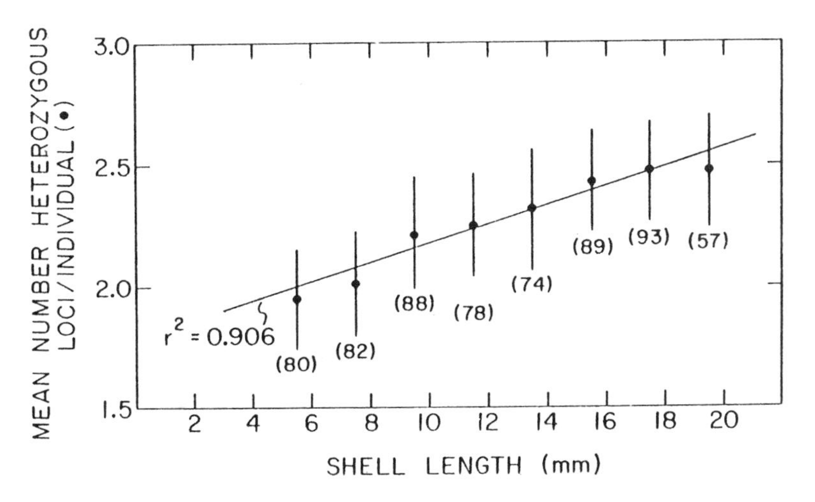

Figure 7.6 The relationship between shell length and allozyme heterozygosity in blue mussels, *Mytilus edulis*. Because all individuals are approximately the same age, differences in size were caused by differences in growth rate. The correlation between mean size and mean heterozygosity is .95. From Koehn and Gaffney 1984.

ity. In both experiments, the growth rates of the glutamate-oxaloacetate transaminase genotypes differed, but the highest rate was exhibited by the most common homozygote rather than by a heterozygote. Further studies of growth rates in oysters have consistently revealed faster growth rates in highly heterozygous individuals (Singh 1982; Singh and Green 1984).

Another marine pelecypod, the blue mussel, exhibits a positive correlation between individual heterozygosity and growth rate (Koehn and Gaffney 1984). Blue mussels colonized a raft anchored in Long Island Sound in June 1983 and were collected approximately three months later. The shell length at three months of age varied from less than 6 mm to almost 20 mm, and the variation in size, caused by the variation in growth rate, was positively correlated with individual heterozygosity (figure 7.6). Correlations between heterozygosity and size for individual loci varied from .36 to .92, with two of the five correlations reaching statistical significance.

Allozyme heterozygosity is related to many aspects of growth in white-tailed deer, *Odocoileus virginianus*, although the influences of heterozygosity appear to vary among years, probably in response to environmental quality. The frequency of twinning, along with fetal growth rate, increases with heterozygosity (Cothran et al. 1983). The incidence of spiked antlers—abnormal, unbranched antlers—decreases with age, body weight, and heterozygosity (Scribner, Smith, and Johns 1984). Body weight, kidney fat, the number of antler points, and main beam length/body weight all increase with heterozygosity (Scribner and Smith 1990; Smith et al. 1982). Within an age class, 10 to 15% of the differences in main beam length, beam diameter, inside beam spread, number of antler points, and incidence of spiking are related to allozyme heterozygosity (Scribner and Smith 1990). The mean of the Boone and Crockett score, which quantifies the size and shape of deer horns, was 17 to 27% higher in highly heterozygous bucks.

To further examine the influence of enzyme heterozygosity on fetal growth in white-tailed deer, Leberg, Smith, and Rhodes (1990) replicated the study by Cothran and colleagues (1983) and extended the analysis to compare the growth rates of twins. Just as in the first study, the growth rates of nontwin fetal deer increased with their heterozygosities. But because heterozygosity was not related to fetal growth rates in twins, Leberg and associates concluded that the protein loci were not responsible for the increase in growth rate with heterozygosity in single fetuses. However, other interpretations are consistent with these data. Because twins grow in a poorer, more crowded environment than singletons do, differences in the fetal environment may explain the lack of a relationship between growth rate and heterozygosity in twins.

The growth rate of tiger salamanders, *Ambystoma tigrinum*, increases with allozyme heterozygosity in both natural environments and the laboratory (Pierce and Mitton 1982). Although growth rate increases with heterozygosity in young larvae, no relationship was seen in samples of older larvae. The loss of the correlation with age may have been caused by differential metamorphosis, removing the larger, more heterozygous individuals from the pond (Carter 1996). Full sibs from a single-pair cross conducted in the laboratory were divided randomly into four aquaria. Significant, positive correlations between heterozygosity and growth rate were observed in two of the four aquaria. Oxygen consumption at rest decreases with heterozygosity (Mitton, Carey, and Kocher 1986), and this relationship probably offers an energy advantage to highly heterozygous individuals that is reflected in their superior growth rates.

Allozyme heterozygosity is related to development in the spotted chorus frog, *Pseudacris clarkii* (Whitehurst and Pierce 1991). Eggs were collected from an ephemeral pond in Texas and allowed to hatch in the laboratory, and juveniles were placed in four artificial ponds. Size was measured early in their development and at the metamorphic climax, when their front legs emerged. Individual heterozygosity was not related to growth rate, but the time to metamorphic climax shortened with individual heterozygosity. Heterozygotes metamorphosed significantly earlier than did homozygotes at alcohol dehydrogenase, glucose phosphate isomerase, and xanthine dehydrogenase. Heterozygotes at alcohol dehydrogenase metamorphosed 11 days earlier than did homozygotes (65.8 days versus 76.4 days). Time to metamorphosis is an important component of fitness for amphibians that develop in ephemeral ponds.

Relationships between allozyme heterozygosity and growth rate are clearly dependent on environmental conditions (Diehl 1988), and under moisture stress, growth rate increases with heterozygosity in earthworms (Diehl and Audo 1995). Juvenile earthworms, *Eisenia fetida*, were collected from a large laboratory breeding population at the age of 2 weeks and were raised in individual plastic cups (75 ml) at a temperature of 25°C under relatively dry conditions (3 ml H_2O/g dry neutral peat moss). Each individual was weighed at the beginning of the experiment and at the end of each week for four weeks. Eight polymorphic allozyme loci were genotyped for each individual in the experiment. For the statistical analysis, worms were divided into two groups—those heterozygous for ≤3 loci, and those heterozygous for ≥4 loci. Weight increased exponentially, and at each weighing the more heterozygous group was significantly heavier than the less heterozygous group. By the end of the experiment, the highly heterozygous group was 50% heavier (106.1 ± 6.0 mg versus 71.1 ± 4.6 mg) than the less heterozygous group.

The first report of a positive relationship between growth rate and allozyme heterozygosity in a plant was reported for quaking aspen, *Populus tremuloides* (Mitton and

Grant 1980). Quaking aspen is a dioecious species known in the Rocky Mountains for its large and ancient clones (Mitton and Grant 1995). Increment cores were taken from five of the largest ramets in each of 104 clones in the Front Range of Colorado, and the genotype of each clone was identified at three enzyme loci. Growth rate decreased with elevation and with the age of the ramet, and it increased with heterozygosity at the three enzyme loci. Similar results were obtained in a study of the growth rates of aspen in Waterton Lakes National Park (Jelinski 1993). From each of 156 clones, cores were extracted from five of the largest ramets, and genotypes were identified at 14 polymorphic enzyme loci. The individual heterozygosity of these clones varied from zero to eight heterozygous loci, and the growth rate increased with heterozygosity. The growth rate of the most heterozygous group exceeded that of the most homozygous group by 35%.

Heterozygosity and Fecundity

Allozyme genotypes are related in a complex way to cone production in ponderosa pine (Linhart and Mitton 1985). A six-year study of reproduction in a stand of 217 ponderosa pines provided the data to test for a relationship between heterozygosity and both pollen and seed production. Each year, male reproduction was estimated by counting the number of branch tips with male cones, and female reproduction was assessed by counting all the mature cones on a tree (Linhart et al. 1981). Individual heterozygosity was measured with seven polymorphic protein loci, and the ages and relative growth rates were calculated from the diameter of the tree and the number of annual rings from a core removed from each tree. No relationship was detected between genotype and pollen production. The mean of female cone production was independent of heterozygosity, but the variance of female cone production among individuals decreased with individual heterozygosity (table 7.6).

The heterozygosity of three enzyme polymorphisms is related to fecundity in *Drosophila melanogaster* (Serradilla and Ayala 1983a,b). Each of three enzyme polymorphisms exhibited strong overdominance (table 7.7) in females but not in males (Serradilla and Ayala 1983a). Females heterozygous for alpha glycerophosphate dehydrogenase, alcohol dehydrogenase, and acid phosphatase produced averages of 37%, 47%, and 46% more eggs per week than did homozygous females. There was a striking interaction between male and female genotypes, so that matings of dissimilar homozygous genotypes (producing the most heterozygous offspring) displayed the highest fecundities (Serradilla and Ayala 1983b).

Table 7.6 Variance of female cone reproduction declines with heterozygosity in ponderosa pine, *Pinus ponderosa*

	No. of Heterozygous Loci			
	0,1	2	3	4,5
No. of trees	39	69	44	20
Variance	88,348	23,786	24,333	18,352

Note: Data from Linhart and Mitton 1985. Variances are statistically heterogeneous ($P < .001$) among heterozygosity classes.

Table 7.7 Fecundity of heterozygotes exceeds the fecundities of homozygotes in *Drosophila melanogaster*

Enzyme	Genotype	No. of Eggs + SE
α-glycerophosphate Dehydrogenase	*F/F*	101.6 ± 3.0
	F/S	138.9 ± 3.0
	S/S	101.8 ± 3.0
Alcohol Dehydrogenase	*F/F*	100.0 ± 3.4
	F/S	149.3 ± 2.5
	S/S	102.5 ± 3.0
Acid Phosphatase	*F/F*	102.4 ± 3.3
	F/S	160.7 ± 3.8
	S/S	118.0 ± 3.8

Note: For each locus, fecundities are heterogeneous among female genotypes ($P < .001$). Data from Serradilla and Ayala 1983b.

Fecundity is related to individual heterozygosity in the blue mussel. The relationship is not seen throughout the life cycle, for young mussels put more energy into somatic tissue, whereas older and larger mussels put most of their energy into gamete production. In a sample restricted to those mussels in which gamete production exceeded somatic tissue production, fecundity increased with heterozygosity (Rodhouse et al. 1986).

Plant size and fecundity increase with allozyme heterozygosity in tetrasomic orchardgrass, *Dactylis glomerata* (Tomekpe and Lumaret 1991). From a natural population near Montpelier, in southern France, 448 adult plants were taken from the field and placed in an experimental garden. The leaves and roots of all plants were trimmed to the same size, and they were allowed to grow and set seed. Leaf weight and the number of panicles were used to estimate growth rate and fecundity. Genotypes were examined at six polymorphic loci, and at each locus, individuals were placed into one of five heterozygosity classes: monoallelic ($a^i a^i a^i a^i$), simplex diallelic ($a^i a^i a^i a^j$), duplex diallelic ($a^i a^i a^j a^j$), triallelic ($a^i a^i a^j a^k$), and tetraallelic ($a^i a^j a^k a^l$). Both leaf weight and the number of panicles increased with heterozygosity (figure 7.7). Across the range of heterozygosity in this experimental population, the number of panicles increased from 9 in the most homozygous group to 27 in the most heterozygous group.

A long-term demographic study of plantain, *Plantago lanceolata*, showed that the level of reproduction increased with allozyme heterozygosity. Demography was studied in a population of marked individuals in a hayfield in the Netherlands (Wolff and Haeck 1990). Heterozygosity was estimated with ten polymorphic allozyme loci. The number of leaves, the percentage of plants with scapes, and the number of scapes increased with allozyme heterozygosity. In heterozygosity classes 0 to 4 or more, the average number of scapes was 0.91, 3.25, 3.57, 5.36, and 5.60, respectively.

The fecundity of *Colias* butterflies varies with the PGI genotype, which also influences metabolism, flight capacity, male mating success, and viability (see "PGI in *Colias* Butterflies," chapter 4). Studies of fecundity in flight cages revealed that the average number of eggs deposited per day by *34* heterozygotes and *33* and *44* homozygotes was 16, 12, and 6, respectively (Watt 1992). Note that this is more than a twofold fitness differential among the genotypes at a locus.

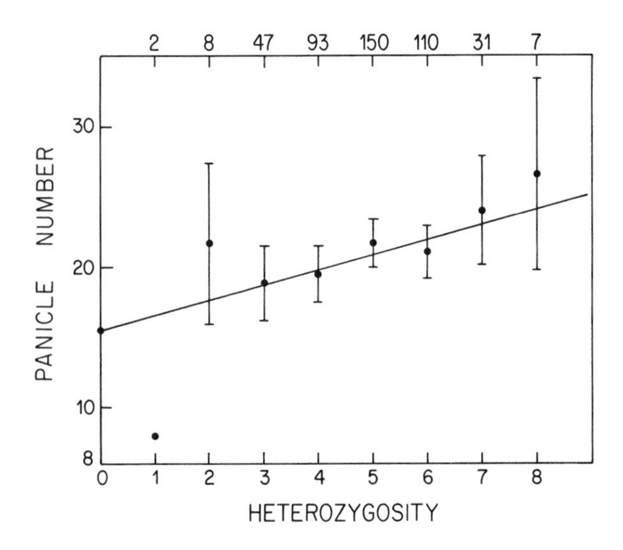

Figure 7.7 The number of panicles increased with heterozygosity in 448 individuals of orchardgrass, *Dactylis glomerata*, in a common garden near Montpelier, France. From Tomekpe and Lumaret 1991.

The production of overwintering cysts in brine shrimp, *Artemia franciscana*, increases with allozyme heterozygosity (Gajardo and Beardmore 1989). Brine shrimp produce either swimming larvae (nauplii) or encysted embryos. The encysted embryos are energetically more expensive to produce than are nauplii, and cysts are the only stage in their life cycle in which they can overwinter in harsh climates. The lifetime fecundity of brine shrimp was estimated for females collected as cysts from the Great Salt Lake in Utah and raised in the laboratory. Each female was genotyped for six polymorphic allozyme loci, and females were grouped into heterozygosity classes of 0, 1, 2, and 3 or more. Across this range of heterozygosity, lifetime fecundity increased from 580 to 900. The increase in fecundity was due primarily to the production of cysts, which varied from 61 in the least heterozygous group to 527 in the most heterozygous group (figure 7.8).

Reproductive success increases with heterozygosity in the green treefrog, *Hyla cinerea* (McAlpine 1993). Eighty-two amplectant pairs and 84 calling males were collected at a single site at the Savannah River Ecology Laboratory near Aiken, South Carolina. The amplectant pairs were placed in buckets in the laboratory, and after the eggs were deposited, the clutch size and the number of hatched offspring were recorded for each pair. Genotypes were obtained for eight polymorphic loci for all parents and the calling males. Both clutch size and the number of hatched offspring increased with allozyme heterozygosity. These relationships were not linear; for example, the number of hatched offspring increased with heterozygosity from about 1,100 to almost 1,600 but then dropped in the most heterozygous group. Heterozygosity was not related to body size in either males or females.

Heterozygosity and Metabolism

Many of the enzymes used by population biologists to estimate the heterozygosity of populations and of individuals are in the main corridor of metabolism: glycolysis, the pentose shunt, and the citric acid cycle. If these metabolic loci are directly responsible

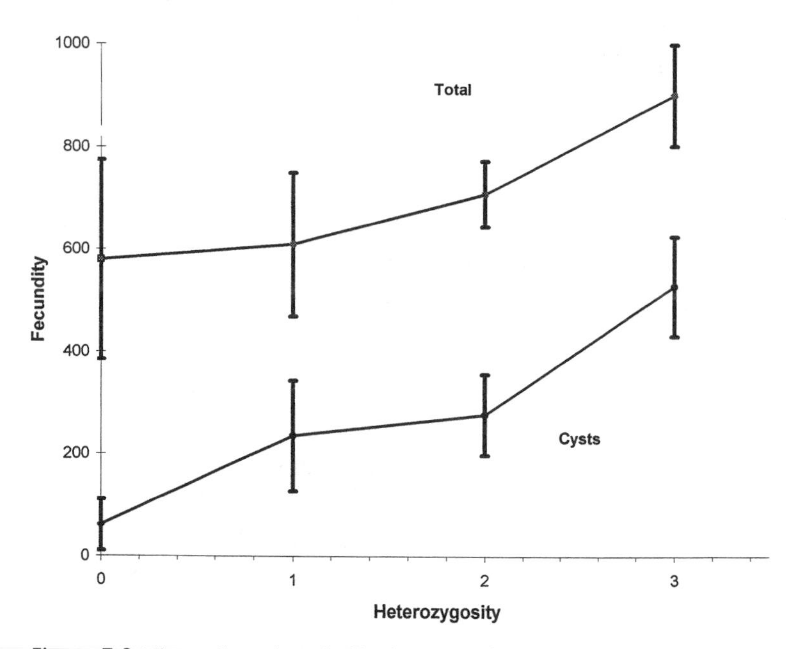

Figure 7.8 The total number of offspring (nauplii and cysts) and the number of cysts produced by brine shrimp, *Artemia franciscana*, grouped by heterozygosity at six allozyme loci. In temperate environments, cysts are the overwintering stage. From Gajardo and Beardmore 1989.

for the associations summarized for viability, developmental stability, and growth rate, we would predict that the enzyme loci would also be associated with some aspects of metabolism, such as those measured by physiological ecologists.

The first clear demonstration of an association between individual heterozygosity and some aspect of metabolism was the study of oxygen consumption and heterozygosity in the American oyster, *C. virginica* (Koehn and Shumway 1982). The reports of increases in growth rate with protein heterozygosity in the American oyster (Singh 1982; Singh and Zouros 1978; Zouros, Singh, and Miles 1980) prompted a more physiological study of the phenomenon. Animals approximately one year old were held in the laboratory for two weeks at 10°C and 2.8% salinity. Oxygen consumption was measured with an oxygen probe in a closed system at this temperature and salinity. The oysters were then transferred to 30°C and 1.4% salinity for 24 hours, and oxygen consumption was measured under these conditions. The first set of temperature and salinity was considered salubrious for oysters, but the second set imposed a stress. After oxygen consumption had been measured, the genotype of each animal was determined for phosphoglucose isomerase, phosphoglucomutase, two alpha-aminoacyl peptide hydrolases, and esterase. Under both the control and the stress conditions, oxygen consumption decreased dramatically with the number of heterozygous loci (figure 7.9). The number of heterozygous loci explained 66% of the variation in oxygen consumption under salubrious conditions and 57% of the variation in oxygen consumption during stress. Those individuals in the most homozygous class consumed approximately twice as much oxygen as did those in the most heterozygous class. These results are consis-

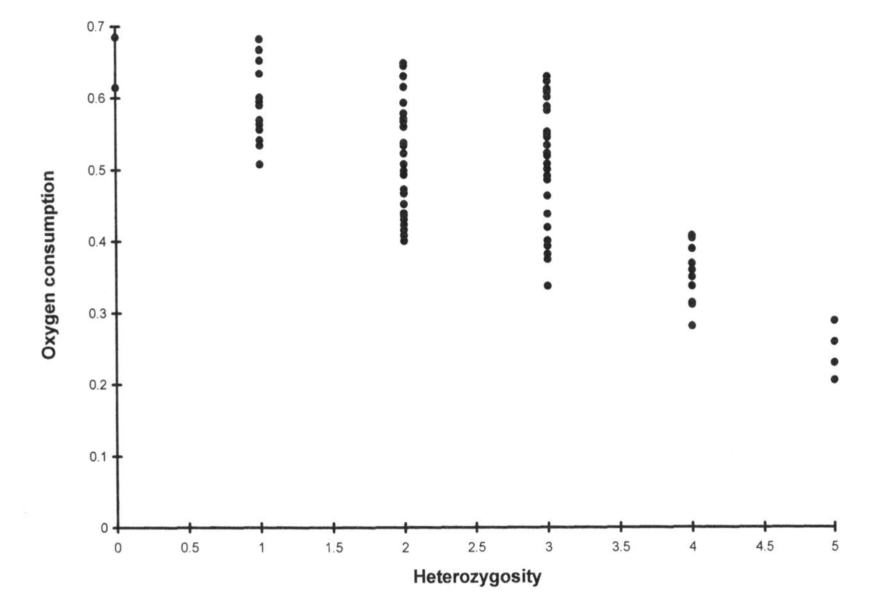

Figure 7.9 Oxygen consumption as a function of individual heterozygosity in the American oyster, *Crassostrea virginica*. Oxygen consumption was standardized by mass and presented as ml per gram per hour. These data were taken under stress conditions of high temperature and low salinity. Note that the oxygen consumption rate of the most homozygous group is approximately twice that of the most heterozygous group. From Koehn and Shumway 1982.

tent with the observations regarding the higher growth rate of highly heterozygous individuals: individuals consuming less oxygen, and therefore respiring at a lower rate, would have greater amounts of energy to invest in growth.

This first observation has now been replicated in a wide variety of poikilotherms; resting oxygen consumption decreases with multilocus heterozygosity in sow bugs, coot clams, oysters, blue mussels, trout, and salamanders (Danzmann, Ferguson, and Allendorf 1986, 1987; Diehl, Gaffney, and Koehn 1986; Diehl et al. 1985; Garton, Koehn, and Scott 1984; Hawkins, Bayne, and Day 1986; Hawkins et al. 1989; Mitton et al. 1986; Mitton, Carter, and DiGiacomo 1997). These studies usually report the lower resting oxygen consumption of highly heterozygous genotypes as evidence of greater physiological efficiency and lower routine metabolic costs in highly heterozygous individuals.

Although the mechanism providing greater physiological efficiency of heterozygotes has not yet been identified, we have gained greater insight into the problem through studies of protein turnover rate (Hawkins, Bayne, and Day 1986; Hawkins et al. 1989). The half-life of soluble enzymes is on the order of 16 hours, so the majority of the thousands of enzymes needed to control metabolism must be replaced each day. Consequently, a substantial portion of the basal metabolic cost, perhaps 20% to 40%, is spent breaking down damaged proteins and synthesizing new proteins. The protein turnover rate in the blue mussel decreases with enzyme heterozygosity, with the turnover rate of the most homozygous class exceeding that of the heterozygous class by 25%. The energy saved in heterozygous individuals can then be invested in growth or reproduction.

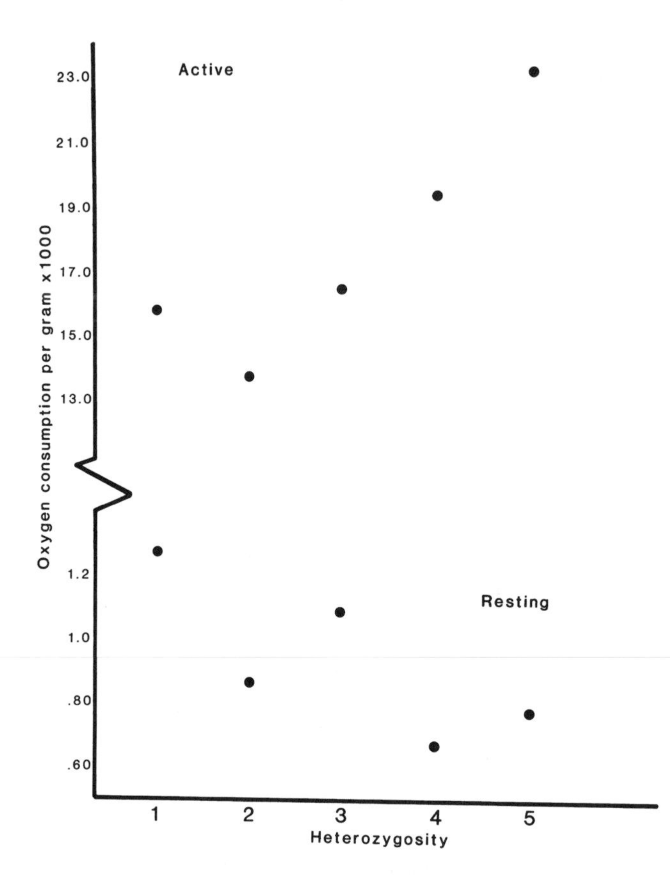

Figure 7.10 Oxygen consumption of the tiger salamander, *Ambystoma tigrinum*, as a function of heterozygosity of allozyme loci. Oxygen consumption at rest declines with allozyme heterozygosity, as it does in other animals (e.g., figure 7.9). But when salamanders are forced to exercise, their oxygen consumption increases with heterozygosity. The scope for activity, defined as the difference between these functions, increases with heterozygosity. Heterozygosity is based on a total of 8 polymorphic loci from a survey of 25 enzyme loci. From Mitton, Carey, and Kocher 1986.

Individual heterozygosity is also related to oxygen consumption of the tiger salamander, *Ambystoma tigrinum* (Mitton et al. 1986). Just as in the oysters, the growth rate of these salamanders increased with allozyme heterozygosity (Pierce and Mitton 1982). At rest, oxygen consumption decreased with heterozygosity, as it does in oysters (figure 7.9). But when the salamanders were forced to exercise, oxygen consumption increased with heterozygosity (figure 7.10). Thus, the scope for activity—defined as the difference between active and resting oxygen consumption—increased with individual heterozygosity. These data suggest that highly heterozygous individuals rest more quietly than homozygous individuals do but that when the occasion demands, they are capable of more strenuous or more prolonged exercise.

Individual heterozygosity is also related to the rate of weight loss during starvation in the American oyster (Rodhouse and Gaffney 1984). For this experiment, oysters were collected in the Nissequogue estuary in Long Island Sound and starved for 42 days at 20°C and 2.8% salinity. The genotype of each individual was determined for six polymorphic loci, the five loci used in the study of oxygen consumption (Koehn and Shumway 1982) and one aminopeptidase. The rate of weight loss was heavily dependent on size, but there was also a significant relationship to heterozygosity, as weight loss declined with individual heterozygosity.

The scope for growth—defined as the number of calories remaining after metabolic

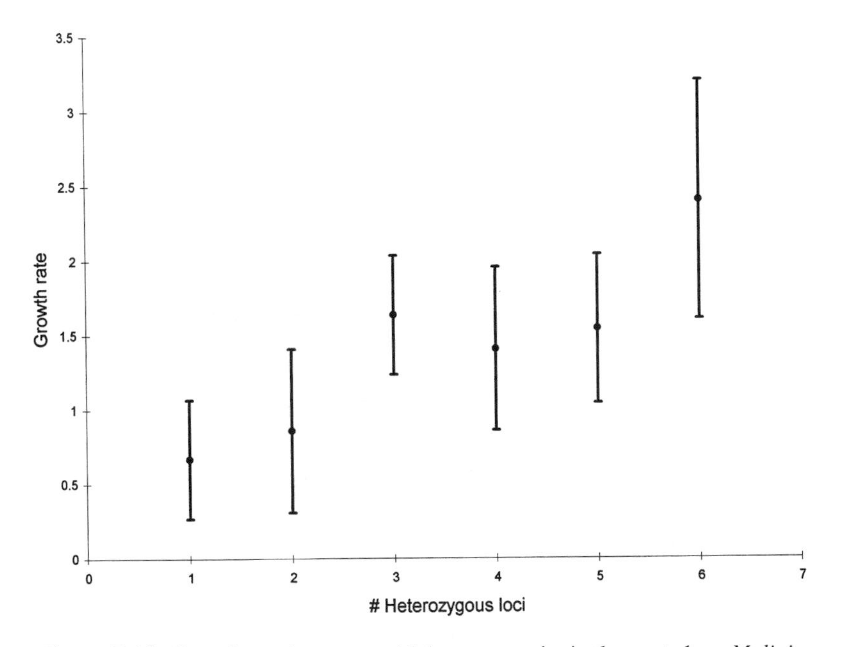

Figure 7.11 Growth rate increases with heterozygosity in the coot clam, *Mulinia lateralis*. From Garton, Koehn, and Scott 1986.

demands have been met—increases with individual heterozygosity in the marine snail *Thais haemastoma* (Garton 1984). Energy budgets were constructed from estimates of rates of ingestion, absorption efficiencies, oxygen consumption, and excretion, and individual heterozygosity was estimated with six polymorphic loci. The greater scope for growth in highly heterozygous individuals was partly attributable to higher feeding rates and partly attributable to lower routine metabolic maintenance costs, measured as energy lost per unit weight. Individual heterozygosity explained about 15% of the variation in the energy budgets.

A study of energetics and growth in the coot clam, *Mulinia lateralis*, revealed a very tight association between individual heterozygosity and metabolism (Garton, Koehn, and Scott 1984). Individual heterozygosity was measured with six polymorphic loci, and energy budgets were constructed from estimates of growth rates, rates of oxygen consumption, ammonia excretion, and clearance rates. Individual heterozygosity was associated with observed growth rates, and this association was driven primarily by a tight relationship between individual heterozygosity and routine metabolic costs (figure 7.11). Routine metabolic costs decrease with increasing enzyme heterozygosity, leaving more energy to be invested in growth. When standardized for differences in feeding rates, the routine metabolic costs explained 97% of the variation in growth rates in this study.

Allozyme heterozygosity of the European mussel, *Dreissenia polymorpha*, is positively correlated with shell length, but mass-specific oxygen consumption is not related to heterozygosity (Garton and Haag 1991).

Heterozygosity at three allozyme loci was related to lipid and fat levels in overwintering house sparrows (Fleischer and Murphy 1992). The birds were collected in February in Kansas; numerous morphometric measures were taken; the amount of lipid in the pectoralis muscle was measured; and fat from the flank, rump, and furcula was measured. In both males and females, the amount of lipid in the pectoralis muscle increased with heterozygosity. Although the total amount of fat also increased with heterozygosity, this relationship was observed only in males. Higher lipid and fat levels in winter indicate that highly heterozygous birds have an advantage, a result consistent with the superior developmental stability and the greater dispersal distances of highly heterozygous house sparrows (Fleischer et al. 1983; Fleischer, Lowther, and Johnston 1984).

Extensive studies by Wilson and his colleagues at the Welsh Plant Breeding Station in Aberswyth demonstrated substantial differences in the rates of dark respiration among individuals of perennial ryegrass, *Lolium perenne*. They reported two estimates of the heritability of dark respiration, which indicate that 50% or more of variation in the rate of dark respiration is attributable to genetic variation (Wilson 1981). Further evidence of the genetic basis for variability in the rate of respiration is found in the response of respiration rate to artificial selection. Lines selected for high respiration and for low respiration diverged in just two generations of selection (Day et al. 1985; Wilson 1975, 1981). In perennial ryegrass, dark respiration is inversely proportional to production, or growth, and so plants selected for low levels of dark respiration were more productive than a random sample of plants (Wilson and Jones 1982). Wilson suggested that the differences seen in productivity were the result of differential maintenance costs between the lines. The control of this variation in respiration rate in ryegrass is not in the mitochondria but appears to reside in glycolysis (Day et al. 1985).

Studies reveal associations between enzyme genotypes and respiration in perennial ryegrass (Rainey, Mitton, and Monson 1987; Rainey et al. 1990; Rainey-Foreman, and Mitton 1995). Data are most extensive for 6-phosphogluconate dehydrogenase (6PGD), which is in the pentose shunt. Enzyme kinetic studies of 6PGD revealed biochemical differences among genotypes (Rainey-Foreman and Mitton 1995). At 35°C, both Michaelis constants and V_{max} differed among genotypes, producing dramatic differences in V_{max}/K_m among genotypes. The Q_{10} of dark respiration, measured between 20°C and 35°C, differed by 30% among the 6PGD genotypes (Rainey et al. 1987). The biological relevance of this difference was tested by exposing ryegrass to a temperature stress for five days and estimating the tolerance of the genotypes to the prolonged stress. As expected, the genotypes with the lower values of Q_{10} survived this specific stress in better condition (Rainey et al. 1987). These data are consistent with the patterns of viability differentials seen in annual ryegrass (table 7.1).

Respiration rates differ among 6PGD genotypes at 35°C (Rainey et al. 1990). The variation in respiration rate is consistent with differences in V_{max}/K_m, with the *22* homozygote having a dramatic advantage over the *11* homozygote. In addition, the flux through the pentose shunt differs among 6PGD genotypes (Rainey-Foreman and Mitton 1995). The flux through the pentose shunt is highest in the *22* homozygote, the genotype with the highest value of V_{max}/K_m. The data on enzyme kinetics, flux through glycolysis, respiration, and viability suggest that genetic variation at a single locus in ryegrass substantially contributes to variation in fitness.

Stressful Environments

Although many studies report associations between protein heterozygosity and components of fitness such as growth rate and developmental homeostasis, numerous other, thorough studies found no relationship between heterozygosity and components of fitness (e.g., Bongarten, Wheeler, and Jech 1985; Booth, Woodruff, and Gould 1990; McAndrew, Ward, and Beardmore 1986; Ward et al. 1985). Relationships between enzyme genotypes and developmental homeostasis are not constant because animals develop in heterogeneous environments and their energy budgets vary with both genotype and environmental conditions. Stressful environments accentuate differences among genotypes and increase the likelihood of detecting overdominance (Mitton and Grant 1984; Parsons 1971, 1973, 1987, 1990).

For example, Koehn and Shumway (1982) found that oxygen consumption decreased with protein heterozygosity in oysters under normal conditions, but they also discovered that the relationship was much stronger when the oysters were kept at a combination of temperature and salinity that imposed a stress. Similarly, allozyme genotypes are independent of growth in earthworms kept at optimal conditions, but genotypes are correlated with growth under stress conditions (Diehl 1989). The feeding efficiency of *Peromyscus polionotus* is independent of enzyme genotype when the quality of food is high, but feed efficiency increases with heterozygosity when the food quality is low (Teska, Smith, and Novak 1990). The relationship between allozyme heterozygosity and growth in the coot clam, *Mulinia lateralis*, is critically dependent on the environment (Scott and Koehn 1990). Growth was independent of genotype in both the control and the highly stressful environments, but growth increased with heterozygosity in two environments with moderate degrees of stress. Growth rates of blue mussels increase with heterozygosity when mussels are grown at high densities, but not when they are grown at low densities (Gentili and Beaumont 1988).

A genetic study of growth in the Appalachian land snail, *Mesodon normalis*, clearly demonstrated that the relationship between heterozygosity and growth is dependent on environmental conditions (Stiven 1995). Adult females were collected from two natural populations near Highlands, North Carolina, and allowed to deposit egg clutches in the laboratory. After hatching, each egg clutch was placed in a plastic chamber lined with moist paper toweling and covered with a thin layer of soil and leaf litter. All snails were fed lettuce dusted with calcium carbonate. The chambers of the control group were cleaned weekly, but the litter of the stressed was changed at longer intervals, allowing water and mucus to accumulate. The conditions in the stress environment reduced growth rate by 65% and increased mortality by more than 300%. Allozyme genotypes were obtained for five polymorphic loci for the 192 snails surviving the stress environment and the 373 snails surviving in the control treatment. Growth rate increased with heterozygosity in the stress environment but not in the control treatment. In the stress environment, the most heterozygous class grew 40% faster than did the least homozygous class. Multiple regression was used to test the effects of individual loci, and the growth rate increased with heterozygosity at an aminopeptidase locus ($P < .001$) and PGI ($P < .05$), but the contributions of the remaining loci did not reach statistical significance. These results are consistent with the study by Koehn, Diehl, and Scott (1988), which revealed that the polymorphisms most directly related to growth rate were those involved in protein recycling.

Resistance of conifers to air pollution

Deteriorating air quality is challenging the forests of Europe, causing either mortality or declining vigor in several species (Giannini 1991; Scholz, Gregorius, and Rudin

1989). Comparisons of resistant and susceptible trees from the same stands suggest that heterozygous genotypes confer resistance to airborne pollutants (Bergmann and Scholz 1984, 1987, 1989; Geburek et al. 1987; Muller-Stark 1985; Scholz and Bergmann 1984).

Resistance to air pollution in Norway spruce, *Picea abies*, appears to be influenced by heterozygosity at the locus coding for phosphoenol pyruvate carboxylase (PEPC). PEPC is involved in a CO_2 fixation system, and it contributes to carbohydrate turnover in guard cells and to the stomatal movements of needles. Demographic studies first generated the hypothesis that PEPC heterozygotes were resistant to air pollution, primarily in the form of SO_2 (Bergmann and Scholz 1989). Two alleles segregate at PEPC, and the genotypic frequencies of mature spruces and natural regenerated seedlings in relatively clean sites do not differ. But in sites receiving high loads of SO_2 and other airborne pollutants in Germany, genotypic frequencies in seedlings differed from those in the mature trees around them. Genotypic frequencies differ at the locus coding for PEPC, but not at any of the other polymorphisms monitored in the study. Whereas the frequencies of heterozygotes in mature trees (the potential parents of the naturally regenerated understory of seedlings) was 17% and 22%, the frequencies in the seedlings was 40% and 36%. Furthermore, the other allozyme polymorphisms revealed F_{is} to be approximately 0.10, indicating a small amount of selfing or inbreeding. But F_{is} at PEPC was approximately −0.20, suggesting that natural selection had favored heterozygotes at this locus. A subsequent kinetic study of PEPC's enzyme products disclosed biochemical differences that might have been the target of selection (Rothe and Bergmann 1995). When kinetic studies used phosphoenol pyruvate as the substrate, the heterozygous enzyme was different from the two homozygotes for K_m, V_{max}, and V_{max}/K_m, and no differences were detected between the homozygotes. V_{max}/K_m is 60% greater in the heterozygote than the mean of the homozygotes, and this enhanced efficiency may be expressed as greater resistance to airborne SO_2.

Resistance to herbivores

Allozyme heterozygosity is associated with resistance to herbivores in pinyon pine, *Pinus edulis* (Mopper et al. 1991). Pinyon pines on the cinder soils around Sunset Crater, near Flagstaff, Arizona, experience chronic water and nutrient stress and also sustain higher densities of herbivores than do pines living nearby on normal soils (Mopper et al. 1991; Mopper and Whitham 1986; Whitham and Mopper 1985). However, the impact of the stress and herbivory is not uniform among the trees on the cinder soils. Some trees suffer little or no damage from herbivores, whereas others are trimmed so regularly by the stem moth, *Dioryctria albovitella*, that they assume a different growth form (Mopper and Whitham 1986; Whitham and Mopper 1985). By defoliating the tips of branches, the stem moths increase internal branching, thereby producing an atypical, densely packed, closely trimmed growth form that records chronic herbivory. Resistant trees were significantly more heterozygous than susceptible trees at two of four allozyme loci (Mopper et al. 1991; figure 7.12). In addition, in both susceptible and resistant trees, older trees were more heterozygous than younger trees, revealing viability differentials favoring heterozygotes.

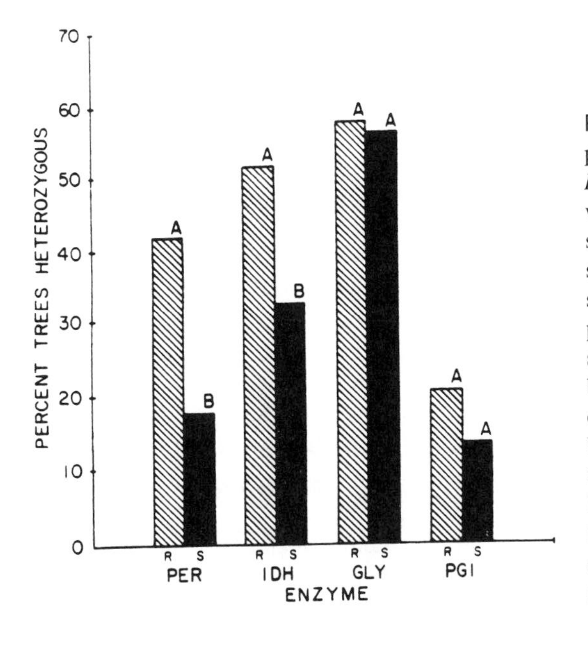

Figure 7.12 Resistance of pinyon pine, *Pinus edulis*, to the tip moth, *Dioryctria albovitella*, increases with allozyme heterozygosity. The study was conducted in a natural stand on the lava soils around Sunset Crater, Flagstaff, AZ, which impose both nutrient and water stresses and allow the levels of herbivores to attain unusually high levels. Resistant trees, R, are indicated by hatched bars, and susceptible trees, S, are indicated by closed bars. The letters above the bars designate statistically significant differences between groups. From Mopper et al. (1991).

Resistance to parasites

Genetic variation at the locus coding for adenosine deaminase (Ada) influences both parasite loads and survival in Soay sheep, *Ovis aries* (Albon et al. 1993; Gulland et al. 1993). The Soay sheep is a primitive domestic sheep, and a free-ranging population has existed in the St. Kilda Archipelago, Scotland, since before written history. The sheep are parasitized by gastrointestinal (*Teladorsagia* spp.) and respiratory tract (*Dictyocaulus filaria*) nematodes. Since 1988, the nematodes have been assayed from feces. Antibiotic treatments have demonstrated that the intestinal nematodes contribute to the probability of death. The population cycles approximately every three years, expanding in successive summers until the population exceeds the winter carrying capacity in the third or fourth winter, when up to 70% of the population dies. Ada plays a role in maintaining cellular immunity through the major histocompatibility complex, and, consequently, the Ada genotypes are associated with nematode egg counts. *Ada-FF* homozygotes consistently had the highest, and *FS* heterozygotes consistently had the lowest parasite loads. During population crashes in 1986, 1989, and 1992, mortality was highest in the *FF* homozygotes and lowest in the *FS* heterozygotes. Estimates of relative fitnesses for the *FF*, *FS*, and *SS* genotypes were 0.59, 1.00, and 0.83, respectively. These estimated fitnesses predict equilibrium frequencies very close to the frequencies observed in the population.

Haptoglobin genotypes of the brown hare, *Lepus europaeus*, are associated with differential susceptibility to a variety of parasites (Markowski et al. 1990a,b). Brown hares were collected from five hunting areas in Poland, and tissue samples of the rectum, duodenum, gall bladder, and lungs of 145 hares were examined for microorganisms of the genera *Mycobacterium*, *Staphylococcs*, *Salmonella*, and *Candida*. When the hares were classified into uninfected and infected groups, the uninfected group had an excess of heterozygotes ($X^2 = 24.3$, $P < .001$), but the infected group did not deviate from equilibrium expectations. Among populations, the proportion of haptoglobin heterozygotes was correlated with the proportion of animals free of coccid infections ($r_s = .61, N = 12, P < .05$).

Allozyme genotypes of the alpine marmot, *Marmota marmota marmota*, are related to differential susceptibility to intestinal parasites (Preleuthner et al. 1995). An extensive allozyme survey of protein variation revealed polymorphisms at only superoxide dismutase (Sod) and peptidase (Pep) in populations in Austria. The restricted genetic variation is a consequence of repeated bottlenecks associated with reintroductions. Examination of the gastrointestinal tracts of 162 marmots revealed that 90% were infested with the cestode *Ctenotaenia marmotae* (range 1–1,273 in infected marmots) and 62% were infested with the nematode *Citellina alpina* (range 1–1,397 in infected marmots). Parasite loads were not related to Sod genotypes but were related to Pep genotypes. More than 70% of the *ss* homozygotes were free of nematodes, but approximately 40% of the other genotypes were free of nematodes ($P < .01$). Approximately 30% of heterozygotes were free of cestodes, but less than 10% of the homozygotes were free of cestodes ($P < .01$).

The transferrin locus in pigeons segregates two common alleles, *A* and *B*, and the hatchability of eggs varies among transferrin genotypes (Frelinger 1972). A common source of mortality of pigeon eggs is microbial growth. Transferrin inhibits a wide variety of iron-dependent microorganisms, including *Saccharomyces cerevisiae*, *Staphlococcus aureus*, *Candida albicans*, *Shigella dysenteria*, *Pasturella septica*, *Pseudomonas*, *Clostridium welchii*, *Lysteria moncytogens*, and *Salmonella typhimurium* (Kochan, Golden, and Bukovic 1968). The transferrin in an egg is produced by the mother, not the squab (Frelinger 1971). Although the number of eggs produced did not differ among genotypes, the percentage of the eggs hatched by females of transferrin genotypes *AA*, *BB*, and *AB* was 46%, 52%, and 67%, respectively ($P < .02$).

To demonstrate the influence of transferrin on microbial growth, yeast was grown in a medium supplemented with purified transferrin. The growth rates of yeast in a medium supplemented with the purified transferrin from maternal genotypes *AA*, *BB*, and *AB* were 0.34, 0.30, and 0.14 ($P < .001$). Thus, the hatchability of eggs is inversely related to the microbial growth rate, and both are related to maternal transferrin genotype in pigeons. The hatchability of maternal heterozygotes exceeds that of homozygotes by approximately 30%. One of the consequences of this overdominance is that the frequencies of transferrin alleles differ very little among populations of pigeons, despite numerous bottlenecks associated with introductions and selective breeding.

Allozyme heterozygosity is associated with resistance to bacterial gill disease in rainbow trout, *Oncorhynchus mykiss* (Ferguson and Drahushchak 1990). Individually tagged fish from both a pooled gamete cross (25 males × 25 females) and from 12 full-sib families were raised in a common tank for 11 months. Approximately one month before the accidental exposure to the bacteria, probably *Flavobacterium* spp., the fork length of each fish was measured. The genotypes of 160 victims and 213 survivors were examined at nine polymorphic loci. The surviving fish were larger before exposure (13.37 cm versus 12.89 cm) and more heterozygous (3.23 versus 2.88 heterozygous loci) than those that succumbed to the infection.

Allozyme genotypes are associated with infection by the trematode, *Halipegus occidualis*, in the snail *Helisoma anceps* (Mulvey et al. 1987). Snails were taken from a farm pond in the Piedmont area of North Carolina and were examined for infection by trematodes. Patent infections result in castration of the host. The genotypes of each snail were examined for six polymorphic allozyme loci. Infected and uninfected snails differed in allelic frequencies for an esterase locus and a leucine aminopeptidase locus. Average heterozygosity was higher in the uninfected snails than in the infected snails (0.054 versus 0.047).

Resistance of insects to pesticides

A few enzymes, most notably esterases, modify the molecular structure of organophosphate pesticides, rendering them less toxic to insects. A wide variety of insects—including those in the genera *Culex*, *Simulium*, *Aedes*, *Tribolium*, *Aphis*, *Myzus*, *Dermestes*, and *Musca*—have evolved a resistance to organophosphate insecticides by amplifying one or a few enzymes, usually esterases (Oppenoorth 1985). For example, resistance to organophosphate pesticides in the mosquito *Culex quinquefasciatus* is provided by a 250-fold amplification of the gene coding for a specific esterase (Mouches et al. 1986, 1987; Raymond et al. 1989). Both resistance and amplification are eroded if the populations are heterogeneous for copy number and are not challenged by the insecticide (Raymond et al. 1993). However, if the population is homozygous for the number of amplified genes, the amplification is stable for at least 60 generations. *Culex pipiens* resistant to organophosphate insecticides are now distributed around the world. A DNA restriction analysis of resistant *C. pipiens* from French Guiana, Venezuela, Puerto Rico, California, and China revealed that they share the amplification of the same esterase haplotype, strongly suggesting that the amplification arose once and then spread rapidly (Qiao and Raymond 1995). A heritable amplification of a cholinesterase was documented in humans chronically exposed to organophosphate insecticides (Prody et al. 1989).

Resistance to heavy metal pollution

Studies of fishes and mollusks reveal significant differentiation between control sites and sites contaminated by heavy metal pollution. The majority of the reports indicate that glucose phosphate isomerase is differentiated between control and pollution sites. It is difficult to extract further generalities from these studies because selection induced by heavy metal pollution may favor heterozygotes (Roak and Brown 1996) or homozygotes (Chagnon and Guttman 1989a,b; Nevo et al. 1986; Nevo, Shimony, and Libney 1978). In vitro assays of enzyme activities of mosquitofish, *Gambusia affinis*, demonstrated that enzyme activity was influenced by low levels of copper but not by low levels of cadmium (Chagnon and Guttman 1989a) and that these assays were consistent with studies reporting the superior survival of homozygotes during exposure to toxic levels of the metals (Chagnon and Guttman 1989b). The genetic variabilities of *G. affinis*, *Pimephales notatus*, and *Fundulus notatus* were compared in two creeks in southeastern Kansas, one creek relatively undisturbed (zinc concentrations 15 to 20 ppb) and the other polluted by mine tailings (300 to 600 ppb). All three species exhibited significant differences between the creeks, and in each case heterozygosity was higher in the polluted creek.

Survival analysis (Dixon and Newman 1991), or failure time analysis, was used to examine the response of mosquitofish, *Gambusia affinis* and *G. holbrooki*, to inorganic mercury and arsenate. In these experiments, fish acclimated to laboratory conditions were exposed to toxic levels of mercury or arsenate, and performance was measured in time to death (TTD). Exposure to toxic levels of inorganic mercury was examined with eight polymorphic allozyme loci, and significant differences among genotypes were found at three (isocitrate dehydrogenase, malate dehydrogenase, and glucose phosphate isomerase) of the eight loci, with heterozygotes favored at each locus. In addition, TTD increased significantly with individual heterozygosity (Diamond et al. 1989). Distinctly different results were obtained with exposure to arsenate: homozygous genotypes were favored at fumarate dehydrogenase and glucose phosphate isomerase, and TTD decreased with individual heterozygosity (Newman et al. 1989). A study using *G. holbrooki* revealed interactions between genotypes and mercury levels; TTD differed among glucose phosphate isomerase genotypes at low levels of mercury but not at medium or high levels (Heagler et al. 1993). This series of studies clearly demonstrated that genotypic responses to pollution are dependent on the severity and type of pollution.

Three pairs of mollusks were used to test the hypothesis that resistance to various forms of pollution would be greater in species with high levels of genetic variation (Lavie and Nevo 1981, 1986; Nevo et al. 1986; Noy, Lavie, and Nevo 1987). All these species are abundant and widespread. Each pair consisted of one narrow-niche species with low genetic diversity and one broad-niche species with higher genetic diversity. *Monodonta turbinata* and *M. turbiformis, Littorina punctanta* and *L. neritoides*, and *Cerithium scabridum* and *C. rupestre* were exposed to copper, zinc, lead, cadmium, detergent, and oil. In virtually every test, the species with the higher level of genetic variation fared better in the exposure to the pollutant. Laboratory studies of the survival of genotypes exposed to inorganic mercury pollution were consistent with the differentiation of genotypes between control and polluted sites (Nevo, Lavie, and Noy 1987).

Empirical Data and the Adaptive Distance Model

The adaptive distance model (Smouse 1986; chapter 6) was used in analyses of annual ryegrass, pitch pine, loblolly pine, killifish, and cod. Most of these studies found that the empirical data showed some agreement with the expectations of this model. For example, when relative viabilities were heterogeneous in annual ryegrass, heterozygotes had higher fitnesses than did homozygotes (table 7.1), and among the homozygous genotypes, fitnesses declined with the frequencies of their alleles (figure 7.4).

The adaptive distance model was applied to empirical data on pitch pine to make inferences about the role of protein polymorphisms in growth (Bush, Smouse, and Ledig 1987). The radial growth rate of pitch pine increased with the heterozygosity of proteins, although this relationship was observed only in mature stands. The initial report (Ledig, Guries, and Bonefield 1983) presented growth rate as a function of heterozygosity class, with regressions run on class means rather than on individuals. When the regression was run with individuals, the positive relationship between growth and heterozygosity was no longer statistically significant (Bush et al. 1987). The adaptive distance model was then applied to these data to determine whether a higher proportion of the variation in growth rate might be explained. Using multiple regression, the researchers tested the relationship of an age-standardized growth rate to adaptive distance at eight polymorphic loci. This analysis explained higher proportions of the variances in growth rates than did linear regressions on the number of heterozygous loci.

The researchers concluded that the data were not consistent with selection solely against selfed genotypes but that specific genotypes played a role in the growth of pitch pine.

Analyses of genetic variation in killifish

The common killifish, *Fundulus heteroclitus*, is abundant in the marshes and bays of the North Atlantic from the Matanzas River in Florida to Newfoundland. It has abundant genetic variation (Mitton and Koehn 1975), and at least some of this genetic variation helps adapt populations to variations in temperature (Mitton and Koehn 1975; Powers 1987). Two polymorphisms, serum esterase (EST) and phosphoglucomutase-1 (PGM), were chosen for this analysis for their high levels of genetic variation. Dilocus genotypes were available from 574 fish collected in 1971 and 1972 from Flax Pond, Long Island, New York. The sample was broken into 107 reproductively immature individuals and 467 reproductively mature individuals on the basis of size (Mitton and Koehn 1975).

With the assumption that the sampled population was at equilibrium, the transition from young to mature individuals yielded both single-locus and dilocus fitnesses (table 7.8). Fitness differentials at the PGM locus were moderate, with the heterozygote having the highest viability. The heterozygote at the EST locus also exhibited the highest viability. The common homozygote had a relative viability of 0.94, and the less common homozygote had a relative viability of 0.46. These single-locus fitnesses and the allelic frequencies in the mature group were used to predict (Smouse 1986) dilocus fitnesses (table 7.8). The observed and predicted frequencies were in good agreement (r = .91, $P < .001$). The double heterozygote had the highest relative fitness, as predicted. The most common (*AABB*) and the least common (*BBCC*) double homozygotes were predicted to have fitnesses of 0.90 and 0.76, and their fitnesses were estimated to be 0.84 and 0.56. The values of D' and D'_z in the 467 adult fish were 0.038 and 0.016, respectively. The adult fish showed no evidence of linkage disequilibrium or zygotic disequilibrium. Thus the genetic differences between young and adult fish revealed substantial fitness differentials, and the fitness differentials appeared to be consistent with the adaptive distance model.

Analyses of genetic variation in cod

A study of viability of Atlantic cod, *Gadus morhua*, was conducted with a sample of young cod collected with a seine at Trondheimsfjorden, Norway (Mork and Sundnes 1985). Unfortunately, the fish refused to eat in captivity, thus turning the planned study of feeding and growth into a study of viability. The experiment was terminated when only about one-third of the fish were still alive. The genotypes of all the fish were obtained for lactate dehydrogenase (LDH) and phosphoglucose isomerase (PGI) (table 7.9), and the observed fitnesses were estimated from the genotypic distributions of surviving and deceased cod (data provided by J. Mork).

Selection was intense in the starving cod. The LDH heterozygote had the highest relative fitness, and the fitnesses of both homozygotes were 0.60. The PGI heterozygote also had the highest relative fitness, and the fitnesses of the common and rare homozygotes were 0.95 and 0.69, respectively.

Table 7.8 Genotypic frequencies and fitnesses for PGM and serum esterase in killifish from Flax Pond, Long Island, New York

A. Single locus

	PGM Genotypes		
	BB	*BC*	*CC*
Population samples			
Young	36	48	23
Mature	153	214	100
Fitnesses	0.95	1.00	0.97

	EST Genotypes		
	AA	*AB*	*BB*
Population samples			
Young	62	35	10
Mature	278	167	22
Fitnesses	0.94	1.00	0.46

B. Dilocus fitnesses, observed and (expected)

	PGM Genotypes		
EST Genotypes	*BB*	*BC*	*CC*
AA	0.84	0.89	0.84
	(0.90)	(0.93)	(0.90)
AB	0.85	1.00	0.86
	(0.97)	(1.00)	(0.96)
BB	0.51	0.34	0.56
	(0.77)	(0.79)	(0.76)

Note: Data from Mitton 1993. r, observed and expected fitnesses = .91; $P < .001$.

The correlation between observed dilocus fitnesses and fitnesses predicted by the adaptive distance model was $r = .69$ ($P < .05$). The double heterozygote was predicted to have the highest fitness, but it was the second highest, with a fitness of 0.96. The fitness of the rare double homozygote (70/70 135/135) was predicted to be 0.51, but not a single individual of this genotype survived, so the fitness was estimated to be zero.

Implications for the Overdominance and Dominance Hypotheses

Because these data are consistent with the adaptive distance model, selection appears not only to favor the most heterozygous genotype but also to discriminate among homozygous genotypes, favoring common homozygotes over rare homozygotes. Thus, these analyses suggest that selection is not simply eliminating inbred individuals but is also discriminating among genotypes in the outcrossed portion of the population. Furthermore, these data demonstrate intense selection in natural populations of ryegrass (table 7.1) and killifish (table 7.8) and a laboratory experiment using cod (table 7.9). Ob-

Table 7.9 Genotypic frequencies and relative fitnesses in cod, *Gadus morhua*, in a laboratory study in Trondheimsfjorden, Norway

A. Single locus

	LDH Genotypes		
	100/100	100/70	70/70
Population samples			
Dead	72	68	27
Alive	24	48	9
Fitnesses	0.60	1.00	0.60

	PGI Genotypes		
	100/100	100/135	135/135
Population samples			
Dead	77	74	16
Alive	37	39	5
Fitnesses	0.95	1.00	0.69

B. Dilocus fitnesses, observed and (expected)

	PGI Genotypes		
LDH genotypes	100/100	100/135	135/135
100/100	0.43	0.70	0.37
	(0.68)	(0.73)	(0.63)
100/70	0.90	0.96	1.00
	(0.94)	(1.00)	(0.86)
70/70	0.95	0.50	0.00
	(0.56)	(0.59)	(0.51)

Note: Data appeared in Mork and Sundnes (1985), and additional information needed to calculate dilocus fitnesses was supplied by J. Mork. The correlation between observed fitnesses and fitnesses predicted from the adaptive distance model is $r = .69$ ($P < .05$). From Mitton (1993a).

served dilocus fitnesses varied from 1.00 to 0.34 in killifish and from 1.00 to zero in cod. The average observed fitnesses of doubly heterozygous, singly heterozygous, and doubly homozygous genotypes in cod were 0.96, 0.78, and 0.44, respectively. If these fitness differentials could be extended to more loci, the decline in fitness with multilocus homozygosity could easily produce severe inbreeding depression. To the extent that these analyses are relevant to hypotheses concerning the mechanisms underlying inbreeding depression and heterosis, the data are consistent with the overdominance hypothesis. These data certainly cannot deny the role of deleterious alleles in the continuum between inbreeding depression and heterosis, but they do show the role of some form of overdominance.

Negative Results

This summary of empirical studies has emphasized the studies that reported positive results. But negative results have been found as well. Hartl et al. (1991) found no relationship between protein variability and the developmental homeostasis measured by antler shape in red deer, *Cervus elaphus*. Although researchers in several studies (Chesser and Smith 1987; Cothran et al. 1983; Leberg et al. 1990) reported that the growth of white-tailed deer increased with heterozygosity, a study of 21,264 deer over a 22-year period uncovered no significant associations between heterozygosity and the asymptotic weight (Leberg, Smith, and Brisbin 1992). Bottini and colleagues (1979) found an association between birth weight and heterozygosity in babies born in Rome and New Haven, but no relationship was found in a larger sample of children born in England (Ward et al. 1985). Livshits and Kobyliansky (Livshits and Kobyliansky 1985; Kobyliansky and Livshits 1985) reported that morphological variability declined with heterozygosity in human populations, but Livshits and Smouse (1994) reported no relationship between asymmetry and heterozygosity in their study of elderly Israelis. McAndrew and associates (1986) found no relationship between multilocus heterozygosity and growth rate in an extensive study of plaice, *Pleuronectes platessa*.

Interpretations of Empirical Observations

Positive associations between individual heterozygosity and components of fitness such as growth and developmental stability were reported in annual plants, forest trees, marine mollusks, fishes, *Drosophila*, butterflies, birds, deer, and humans (reviewed in Mitton 1993a,b; Mitton and Grant 1984). Clearly, the phenomenon has a broad systematic basis—it is general, although not universal. Biologists have long recognized a link between heterozygosity and measures of viability, developmental stability, and vigor (Lerner 1954), but that intuition was gained from experience with inbred strains and the highly heterozygous progeny of crosses between inbred strains. Most biologists did not expect to see the same relationships in populations of sexually reproducing organisms. The delightfully perplexing aspect of this phenomenon is that individual heterozygosity can be measured with a single enzyme polymorphism or, more typically, with 4 to 12 polymorphisms. Why do we observe such general and dramatic differences in viability, growth, development, and physiology when we categorize individuals with so few loci?

Generally, when geneticists grapple with the interpretation of associations between components of fitness such as viability or growth and individual heterozygosity, they point to three possible explanations (Mitton and Grant 1984): (1) the enzyme loci estimate genomic heterozygosity or perhaps estimate heterozygosity of large segments of chromosomes; (2) polymorphic markers identify individuals with different levels of inbreeding; and (3) the enzymes, or closely linked genes, directly affect the associated measure of fitness. I will discuss each of these, with the enzymes and linked loci (item 3) treated as separate possibilities.

Genomic or chromosomal heterozygosity

A small number of loci, such as the 1 to 12 polymorphic loci typically employed to measure individual heterozygosity, cannot reliably rank individuals in a population for levels of heterozygosity at hundreds or thousands of polymorphic loci (Chakraborty 1981; Mitton and Pierce 1980). Computer simulations demonstrated that a small number of loci (say 20) might reasonably reflect the heterozygosity of a larger set (perhaps 100) of loci (Mitton and Pierce 1980), but such a relationship is due largely to the correlation between a part and the whole. As the set of sampled loci becomes a smaller and smaller fraction of the entire set of polymorphic loci, the correlation tends toward zero.

If a small set of loci cannot measure the heterozygosity of the entire genome, could they measure the heterozygosity of a large segment of the chromosome(s) in which the loci are embedded? After all, genes come packaged in chromosomes, and even though we can record the genotypes of a few loci, those markers are linked to many other loci. The important point here is that linkage by itself is not sufficient. For a marker locus to reflect the genotype of other loci, it must be in strong linkage disequilibrium with those loci. Remember that for neutral loci at equilibrium in large panmictic populations, we expect even tightly linked loci to be independent of one another—the genotypes at a locus would hold no information concerning genotypes at an adjacent locus. Under some models of fitness determination, we expect linkage disequilibrium to become a common characteristic of the genome (Franklin and Lewontin 1970; Lewontin 1964). Truncation selection on the axis of individual heterozygosity does not generate linkage disequilibrium (figure 7.3), however, and the linkage disequilibrium generated by stochastic processes in finite populations is broken down much more quickly than is predicted by neutral models (Clegg 1978; Clegg, Kidwell, and Horch 1980). In general, surveys for linkage disequilibrium find genetic organization within and adjacent to inversions in *Drosophila* and throughout the genome of inbreeding species, but otherwise it is rare (Hedrick, Jain, and Holden 1978; Schaeffer and Miller 1993). Linkage disequilibrium is not a common characteristic of the genome of sexually reproducing, outcrossed species.

Heterozygosity as an indication of level of inbreeding

Consider those species with mixed mating systems, such as ponderosa pine or annual ryegrass, in which some of the seeds are produced by outcrossing to unrelated individuals and some of the seeds are produced by selfing (for simplicity, we disregard outcrossing to related individuals). Random outcrossing produces a distribution of individual heterozygosity, such as the distribution presented for the killifish (table 7.1). This distribution of individual heterozygosity is surprisingly restricted (Wills 1981). In a population with 2,000 polymorphic loci, all segregating two alleles each with a frequency of .50, the mean individual heterozygosity is 1,000, and 95% of the population falls between 955 and 1,045 heterozygous loci. Selfing reduces the heterozygosity of all loci by 50%, so if some of the adults in this population produced offspring by selfing, the mean heterozygosity of their offspring would be 500, far outside the range of individual heterozygosity produced by outcrossing.

Under these conditions, the mating system has a greater effect on individual heterozygosity than do the vagaries of recombination and syngamy. Thus, in a species with a mixed mating system, it is conceivable that a handful of polymorphic loci could indicate levels of heterozygosity throughout the genome. An individual predominantly homozygous for a set of marker loci might have been produced by selfing; if so, the remainder of that individual's genome would also have a low level of heterozygosity relative to the individuals produced by outcrossing. Thus, individual heterozygosity in a set of marker loci can predict individual heterozygosity in populations in which individuals exhibit extreme variation in their level of inbreeding. This is the interpretation favored by Ledig, Guries, and Bonefield in their 1983 discussion of positive correlations between protein heterozygosity and growth rate in pitch pine, *Pinus rigida*. This is a plausible explanation in species with mixed mating systems, but it does not suffice for obligately outcrossing species in which the range of F is typically much more restricted.

Consider now a population of an obligately outcrossing species in which the matings vary from those between unrelated individuals to those between first cousins, so that the F of progeny varies from zero to .062. The mean heterozygosity of individuals with an $F = .062$ is 938. The ranges of individual heterozygosity of individuals with $F = 0$ and $F = .062$ now broadly overlap, and the ability of a small set of polymorphic markers to predict individual heterozygosity of the entire genome is largely lost.

Tightly linked genes

Perhaps the relationships between heterozygosity and components of fitness can be attributed to genes very tightly linked to the enzyme markers. Once again, linkage alone does not suffice—the enzymes must be in strong linkage disequilibrium with the linked loci to see a correlated effect through neutral loci. But it would be absurd to propose that linkage disequilibrium *never* occurs; it is well known in antigen–antibody systems such as the Rh system and between the S and MN systems in man. Leary, Allendorf, and Knudsen (1984b) cite *cis-trans* modifiers as a possible mechanism to explain the associations between protein heterozygosity and developmental stability in trout.

Let us look again at the associations between protein genotypes and growth rate in oysters (Singh and Zouros 1978; Zouros, Singh, and Miles 1980) and in mussels (Koehn and Gaffney 1984). Each locus appears to contribute something to the variation in growth rate. Although not all associations are statistically significant, the heterozygosity of all loci (except at glutamate oxaloacetate transaminase in oysters, in which a homozygote is favored) appears to be related to an enhanced growth rate. If we favor the hypothesis relying on linked genes, we must propose an unseen locus that influences the growth rate and which is linked and in strong linkage disequilibrium with each of the protein polymorphisms used in these studies.

If the associations between allozyme heterozygosity and components of fitness can be attributed to linkage disequilibrium between the allozyme loci and loci that directly influence components of fitness, then any set of marker loci should reveal these correlations. Heterozygosity at loci controlling pelage color and pattern in cats, mice, and butterflies would also be expected to be correlated with viability, growth, and fecundity. Similarly, DNA markers, such as nuclear RFLPs and VNTRs, would also be expected to be in linkage disequilibrium with loci that directly influence components of fitness and to be correlated with components of fitness. An empirical study of genetic variation and growth in scallops, *Placopecten magellanicus*, demonstrated that protein heterozygosity was correlated with growth rate but that the heterozygosity of DNA markers was not (Pogson and Zouros 1994; Zouros and Pogson 1993).

The direct influence of enzymes on metabolism

Because enzymes catalyze metabolic reactions, they control flux in metabolic pathways. Differential fluxes have been measured as a consequence of different genotypes at an enzyme locus (Cavener and Clegg 1981; Eanes 1984; Paynter et al. 1991; Rainey-Foreman and Mitton 1995; Silva et al. 1989; Zamer and Hoffmann 1989). Furthermore, individual protein polymorphisms can and do influence whole animal physiology (DiMichele and Powers 1982a,b; Frelinger 1972; Koehn, Newell, and Immerman 1980; Koehn, Zera, and Hall 1983; Mane, Tompkin, and Richmond 1983; Powers, DiMichele, and Place 1983; Powers et al. 1994; Van Delden 1982; Watt 1977, 1983; Zamer and Hoffmann 1989; see chapter 4). It is more parsimonious to attribute physiological effects to polymorphic metabolic enzymes than to unseen loci with unknown functions, linked to and in strong linkage disequilibrium with every protein polymorphism.

We can learn about the role of protein polymorphisms in growth, development, and reproduction by considering the balanced energy equation presented by Bayne and Newell (1983): $P_g + P_r = C{*}AE - (R_{\mathrm{m}} + R_r)$, where P_g and P_r are the production of gametes and somatic tissues, C is the consumption of energy, AE is the efficiency of absorption of consumed energy, R_{m} is the metabolic cost of maintenance, and R_r includes all the costs of activity, such as seeking food and mates and defending territories. Thus the terms on the left side of the equation represent growth and reproduction. On the right side of the equation, the first term represents the absorbed energy (A), and the second term represents costs. Differences among individuals in any of the variables on the right side of the equation could contribute to variation among individuals for P_g and P_r.

R_{m}, which is typically estimated as oxygen consumption at rest and in the absence of food, is related to enzyme heterozygosity. R_{m} decreases with allozyme heterozygosity in a wide range of invertebrates and vertebrates (summarized earlier). Although variation in R_{m} is typically used to explain differences in growth rate (Hawkins, Bayne, and Day 1986; Hawkins et al. 1989; Koehn and Bayne 1988; Koehn, Diehl, and Scott 1988; Koehn and Shumway 1982; Mitton et al. 1986), the same reasoning shows how enzyme polymorphisms can contribute to variation in developmental homeostasis and fecundity. Just as energy is needed for growth, energy is needed for normal development and for reproduction.

The balanced energy equation helps explain why some studies report associations between protein heterozygosity and components of fitness, whereas other studies report no associations, and still other studies report associations with some loci but not with others. For example, when animals are developing under optimal conditions with abundant food, subtle differences in R_m do not produce important variations in P_g, so all individuals appear to have high levels of developmental homeostasis. It is when P_g is lowered by some form of stress (insufficient food, disease, high or low temperature, etc.) that differences in R_m become important and measurable as variations in morphological variance or fluctuating asymmetry.

One of the critical issues in this line of reasoning is whether the products of a single gene can have a sufficient impact on physiology to produce the observed differences in developmental stability between heterozygotes and homozygotes at a single locus. Two observations suggest that this is possible. Koehn (1991) calculated the metabolic cost of protein turnover for a range of proteins and demonstrated that enzymes with moderate turnover rates, maintained in high concentration in muscle, could cost up to several percent of the mass-specific energy demands for maintenance metabolism (R_m). Because minor changes in the amino acid sequence can dramatically change the half-life of a protein (Rogers, Wells, and Rechsteiner 1986; Tobias et al. 1991), alternative genotypes at a locus can differ in their energy costs.

Can the associations between individual heterozygosity and viability, growth rate, and developmental stability arise as simple consequences of the control of flux in metabolic pathways by polymorphic enzymes? Haldane (1954) was the first to propose that heterozygotes at enzyme loci would be more efficient than homozygotes at controlling flux in metabolic pathways. This idea has been elaborated by others, each with different perspectives on the problem (Berger 1976; Fincham 1972; Johnson 1974; Milkman 1967; Mitton and Koehn 1985; Clark and Koehn 1992). One way to envision superior control of metabolic flux is to assume that the genotypes have different biochemical properties under different conditions of temperature or pH and also that these environments fluctuate between conditions favoring first one homozygote and then another. This model was developed most extensively by Gillespie (1973, 1976, 1978a,b, 1991), and it is presented in a simple form in table 3.1, in which the various environmental conditions produce selective events in the life cycle. Note that the heterozygote has intermediate fitness in each of the environments but that it has the highest fitness when the genotypes experience both environments.

Summary

Empirical data regarding allozyme polymorphisms are often, but certainly not always, consistent with the theoretical expectation that components of fitness increase with heterozygosity. Studies comparing two portions of a life cycle often reveal viability differentials favoring heterozygous individuals. In populations, morphological variation and fluctuating asymmetry decrease with allozyme heterozygosity, indicating that developmental stability increases with heterozygosity. Growth rates and fecundity also increase with allozyme heterozygosity. All these relationships are easier to detect, or are more apparent, under a moderate degree of stress.

Physiological variation is associated with allozyme heterozygosity, and some of this variation is consistent with the advantages in viability, development, growth rate, and fecundity accruing to the highly heterozygous individuals. Routine metabolic costs, estimated from oxygen consumption in resting animals, decline with allozyme heterozygosity. The rate of protein cycling falls with allozyme heterozygosity, and this cost may explain the associations between resting oxygen consumption and heterozygosity.

Analyses of empirical data using the adaptive distance model suggest that some of the correlations between heterozygosity and components of fitness can be attributed to allozyme loci, or loci in strong linkage disequilibrium with them.

8

Female Choice and Male Fitness

Whether mate choice could be based mainly on genetic quality of the potential mate has been a puzzle to evolutionary biologists. Population genetic theory predicts that any balanced polymorphism for a selected trait ends with zero heritability of fitness, so that no one mate is better for "good genes" than any other.

W. D. Hamilton and M. Zuk,
"Heritable True Fitness and Bright Birds" (1982)

There are both theoretical and observational grounds for thinking that there will be a low parent–offspring correlation for fitness (in effect, additive genetic variance for fitness is rapidly depleted . . .). Therefore, a female who selects as a mate a male of high fitness does not increase the expected fitness of her own offspring.

J. Maynard-Smith,
The Evolution of Sex (1978)

Contrary to the expectations of the lek paradox, selection on sexual traits has not caused an exhaustion of additive genetic variance. If anything, the reverse appears to be the case.

A. Pomiankowski and A. P. Møller,
"A resolution of the lek paradox" (1995)

Sexual selection (Darwin 1871) may be defined as two forms of natural selection driven by the disparate sizes of male and female gametes and the consequent differential parental investment in offspring (Thornhill and Alcock 1983; Williams 1975; Williams and Mitton 1973). Females are usually the limiting sex in reproduction, for they invest more energy in each gamete and they supply the majority of the parental effort. Because males often commit only some tiny cells and the energy needed for courtship and copulation, they are usually cavalier in their engagement in sex. Because females lose a much greater investment in energy when an offspring is lost, they generally are coy. In the most common form of sexual selection, males compete with one another for access to females.

This *intrasexual selection* is the force behind the evolution of a broad spectrum of attributes used in either battles or ritualized combat. Examples of these attributes are the horns of deer, the relatively large body size of the male mountain gorilla, and the colorful epaulets of the red-winged blackbird. Under breeding systems dominated by female choice, however, females shop to find males protecting the best breeding territories, offering the most aid, or possessing the genes that will give their progeny the greatest advantage. This female choice is referred to as *intersexual* or *epigamic selection*. This brief introduction does not do justice to the fascinating history of the theory of the evolution of sexual selection (Bradbury and Andersson 1987; Bull 1983; Charnov 1982; Ghiselin 1974; Maynard-Smith 1978; Williams 1975; Brown 1997) or the evolution of the rich diversity of mating systems (Batten 1992; Cronin 1991; Emlen and Oring 1977; Møller 1994a; Thornhill and Alcock 1983).

Although it is clear that epigamic selection works through the female choice of superior feeding territories and nuptial gifts (Thornhill and Alcock 1983), theoreticians argue that female choice based solely on male genetic quality should be nonexistent or rare. There are several reasons for this expectation. First, for females to choose among potential mates, incentives or penalties must be incumbent on the choice. But as indicated by the quotations at the beginning of this chapter, the heritability of fitness is expected to be reduced to zero as a genetic system approaches equilibrium. When the most fit male is homozygous at one or more loci, unrelenting directional selection erodes the genetic variability, leaving all males the same and therefore providing no incentive for females to continue this selection. On the other hand, if the most fit males are heterozygous at one or more loci, the additive genetic variation, and therefore the heritability, will fall to zero as the genetic system reaches equilibrium.

This point was made explicitly by Partridge (1983) and is illustrated for a set of equilibrium frequencies in table 8.1. It is assumed that females are not aware of their own genotype but that they can detect the genotypes of males. For this example, there are two alleles at a locus, with $p = f(A) = 0.8$ and $q = f(a) = 0.2$. The fitnesses of genotypes *AA*, *Aa*, and *aa* are 0.8, 1.0, and 0.2, so that allelic frequencies are at the equilibrium predicted by $p = s_2 / (s_1 + s_2)$. In this example, the most common female genotype benefits by choosing her mate according to his genotype, whereas all other females are actually penalized for choosing males with the most fit genotype. Females mating randomly and females choosing the most fit males have the same average fitness. Further exploration of models of female choice based solely on male genetic quality were conducted by Lande (1981) and Kirkpatrick (1982). Their simulations of an intermediate phenotype with superior viability revealed no tendency for random mating to be replaced by females' choosing males with high fitness.

The notion that sexual selection would erode genetic variability underlying selected traits was challenged by Pomiankowski and Møller (1995), who argued that heritability was not the best measure of genetic variation for a trait. Although narrow sense heritability is the most widely used index for comparing additive variation in traits, Pomiankowski and Møller pointed out that heritability conflates the effects of additive and residual variance. They proposed, instead, the coefficient of additive genetic variance, $CV_A = \sqrt{V_A}/\overline{X}$, where V_A is the additive genetic variance and X is the mean of the trait value. When they compared data from the literature (59 characters, 27 species) for sex-

Table 8.1 A comparison of random mating and choice of the most fit male parent for a specific set of equilibrium frequencies produced by overdominance

Allelic frequencies $p = f(A) = .8$
$q = f(a) = .2$

Fitness

	Genotypes	
AA	*Aa*	*aa*
.8	1.0	.2

Parents				Offspring Produced by			
				Random Mating		Choice	
Female Genotype	Frequency	Male Genotype	Frequency	Genotype	Progeny Fitness	Genotype	Progeny Fitness
AA	.64	*AA*	.64	*AA*	.8		
		Aa	.32	.5*AA* + .5*Aa*	.9	.5*AA* + .5*Aa*	.9
		aa	.04	*Aa*	1.0		
				Weighted mean	.84		.9
Aa	.32	*AA*	.64	.5*AA* + .5*Aa*	.9		
		Aa	.32	.25*AA* + .5*Aa*=.25*aa*	.75	.25*AA* + .5*Aa* + .25*aa*	.75
		aa	.04	.5*Aa* + .5*aa*	.60		
				Weighted mean	.84		.75
aa	.04	*AA*	.64	*Aa*	1.0		
		Aa	.32	.5*Aa* + .5*aa*	.6	.5*Aa* + .5*aa*	.6
		aa	.04	*aa*	.2		
				Weighted mean	.84		.6
				Grand weighted mean	.84		.84

Note: Data modified from Partridge 1983. Allelic frequencies are at equilibrium, and adult genotypes occur in the frequencies expected under the Hardy–Weinberg law. The weighted means of offspring produced by random mating and by choice of the most fit genotype (the heterozygote) are equivalent.

ually selected traits and nonsexual traits, the found that CV_A's were significantly higher in the sexually selected traits. This empirical result is diametrically opposed to the expectation entrenched in the literature.

Although Partridge's reasoning (table 8.1) is clear, I wonder whether it is biologically correct. Perhaps we should consider not the average fitness of a female but the relative production of the most fit genotypes (Williams 1975; Williams and Mitton 1973). The distinction here is between fitness and fitness potential. Partridge assumed that she was dealing with fitnesses, but it seems to me that these are fitness potentials, for the placement of the threshold of selection alters both the numbers of individuals in the various genotypic classes and their fitnesses. Those individuals falling below the threshold either disappear or fail to breed and thus have fitnesses of zero. If natural selection truncates the axis of individual heterozygosity, it is more important to compare mating strategies according to their production of heterozygous offspring than their mean fitnesses.

Table 8.2 A comparison of the proportion of heterozygous progeny produced by random mating and choice of the most fit (heterozygous) male

Allelic frequencies $p = f(A) = .8$
$q = f(a) = .2$

Fitness

Genotypes		
AA	*Aa*	*aa*
.8	1.0	.2

Parents				Offspring Produced by	
Female Genotype	Frequency	Male Genotype	Frequency	Random Mating	Choice
AA	.64	*AA*	.64	.0	
		Aa	.32	.5	.5
		aa	.04	1.0	
			Weighted mean	.2	.5
Aa	.32	*AA*	.64	.5	
		Aa	.32	.5	.5
		aa	.04	.5	
			Weighted mean	.5	.5
aa	.04	*AA*	.64	1.0	
		Aa	.32	.5	.5
		aa	.04	.0	
			Weighted mean	.8	.5
			Grand weighted mean	.32	.5

Note: Allelic frequencies are at equilibrium determined by overdominance, and adult genotypes occur in the frequencies expected under the Hardy–Weinberg law. Choice of the most fit male enhances the frequency of the most fit genotype over random mating.

A comparison of the production of heterozygous offspring by females practicing random mating versus those choosing heterozygous males once again reveals different consequences for three female genotypes (table 8.2). When the allelic frequencies and fitnesses are the same as in the previous example (table 8.1), the most common females (genotype *AA*) can increase their proportion of heterozygous progeny from 20 to 50% by selecting mates by their genotypes. The two mating systems have identical consequences for heterozygous females—all matings result in 50% heterozygous offspring. Female choice is counterproductive for females homozygous for the rarer allele: they produce 80% heterozygous offspring by randomly mating, but only 50% of their progeny are heterozygous when they choose a heterozygous mate. When these productions of heterozygous progeny are weighted by the frequencies of the female genotypes, females choosing heterozygous mates gain an advantage (table 8.2). If a female is unaware of her genotype but wishes to maximize her production of heterozygous offspring, she should seek a heterozygous mate. Depending on the placement of the threshold of selection, this may also enhance the average fitness of her offspring.

Correlation of Heterozygosity Between Parents and Their Offspring

The choice of a heterozygous mate, or a highly heterozygous mate, helps maintain genetic variation, for heterozygosity is correlated between parents and their offspring. The correlation between the individual heterozygosities of parents and their offspring is positive because at most allelic frequencies, heterozygous parents produce higher proportions of heterozygous progeny than do homozygous parents. Consider a locus segregating two alleles, *A* and *a*, with frequencies *p* and *q*, respectively, in an infinite population with no selection. To simplify the explanation, consider the production of progeny genotypes from the perspective of the maternal parent.

Homozygous mothers produce 50% heterozygous offspring when the allelic frequencies are $p = q = 0.5$, but at all other frequencies, they produce lower proportions of heterozygous offspring. More precisely, the proportion of heterozygous progeny from heterozygous mothers is $pq / 2pq$, and the proportion of heterozygous progeny from homozygous mothers is $pq / (p^2 + q^2)$. As the allelic frequencies become increasingly unequal, the proportion of heterozygous progeny produced by homozygous mothers decreases.

The expected value for the correlation between offspring and parent heterozygosity, r_{op}, can be derived (Mitton et al. 1993) most simply from the formulations of genotypic value, additive genetic variance, and dominance genetic variance (Falconer 1989). Homozygous genotypes are assigned genotypic values of *a* and $-a$, and the heterozygote is assigned a value *d*. When homozygous and heterozygous genotypes are assigned the scores of 0 and 1, respectively, the genotypic values of *a* and *d* become 0 and 1, respectively. The additive genetic variance is $V_A = 2pq\alpha^2$, where $\alpha = a + d(q - p) = q - p$ and therefore $V_A = 2pq(q - p)^2$. The dominance variance is $V_D = (2pqd)^2 = 4\,p^2q^2$. The correlation of offspring heterozygosity with parental heterozygosity, r_{op}, is $r_{op} = 1\,/2V_A\,/\,(V_A + V_D) = pq(q - p)^2/\,[2pq(q - p)^2 + 4p^2q^2] = (q - p)^2/\,[2(q - p)^2 + 4pq]$. Figure 8.1 presents r_{op} as a function of *q* and shows that the correlation is equal to zero only when $p = q = 0.5$ and that it increases to a maximum of $r_{op} = 0.5$ as allelic frequencies become increasingly unequal.

The correlation between the individual heterozygosities of parents and their offspring can be extended to *M* independent loci (Mitton et al., 1993) by summing V_A and V_D across loci, as follows:

$$V_A = V_{A1} + V_{A2} + \ldots V_{Am}$$

and

$$V_D = V_{D1} + V_{D2} + \ldots V_{Dm}$$

Consequently,

$$r_{op} = \sum_{i=1}^{M} p_i q_i (q_i - p_i)^2 / \,[2\sum_{i=1}^{M} p_i q_i (q_i - p_i)^2 + 4\sum_{i=1}^{M} p_i^2 q_i^2]$$

where p_i is assigned to the allele with the higher frequency. A close approximation to this value can be obtained by using the mean allelic frequency in the expression for r_{op} for a single locus.

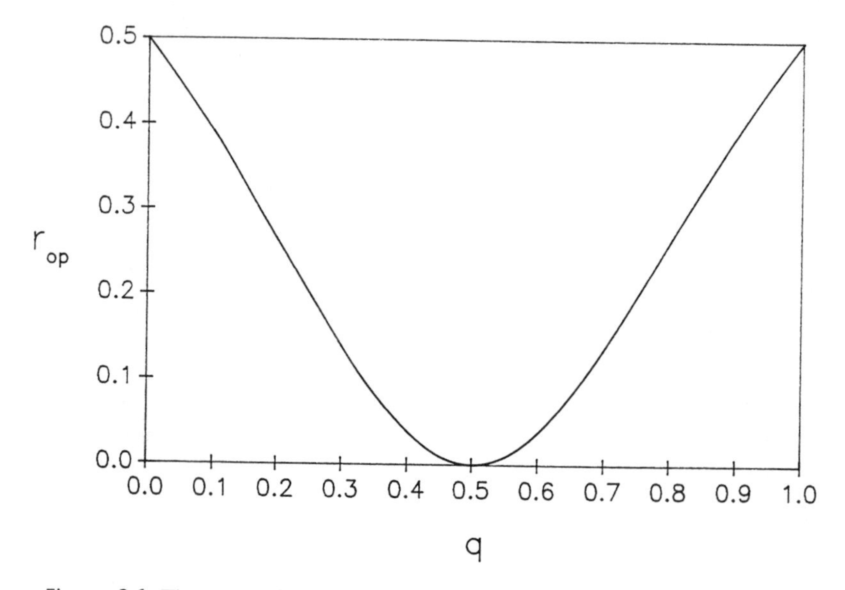

Figure 8.1 The correlation in heterozygosity, r_{op}, between offspring and parents. From Mitton et al. 1993.

Female choice of "good genes"

Many studies report that females select either the larger, brighter, more ebullient males available to them or the males defending desirable territories (Halliday 1987). Kodric-Brown and Brown (1984) argue that sexual selection focuses on characters that reveal true differences in physiological condition, characters that cannot be faked. Males more heterozygous for metabolic enzymes may have greater physiological efficiency and may exhibit superior viability, growth rate, and vigor (chapter 7). The selection for highly heterozygous individuals actively maintains—does not constantly erode—the genetic variation underlying the traits of choice. It is clear that choosing highly heterozygous males maintains the genetic variation underlying the characters selected by females, but the system must also be repeatedly perturbed from equilibrium so that the heritability of fitness can be maintained.

Are there circumstances under which the heritability of these characters can be maintained while under persistent sexual selection? If males are in an evolutionary race with debilitating parasites, frequency-dependent selection might provide a store of variation not eroded by directional selection; that is, females could be choosing the most "resistant" genotypes available (Hamilton 1980; Hamilton and Zuk 1982). In fact, the plumage color of male birds burdened by debilitating parasitic infections pales in comparison with that of healthy males (Zuk 1984). The key here is that the system never rests at equilibrium, and therefore fitness is either continually or recurrently heritable. But Kirkpatrick (1986) and Pomianowski (1987) modeled this mode of evolution of female choice and found it unable to increase the frequency of very rare alleles.

Charlesworth (1988) modeled sexual selection in a fluctuating environment and convincingly demonstrated that the choice of "good genes" can drive the evolution of female choice. His model included a balanced polymorphism and a locus influencing females' choice of male mates. The variation at the selected locus was maintained by environmental switches modeled as fluctuations in selection coefficients over time. A dominant choice allele influences females to mate with either heterozygotes or the most fit genotype available. The "choice" allele is initially rare but increases in frequency under some sets of conditions of linkage between loci, intensity of selection, and frequency of cycling of the environment. Females' choice of good genes can evolve because under weak selection in a fluctuating environment, the expected log fitness of the progeny of heterozygotes is always greater than the population mean fitness. When the choice allele determines the unerring choice of heterozygous mates and the choice locus and selected locus are not linked, the choice allele is always favored. This model does not exhibit the lines of neutral equilibria that dominate the models of sexual selection fashioned by Lande (1981) and Kirkpatrick (1982). The choice allele increases in frequency most quickly when the environment cycles are approximately ten generations long, but it does not increase in frequency when the environment cycles very rapidly, for rapid environmental changes create a negative correlation between parent and offspring fitnesses (Eshel and Hamilton 1984).

Multilocus Simulation of Sexual Selection

I built a multilocus model of sexual selection to explore the evolution of female choice of "good genes" in a finite population. This model departs from other models in that it assigns genotypes to individuals in a relatively small population, on the order of a population of sage grouse (Gibson and Bradbury 1987). Consistent with the theme of this chapter, viability and vigor were influenced by a small number of independently segregating loci. Appendix 2 presents the simulation program.

The simulated life cycle included two challenges—first a bout of viability selection and then a bout of sexual selection. The fitness potential was determined by one to ten independently segregating, overdominant loci that determined the rankings for both viability selection and sexual selection. Single-locus fitness potentials for the overdominant loci were initialized at the outset of each run, and the fitness potentials at the 1 to 10 loci were multiplied together to calculate the multilocus fitness potential. The truncation or threshold selection of the distribution of multilocus fitness potential determined the intensity of selection. The placement of the threshold was determined by the number of adults that the environment could support and the number of progeny produced: an abundance of youngsters was reduced to the carrying capacity of each generation by removing those individuals with the lowest fitness potentials. Generations were discrete. Choosy females mated with the male with the highest fitness potential in the population, or they chose randomly among the males tied for the highest fitness potential; their discrimination was perfect. The remainder of the females in the population chose randomly among the males available to them.

A single locus determined a female's mate choice behavior. Two alleles were at this locus—a recessive allele determining random mating and a dominant allele causing discriminatory mate choice. Simulations generally began with the allele for choice at a low frequency.

Environmental variability was simulated by exchanging the single-locus fitness po-

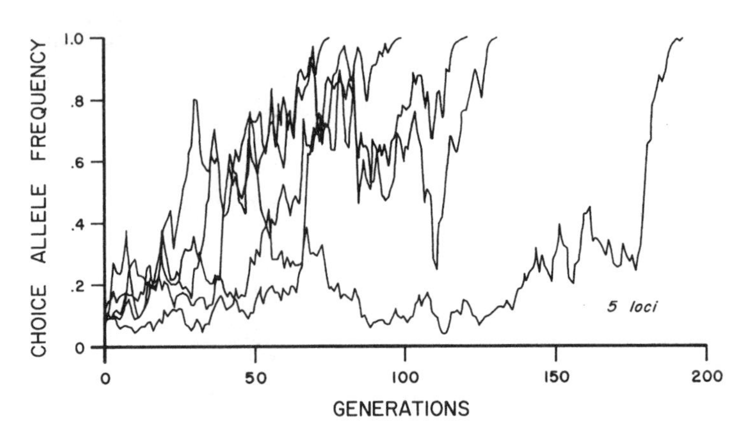

Figure 8.2 The evolution of female choice of "good genes" in a finite population. A dominant allele that causes females to mate with the male with the highest fitness potential increases to fixation. Fitness is determined by five unlinked, overdominant loci, and each generation viability selection reduced 300 progeny to 150 adults.

tentials of the homozygotes at a locus. The periodicity of the environment was specified at the beginning of each simulation.

Some sets of conditions favored the increase in frequency and fixation of the choice allele. Replicates of one set of conditions are presented in figure 8.2. For these simulations, five overdominant loci determine fitness potentials, and the simulations are started with allelic frequencies at each locus of $p = 0.9$ and $q = 0.1$ and genotypic proportions in Hardy–Weinberg equilibrium. Single-locus fitness potentials were 0.9, 1.0, and 0.1 for the *AA*, *Aa*, and *aa* genotypes, respectively, at each of the loci. In each simulation, the frequency of the allele influencing choice was initiated at a frequency of 0.1. The number of adults was held at 150 (with an approximately equal sex ratio), and the number of progeny was 300.

In all replicates of this set of conditions, the frequency of the choice allele rose; in the majority of cases, this frequency was fixed in the population. But many sets of conditions hold the choice allele at low frequencies or allow it to be lost from the population. The key to the evolution of female choice in this model is the continual or repeated departure from equilibrium conditions. If the simulation is modified so that the environment is constant, the choice allele is influenced primarily by genetic drift and usually is lost.

The exact conditions of selection intensity, periodicity of the environment, and number of loci contributing to fitness potential are not critical to our purposes. These all influence the evolution of female choice to some degree, and their influence would be modified by changes in the model intended to make it more biologically realistic. For example, if the dominant allele instilled a proclivity to mate with males with a high fitness potential, rather than unerringly determining a perfect choice, the results of simulations would certainly change. What is important to us is understanding that female choice can evolve relatively quickly in small populations if an allele arises that induces females to mate with vigorous, successful males. Given that this is theoretically possible, we would then want to know whether this type of female choice can be found in natural populations.

Genetic consequences of female choice

A few studies report the enhanced fitness of offspring as a consequence of female choice. The first clear demonstration of female choice that results in the superior genetic quality of their mates was found in a laboratory study of sexual selection in *Drosophila subobscura* (Maynard-Smith 1956). The mating ritual of this species includes a dance in which the female leads and the male follows in a series of sideways jumps. Females offered a choice between inbred and outbred males typically chose the outbred males for their greater agility and endurance. This choice was propitious, as the outbred males supplied more sperm with higher viability, resulting in an immediate fitness advantage. Similarly, the female choice of mates in *Drosophila melanogaster* resulted in greater fitness (Partridge 1980). The competitive abilities of progeny from females allowed to choose their mates were consistently higher than the competitive abilities of progeny from randomly bred females.

Female choice sexual selection in the great tit, *Parus major*, focuses on a character in males that is heritable and is associated with the offspring's fitness (Norris 1990, 1993). The plumage patterns are sexually dimorphic, and females prefer males with a large, black breast stripe, a character not seen in females (Norris 1990). To determine whether the size of the breast stripe was heritable and related to offspring quality, a cross-fostering study was conducted in Bagley and Wytham Woods, near Oxford (Norris 1993). The size of the male offspring's breast stripe was not correlated with the size of their foster fathers' stripe but was correlated with the size of their biological fathers' stripe. The heritability of the stripe's size was $h^2 = 1.44 \pm 0.62$. The size of the biological father's breast stripe was positively correlated with his number of surviving male offspring ($r = .49$, $P < .01$).

Experimental manipulations showed that female black widowbirds, *Euplectes progne*, prefer males with very long tails (Andersson 1982a). Black widowbirds are strikingly dimorphic in both color and form. Reproductive males are black except for a red epaulet on the wing, and females are brown and mottled. The female's tail is approximately 7 cm long, but the male's tail is about 50 cm long and is draped as a keel while the male flies with slow wing beats just above his territory. This unwieldy tail is carried only during the breeding season, and it is clearly part of the flight display.

Male widowbirds defend territories of 0.5 to 3 hectares on the Kinangop Plateau of Kenya. Although females breed with the males controlling the territory around the females' nest, they get no help from the males while building their nests or rearing their offspring. Some males' tails were shortened by cutting 25 cm from the tail feathers, and others were lengthened by gluing the 25-cm feather fragments to intact tail feathers. The tails of nine males were shortened; the tails of nine were lengthened; and the remainder served as manipulated and unmanipulated controls. The number of nests on each territory was monitored before and after the manipulation, so that each male served as his own control. Females clearly preferred the males with the experimentally lengthened tails. The average number of nests on the territories of experimentally manipulated birds was 1.9 for birds with lengthened tails, and 0.4 for birds with shortened tails. Female choice sexual selection in this species focused on an energetically expensive set of feathers used only during the mating season. The elaborate tail hinders flight, but a substantial investment in the tail and defense of a territory are needed to attract breeding females. Energetic investments in the tail and territorial defense cannot be faked and may be used as reliable measures of the male's physiological condition.

Fluctuating asymmetry and sexual selection

Fluctuating asymmetry is measured as small, random deviations from symmetry in otherwise bilaterally symmetrical characters (Palmer and Strobeck 1986). The frequency and amplitude of fluctuations from bilateral symmetry measure the ability to buffer development. A growing set of empirical observations indicate that fluctuating asymmetry is negatively associated with components of fitness and that females use the symmetry of sexually selected characters to assess a male's quality. For example, female zebra finches, *Taeniopygia guttata*, demonstrated a preference for males wearing colored leg bands arranged in symmetrical patterns over males with bands in asymmetric patterns (Swaddle and Cuthill 1994a).

Empirical studies of horses, insects, and birds indicate that performance and competitive ability increase with bilateral symmetry. For example, performance and symmetry are positively correlated in thoroughbred horses (Manning and Ockenden 1994). The performance of 73 flat-racing thoroughbreds was estimated using the official ratings compiled by handicappers. Symmetry was estimated with the fluctuating asymmetry of ten paired characters, four on the forelegs and six on the head. Although all the correlations between ratings and fluctuating asymmetry were negative, the correlation was statistically significant only for the distance from the cheekbone to the mouth ($r = -.33$, $P < .01$). The correlation was stronger with the mean of the ten measures of fluctuating asymmetry ($r = .48$, $P < .001$).

Symmetry is a better predictor of components of fitness than size in the Japanese scorpionfly, *Panorpa japonica* (Thornhill 1992). The size and symmetry of forewings were measured in two natural populations in Japan. The symmetry of mating flies was higher than the symmetry of a random sample of single flies. Males fight over dead arthropods, which they use as food and as nuptial gifts to secure copulations, and the winners of the fights were more symmetric but not necessarily larger than the losers. Asymmetry was negatively correlated with the number of copulations secured by males ($r_s = -.84$, $P < .001$). The survival of both males and females increased with symmetry.

The quintessence of the sexually selected character, the tail of the blue peacock, *Pavo cristatus*, reveals a relationship between fluctuating asymmetry and fitness. The number of eyespots, or ocelli, in a peacock's tail varies from 110 to 170, and symmetry is strongly correlated with the number of ocelli ($r = .84$, $P < .01$) (Manning and Hartley 1991). A male's mating success increases with the number of ocelli in blue peacocks (Petrie, Halliday, and Sanders 1991). Furthermore, a controlled breeding experiment demonstrated the fitness advantages of mating with males with elaborate tails (Petrie 1994). Females were randomly assigned to males, and once they mated, their eggs were incubated by broody chickens and then placed in an incubator. The hatched chicks were raised together, under standard conditions. The chicks' survival increased with the size of their ocelli ($r = .84$, $P < .01$), and the males' growth rate increased ($r = .74$, $P < .05$) with the size of their fathers' ocelli.

Female house sparrows, *Passer domesticus*, use the male's black bib in their choice of mates (Kimball 1995), favoring males with large, symmetric bibs. Asymmetry of the

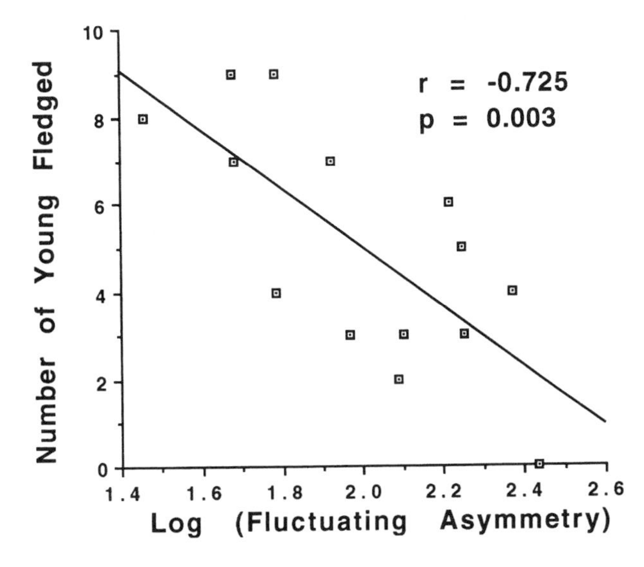

Figure 8.3 Fledgling success in the house sparrow, *Passer domesticus*, as a function of the asymmetry of the male parent's bib. From Kimball 1995.

bib is negatively correlated with the size of the bib, which is the same relationship between the size and symmetry of a sexually selected character as in the tail of the blue peacock (Manning and Hartley, 1991) and the tail of the barn swallow (Møller, 1992, 1994b). Male house sparrows with symmetric bibs began breeding earlier and fledged at a younger age (figure 8.3).

Male mating success increases with symmetry in the desert pupfish, *Cyprinodon pecosensis* (Astrid Kodric-Brown, pers. comm.). Bilateral symmetry was measured with preopercular pores, preorbital scales, pectoral fin rays, and pelvic fin rays from 80 territorial and 63 nonterritorial pupfish from Bottomless Lake State Park, New Mexico. Male reproductive success in Pecos pupfish is dependent on the defense of a breeding territory. Although the mean counts for these characters do not differ between territorial and nonterritorial males, nonterritorial males were substantially more asymmetric than males defending territories (table 8.3).

Table. 8.3 Fluctuating asymmetry at four meristic traits in territorial and nonterritorial Pecos pupfish from Bottomless Lakes State Park, NM.

	Territorial (80)			Nonterritorial (63)		
Traits	*X* ± SD	Range	%FA	*X* ± *SE*	Range	%FA
Preop. pores	5.6 ± 0.7	4–7	44	5.6 ± 0.8	3–7	68
Preorb. pores	3.3 ± 0.8	2–6	19	2.8 ± 0.9	1–6	52
Pectoral fin rays	15.4 ± 0.6	14–16	25	15.2 ± 0.6	14–17	56
Pelvic fin rays	5.8 ± 0.7	5–8	28	5.8 ± 0.8	4–8	76

Note: Data from Astrid Kodric-Brown, personal communication. The means for these traits do not differ between territorial and nonterritorial fish, but the percentage of fluctuating asymmetry is consistently lower in the territorial fish. Male pupfish must defend territories to breed.

Sexual selection acts on both the fluctuating asymmetry and the length of the tail in the barn swallow, *Hirundo rustica* (Møller, 1994a,b). The length and symmetry of the tail influence barn swallows' aerial maneuverability, viability, and reproductive success. Barn swallows are small (20 g) aerial insectivores. They are monogamous, and although pairs occasionally live alone, more frequently they nest in colonies of up to 100 pairs. Sexual dimorphism is slight, with the exception of the outermost tail feathers, which are approximately 20% longer in males. In contrast to the general pattern seen in morphological characters, fluctuating asymmetry of the tail is high, and fluctuating asymmetry is negatively correlated with tail length (Møller 1990) in males but not in females (Møller 1994a). To study the significance of the tail's size and symmetry, Møller manipulated the size and symmetry of the tails of 96 males in a population in Draghede, Denmark, when they arrived in the spring. He cut and/or glued their feathers in all combinations so that the influence of size and symmetry could be analyzed independently. Females preferred males with longer, symmetric tails. Males whose tails had been lengthened and made symmetric found mates earlier, had shorter intervals until the eggs were laid, and fledged more offspring than did males whose tails had been shortened and made asymmetric. The mean number of young fledged by males modified to have long, symmetric tails was 9, and the mean for males manipulated to have short, asymmetric tails was 4 (Møller 1992).

The size and symmetry of barn swallows' tails reflect their physiological condition in the present or recent past (Møller 1994b). The length of the tail reflects environmental conditions, principally rainfall and food availability, during the winter, when the tail feathers grow. The length of the tail generally increases during the first three years as the males gain skill in searching for and capturing insects in flight. Finally, the length of the tail decreases with the load of two ectoparasites—the tropical fowl mite, *Ornithonysus bursa*, and the feather louse, *Hirundoecus malleus*. For a character to honestly advertise the condition of an individual, the energetic cost must increase with the intensity of the signal or with the size and symmetry of the character. Møller (1990) argued that the size and symmetry of the tail reliably reveal the physiological condition of male barn swallows.

A study of the population at Kraghede, Denmark, showed that the tail's fluctuating asymmetry has a genetic basis and is related to several components of fitness (Møller 1994b). Parent–offspring comparisons yielded estimates of heritability to be $h^2 = 0.80 \pm 0.33$ for fathers and sons and $h^2 = 1.88 \pm 0.70$ for mothers and daughters. Males with more asymmetric tails arrived at the breeding site later and began incubating eggs later than did symmetric males. Asymmetric males were also less likely to acquire a mate; the tail asymmetry of single males was more than twice that of mated males. Comparisons of males that died and males that survived revealed that surviving males were more symmetric than the males that died.

Detection of sexual selection at specific loci

Allozyme variation at two loci was used to test for sexual selection in the milkweed beetle, *Tetraopes tetraopthalmus* (Eanes et al. 1977). Beetles were collected from nine natural populations on Long Island, New York, and groups composed of solitary individuals and copulating pairs were compared. Evidence was found for frequency-dependent sexual selection among females but not among males.

Laboratory studies of female choice revealed strong sexual selection in *Drosophila melanogaster* and *D. pseudoobscura* (Brittnacher 1981). The studies exposed 20 females to 10 virgin control males and 10 experimental males, homozygous or heterozygous for inversions on chromosome 2. After 3 hours, the females were isolated, and analyses of the karyotypes of their offspring were used to determine the father's karyotype. Relative to the fitness of heterozygous males (1.0), homozygotes had relative fitnesses of 0.56 and 0.58 in *D. melanogaster* and *D. pseudoobscura*, respectively.

Field studies of the polymorphic African butterfly, *Danaus chrysippus*, revealed sexual selection favoring heterozygotes at a locus controlling color patterns (Smith 1975, 1980, 1981). Color and pattern morphs are determined by three genes, each segregating two alleles; the *C* locus has a major effect on the color pattern. *CC* homozygotes, called *dorippus*, are lightly colored, whereas *cc* homozygotes, called *aegypticus*, are predominantly black and brown. *Cc* heterozygotes, called *transiens*, are usually intermediate, although a small proportion of these cannot be distinguished from *dorippus*. In males, *Cc* genotypes are larger than either of the homozygotes, but overdominance is not detected in females (Smith 1980).

Butterflies were sampled for 17 consecutive months in Dar es Salaam, Tanzania, where *Danaus* mate continuously, completing 13 to 14 generations per year. Butterflies were grouped into those captured as mating pairs and those captured as solitary individuals. Data from many collection dates were collapsed into two environmental periods—the hot, sunny, dry season and the cool, cloudy, rainy season. Comparisons of the morph frequencies in the mating and solitary groups yielded estimates of relative mating success (table 8.4). The heterozygotes are favored in both seasons, and the relative fitnesses of the homozygotes cycle between the seasons. The light-colored *CC* homozygotes enjoy great mating success during the hot, dry season, but they do poorly during the cool, rainy season. Conversely, the dark *cc* homozygotes enjoy great mating success during the cool, rainy season, but they do poorly during the hot, dry season. In each season, the selection coefficients for the disadvantaged homozygotes exceed .80.

Allozyme genotypes have been used to measure male mating success in natural pop-

Table 8.4 Male mating success for three color and pattern phenotypes of the African butterfly *Danaus chrysippus* at Dar es Salaam, Tanzania

		Phenotypes			
Period	Condition	dorippus	transiens	aegyptius	*N*
April–August 1975	mated	3	85	23	111
	unmated	149	307	155	611
	mating success	0.09	1.0	0.60	
September 1974–March 1975	mated	59	76	2	137
	unmated	209	196	47	452
	mating success	0.79	1.0	0.14	
April–August 1975	mated	5	61	26	92
	unmated	114	197	145	456
	mating success	0.18	1.0	0.64	

Note: Data modified from Smith 1981. Dorippus and aegyptius are homozygous genotypes, and transiens is heterozygous. In all study periods, male mating success is heterogeneous among the phenotypes and always favors the heterozygote.

Table 8.5 The proportion of heterozygous genotypes in a random sample of male *Colias* butterflies (flying) and in males with reproductive success (mating). Sexual selection favors heterozygotes at three allozyme loci.

		Percentage of Heterozygotes		
Locus	Date	Flying[a]	Mating[b]	Probability[c]
Colias eurytheme				
PGI	1984	40	77	***
	1985	52	67	*
PGM	1984	56	74	+
	1985	46	55	NS
G6PD	1984	46	85	***
	1985	47	72	**
Colias philodice eriphyle				
PGI	1984	52	74	***
	1985	56	85	***
PGM	1984	44	63	**
	1985	47	63	+
G6PD	1984	33	61	***
	1985	38	52	+
Average		46	69	

Note: Data from Carter and Watt 1988.

[a]Percentage of males in the field heterozygous for the enzyme.

[b]Percentage of males successfully siring broods. These genotypes were inferred from the mother's genotype and the distribution of genotypes in her brood.

[c]Probability that the percentages are the same: + = $P < .10$, *= $P < .05$, ** = $P < .01$, *** = $P < .001$.

ulations of the sulfur butterflies, *Colias eurytheme* and *C. philodice eriphyle*. The data are most extensive for the phosphoglucose isomerase (PGI) locus, for which there are enzyme kinetic data, demographic data, behavioral data, and data on mating success (Watt, 1977, 1983; Watt, Carter, and Blower 1985; Watt, Carter, and Donohue 1986; Watt, Cassin, and Swan 1983; see chapters 4 and 7). Kinetic analyses documented biochemical differences among genotypes and showed that the most common heterozygote is overdominant in the predominant environmental conditions. The heterozygotes have greater viability and fly over a greater range of temperatures than do the homozygotes. But the most remarkable aspect of this story is the relative mating success of the heterozygotes. Because a female uses the sperm from only a single male to fertilize her eggs, inspection of the female's genotype and those of her offspring allows the inference of the paternal genotype. A comparison of the genotypes of males flying in the field with the genotypes revealed by the paternity analysis revealed a strong and consistent male mating advantage for the heterozygotes (table 8.5). In *C. eurytheme*, the average frequency of PGI heterozygotes in flying males was 46%, but among the males with reproductive success, the average frequency of heterozygotes was 72%. In *C. philodice eriphyle*, the average frequency of PGI heterozygotes in flying males was

54%, but the average frequency of heterozygotes in males with reproductive success was 80%.

Similar differentials of mating success were found (Carter and Watt 1988) with the phosphoglucomutase and the glucose 6 phosphate dehydrogenase polymorphisms in *Colias* (table 8.5). At both these polymorphisms, the proportions of heterozygous males were higher in males with mating success than in the total sample of males caught in the field. The pattern of heterozygous advantage for male mating success in the three loci examined was relatively consistent across loci, across species, and over years. The frequency of heterozygous males in natural localities ranged from 33 to 56%, but the frequency of heterozygotes in males siring broods was 52 to 85% (table 8.5). When the data were combined across years, loci, and species, 46% of the available males were heterozygous, but 69% of the mating success was captured by the heterozygotes. If these loci are assumed to be independently assorting and additive in their effects, the relative mating success of the triple heterozygotes is 1.00, and the relative mating success of the triple homozygotes is only 0.38.

Female *Colias* butterflies become choosier as they age, and their more discriminating choice is reflected in greater fitness differentials among the PGI genotypes of males (Watt et al. 1986). This study categorized genotypes by the V_{max} / K_m ratio and contrasted the kinetically favored genotypes with all other genotypes. All the kinetically favored genotypes were heterozygotes, but not all heterozygotes in this multiallelic system were considered to be favored. Female choice is not based on competition among males but is exercised with respect to individual males. When young, females are minimally discriminating in regard to males, but their discrimination increases with age. A courtship bout begins when a flying male approaches a female sitting on a leaf. To get the attention of a female, a male must hover above and behind her. An unmated female may accept a male after as little as 5 seconds of courtship flight, but females that have already mated once or twice typically demand much longer courtship flights. This is not romance; this is an endurance test. This field study showed that in comparison to virgin females, older, choosier females of both *C. eurytheme* and *C. p. eriphyle* mated with higher proportions of kinetically favored genotypes than did virgin females—71% of the mates of females mating for the first time were favored genotypes, but favored genotypes were 92% of the mates of experienced females.

Male mating success increases with heterozygosity in laboratory populations of brine shrimp, *Artemia franciscana* (Zapata, Gajardo, and Beardmore 1990). Eighty sexually mature virgin females and 160 males were placed in an aquarium, and mating pairs were removed once they joined in amplexus. The genotype of each individual was obtained for five allozyme polymorphisms. At single loci, heterozygotes had higher male mating success at two of the loci and lower mating success at one locus. When individual heterozygosity was examined, males' mating success increased with the number of heterozygous loci (table 8.6). In comparison to completely homozygous genotypes, which were arbitrarily assigned a fitness of 1.0, individuals heterozygous for three and four loci had fitnesses of 2.99 and 2.55, respectively.

Horn growth in rams is related to allozyme heterozygosity in Rocky Mountain bighorn sheep, *Ovis canadensis canadensis* (Fitzsimmons, Buskirk, and Smith 1995). The males use their horns in displays and as weapons and shock absorbers in collision fighting (Geist 1966). Horn size is important when males are struggling to establish their position in the breeding hierarchy (Geist 1971; Hogg 1987). Hogg (1987) specu-

Table 8.6 Male mating success as a function of allozyme heterozygosity in the brine shrimp, *Artemia franciscana*

	Individual Heterozygosity				
	0	1	2	3	4
Male mating success	1.00	1.39	1.45	2.99	2.55
±Standard error		±.19*	±.14**	±.95*	±.25**

Note: Data from Zapata, Gajardo, and Beardmore 1990. * and ** indicate values significantly different from 1.00 at the 5% and 1% level, respectively.

lated that by choosing males with large horns, females gain mates of superior genetic quality, for both resistance to disease and superior foraging ability are needed to grow large horns.

Fitzsimmons measured the horns of 113 bighorn rams taken by hunters in Wyoming in 1989 and 1990 and estimated the annual increase in the horns' volume. Samples of tissue from each animal were analyzed with electrophoresis to identify their genotypes at four polymorphic loci. The sample was broken into two groups, those heterozygous for zero to one and those heterozygous for two to four loci. Multilocus heterozygosity accounted for 25% of the variability of growth in years 6 through 8, a time when growth in the highly heterozygous group was significantly higher than that in the less heterozygous group. Growth rates did not differ significantly between the two groups in the first five years of growth, or after year 8, although increases in horn volume in these years favored highly heterozygous rams. Rams typically begin to breed in year 7 or 8, after two or three years of differential growth between the heterozygosity groups. Growth was related to heterozygosity at individual loci as well. The growth of heterozygotes significantly exceeded the growth of homozygotes for transferrin, glyoxyalase, and lactate dehydrogenase. Symmetry of the horns was not related to heterozygosity (S. Buskirk, pers. comm.).

Sexual selection favoring heterozygous males was detected in the marine snail, *Littorina mariae* (Rolán-Alvarez, Zapata, and Alvarez 1995). Snails were collected from the marine alga *Fucus vesiculosus* in the intertidal zone in Muros-Noya Ria, Galicia, Spain. Individuals were characterized as being solitary or copulating, and each one's genotype was identified for nine polymorphic allozyme loci. Reproductive success was estimated as the proportion of copulating individuals for each genotype or heterozygosity class. Although larger females had greater mating success, allozyme heterozygosity had no connection to it. However, mating success did increase with heterozygosity in males, a trend that was strong in young males but not statistically significant in older males. The male mating success of heterozygosity classes 0, 1, 2, and 3 through 5 was 1.00, 1.43, 1.81, and 2.30, respectively.

Summary

A model of female choice sexual selection of "good genes" was presented. This model extended earlier models by incorporating the threshold selection of overdominant loci in fluctuating environments. Threshold selection shifts the emphasis from the average fitness of progeny to the relative production of highly fit progeny. Both the fluctuating environment and the varying intensity of selection keep the system from reaching equilibrium, thereby maintaining the heritability of fitness and the incentive for females to choose males on the basis of their genetic quality. In this system, a new allele that triggers female choice can rise from low frequencies to fixation.

Empirical data consistent with the model were presented. Sexual selection generally favors symmetric males. This observation is consistent with the model because some empirical studies show that symmetry increases with heterozygosity (chapter 7). More direct support for the model is found in sexual selection in butterflies, brine shrimp, and marine snails, which favors heterozygosity for allozyme loci.

9

Patterns Among Species

Summarizing, we might say that the usual explanation that polymorphisms are the consequences of heterozygote superiority is basically wrong. . . . What we should say is that natural selection results in a subset of alleles with the average heterozygote superiority so that *polymorphism and heterosis are a joint consequence of natural selection rather than the cause of each other.*

L. R. Ginzburg, "Why Are Heterozygotes Often Superior in Fitness?" (1979)

Electrophoretic surveys of genetic variation of proteins provide estimates of the levels of genetic variability in a large number of plants (Brown 1979; Hamrick and Godt 1990; Hamrick, Linhart, and Mitton 1979) and animals (Nevo 1978; Nevo, Beiles, and Ben-Shlomo 1984; Powell 1975; Selander 1976). The percentage of loci polymorphic ranges from zero in elephant seals (Bonnell and Selander 1974) and cheetahs (O'Brien et al. 1983, 1985) to 92% in quaking aspen (Cheliak and Dancik 1982) and 100% in the mussel *Modiolus auriculatus* (Nevo et al. 1984). One of the most enduring objectives of electrophoretic studies is to understand the forces that produce differences among species in genetic variability. This chapter briefly summarizes several theories and some data relevant to this objective.

The Niche-Variation Hypothesis

The niche-variation hypothesis was presented (Dobzhansky 1970) as an explanation for the pattern of chromosome inversion frequency seen in *Drosophila willistoni* (Da Cunha et al. 1959) and *Drosophila robusta* (Carson 1959, 1965; Carson and Heed 1964). Populations of *Drososphila willistoni* near the center of their geographic distribution contain 14 to 18 inversions, but geographically peripheral populations contain as few as 1 to 4.

Dobzhansky perceived an axis of environmental heterogeneity parallel to the continuum in inversion heterozygosity between central and marginal populations. Geographically central populations had much more heterogeneous environments than peripheral populations, and Dobzhansky proposed that the inversion multiplicity enabled central populations to exploit their environments more efficiently than could a population with a uniform karyotype. Inherent in this hypothesis was the assumption that in the microhabitats of this complex environment, inversion genotypes would differ in fitness, perhaps with one genotype superior in one microhabitat but inferior in others. The lower levels of inversion polymorphism in peripheral populations were due partly to the simpler environments and partly to directional selection not typical of central populations. Studies of field behavior of *Drosophila persimilis* (Powell and Taylor 1975) and the intertidal snail *Tegula funebralis* (Byers and Mitton 1981) demonstrated that various genotypes perceive environmental heterogeneity on a microgeographic scale and "choose" alternative enviroments.

The pattern of protein variation does not follow the pattern of inversion polymorphism. Rather, the variation in peripheral populations is equal to or greater than the variation in central populations (Soulé 1973). Nevertheless, the niche-variation hypothesis continued to be tested with allozyme data, with mixed results.

Allozyme variation decreases with the number of sympatric congeners in Hawaiian *Drosophila* and in Plethodontid salamanders. These observations may be relevant to the niche-variation hypothesis, for one measure of the niche breadth of a species is the number of closely related sympatric species. A solitary species is assumed to occupy all the appropriate niche space, but a species with several or many sympatric congeners is assumed to be excluded by competition from some portion of the fundamental niche space and restricted to a narrower realized niche. Note the implicit, dubious assumption that the volume of appropriate niche space is a constant.

The first test of the relationship between the number of congeneric species and allozyme variability was conducted with Hawaiian *Drosophila* (Johnson 1973b). Genetic variation was assayed at the same six loci in 48 species of picture wing *Drosophila.* All six loci were scored in 38 species; five were scored in 9 species; and only four loci were scored in 1 species. Genetic variation and the abundance of sympatric species were negatively correlated, with the number of alleles per species decreasing as the number of *Drosophila* on that island increased (figure 9.1).

A similar result was obtained with Plethodontid salamanders (Karlin, Guttman, and Rathbun 1984). Genetic variation was scored at 26 loci in 27 populations of *Desmognathus fuscus*, and the latitude, elevation, and the number of species of salamanders at each site were determined. The greatest species diversity for Plethodontid salamanders was centered in the Blue Ridge Mountains, and the number of species declined in all directions. As the number of sympatric species of salamanders fell, the average heterozygosity within populations increased (figure 9.2). Spatial autocorrelation analysis suggested that the variable that best described the geographic pattern of average heterozygosity was the number of syntopic congeners at each site. If the niche breadths of the *Drosophila* and *Desmognathus* in these studies decreased with the number of closely related, sympatric species, the results of these studies were consistent with the niche variation hypothesis.

The most direct test of the niche-variation hypothesis compared the genetic variability within a group of closely related, sympatric species exhibiting variation in their niche widths. Sabath (1974) sampled 11 species of Drosophilid flies from plots in a 140 × 80 m study plot in Indiana and estimated the niche width of each species by the method devised by Colwell and Futuyma (1971). The heterozygosity of each species was estimated using electrophoresis of six enzyme systems, revealing five to nine loci

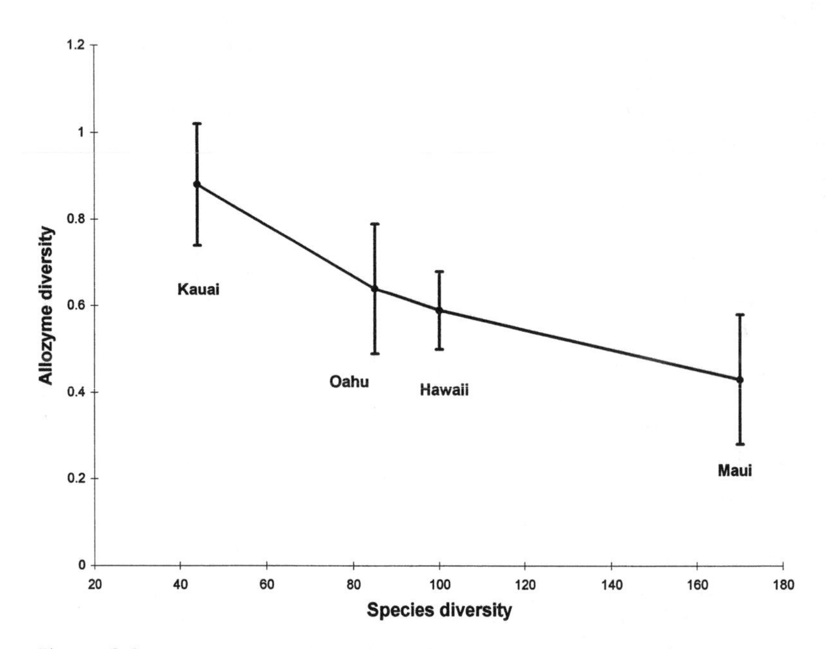

Figure 9.1 Allozyme diversity of Hawaiian *Drosophila* as a function of species diversity. Allozyme diversity is expressed as the number of electromorphs observed on the island divided by the number of species examined on that island. From Johnson 1973b.

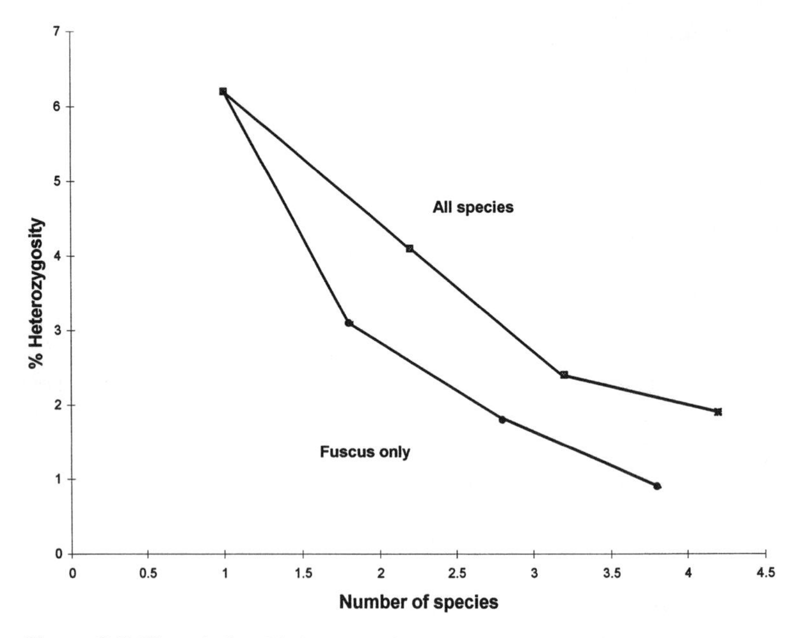

Figure 9.2 The relationship between the mean percentage of heterozygosity and the number of species sympatric with the dusky salamander, *Desmognathus fuscus*. Genetic variation declines in *D. fuscus* and in all species examined with an increase in the number of salamanders at a site. From Karlin, Guttman, and Rathbun 1984.

per species. Protein heterozygosity was not related to niche width. The strength of Sabath's study was in the quantitative estimation of niche breadth among 11 sympatric, closely related species; its weakness was in its estimation of genetic variation, which was based on five to nine loci per species. Fortunately, the same protein systems were surveyed in each species, so the estimates of genetic variation, although narrowly based, were highly comparable.

Enzyme activity is highly dependent on temperature, and therefore the range of temperature experienced by a population may be a measure of niche width that is especially pertinent to the forces acting on enzymes. Somero and Soulé (1974) tested the niche-width hypothesis by comparing the range of temperature experienced by a species and its protein heterozygosity in 13 marine fishes. They found no correlation between these measures. The power of this test was undoubtedly diminished by the use of distantly related, allopatric species.

Another measure of niche breadth, subject to several qualifying disclaimers, is the geographic range of a species. Species with greater geographic ranges, all else being equal (and all else is never really equal), exploit a greater range of environments. Nevo (1978) reported that cosmopolitan species had higher proportions of loci polymorphic than temperate (but not tropical) species did but that levels of heterozygosity were not different among these groups. In a study of genetic variation in 1,111 species of plants and animals, Nevo, Beiles, and Ben-Shlomo (1984, p. 109) concluded, "Protein diversity of species analyzed here is generally positively correlated with broader geographic, climatic, and habitat spectra."

In a multivariate study of genetic variation and life-history variation in plants, plants that differed in the size of their geographic ranges were heterogeneous for their levels of genetic variation (Hamrick, Linhart, and Mitton 1979). The researchers used a polymorphic index, which they defined as $PI = 1 - \Sigma p_i^2$, where p_i is the frequency of the ith allele at a locus. This measure gives the proportion of heterozygosity expected when there is random outcrossing with no selection. For species with endemic, narrow, regional, and widespread distributions, the values of PI were 0.086, 0.158, 0.185, and 0.120, respectively. Of the 35 widely distributed species, 29 were categorized as weedy or early successional, and these tended to be self-pollinated species with relatively low levels of genetic variation. The researchers concluded that non-weedy species with broad geographic distributions generally had high levels of genetic variability.

In a study of genetic variation within and among populations of 122 species of plants, the amount of genetic variation within a species increased with the geographic range (Hamrick 1983). Clearer patterns are seen in more recent surveys of the largest numbers of species (Hamrick and Godt 1990; Nevo et al. 1984). Heterozygosity increases regularly with geographic range in plants and less regularly in animals (table 3.4). In both vertebrates and invertebrates, habitat generalists had significantly higher levels of heterozygosity than did habitat specialists (table 3.4). Closely related species are expected to reveal clearer patterns between levels of genetic variation and the variables that influence variation. For example, Loudenslager (1978) reported that genetic variation in the genus *Peromyscus* increased with the area of geographic distribution (figure 9.3).

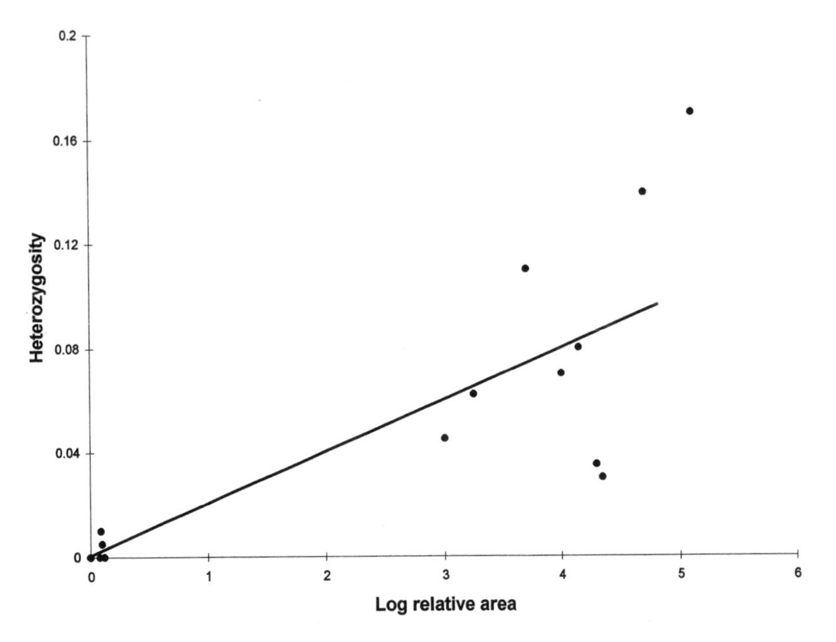

Figure 9.3 Allozyme heterozygosity increases with the size of the geographic distribution in 14 species of *Peromyscus*. From Loudenslager 1978.

Population Size: Neutral Theory

In only those mutations that are not detected by natural selection, the relationship of heterozygosity to population size is $H = 4N_eu/4N_eu + 1$, where H, N_e, and u represent average heterozygosity, effective population size, and the neutral mutation rate, respectively (Kimura and Crow 1964). Electrophoresis is not expected to detect all the neutral mutations, but this equation can be modified (Ohta and Kimura 1973) to represent electrophoretically detectable heterozygosity: $H = 1 - 1\sqrt{(1 + 8N_eu)}$.

Thus, larger populations are expected to contain higher levels of genetic variability than small populations do. This relationship between gene diversity and population size was used to construct a specific test of neutral theory (Nei and Graur 1984). The preceding equations that relate heterozygosity to effective population size and mutation rate rely on the neutral mutation rate. Although the rates of deleterious mutations are approximately constant per generation, the neutral mutation rate is expected to be constant per year (Nei 1975, pp. 31–34) and therefore to be a function of generation time.

For this reason, Nei and Graur (1984) tested the relationship between gene diversity and the product of population size and generation in 77 species of animals. For each species, they estimated gene diversity, *PI*, in the same way that Hamrick, Linhart, and Mitton (1979) defined the *polymorphic index*, which is the heterozygosity expected under random mating and no selection. Recognizing the difficulties of estimating N_e, Nei and Graur estimated the population size of the entire species. The geometric mean effective population size over several or many millennia is actually the variable of interest, and Nei and Graur reasoned that the total size of the species today probably puts an upper bound on the estimate. They found a highly significant correlation ($r = .65$, P

< .001) between gene diversity and the product of the number of individuals and generation time. These and other results of this study were judged to be consistent with predictions of neutral theory. However, these data are also consistent with a prediction based on natural selection (see "Fecundity").

Population Size: Experimental Data

Studies with experimental populations of *Drosophila* suggest that population size increases with heterozygosity for populations limited by space and/or nutrients. For example, following studies of *D. pseudoobscura* that reported differential fitnesses of genotypes carrying the Arrowhead and Chiricahua gene arrangements, Beardmore, Dobzhansky, and Pavlovsky (1960) estimated the productivities of monomorphic and polymorphic populations. Populations homozygous for Arrowhead and polymorphic and homozygous for Chiricahua had equilibrium population sizes of 230, 327, and 261, respectively, with biomasses of 190, 261, and 164 milligrams, respectively. Populations polymorphic for these gene arrangements supported more individuals and more biomass than did homozygous populations.

Laboratory experiments with *Drosophila melanogaster* demonstrated that an increase in genetic variation increases the relative population fitness of laboratory strains. Relative population fitness was defined as either the number of individuals or the biomass maintained in populations limited by both food and space (Carson 1961). A population was allowed to come to equilibrium, and then a single male—the F_1 of a female from the experimental population and the male from another strain—was introduced. Both biomass and size of the experimental population and the control populations were measured regularly. Within three weeks of the introduction, the experimental population began to grow, reaching approximately 3.5 times the size of the control population at 20 weeks. Results were similar in the reciprocal experiment, in which the donor strain and the recipient strain were switched. The introduction of the haploid genome from another strain dramatically increased the number of flies and the biomass of flies maintained in a population vial.

Drosophila serrata from Popondetta, New Guinea, and Sydney, Australia, were placed in population cages to examine the rate of adaptation to artificial conditions as a function of the amount of genetic variability within the population (Ayala 1965). Cages were established with flies from a single locality and with a mixture of flies from both localities. The geographic distance and the ecological differences between the collection localities ensured that the cages with a mixture of flies would have greater genetic variability than would the cages containing flies from a single locality. The size of the mixed-locality populations increased more rapidly than did that of the single-locality populations, and the mixed-locality populations almost always contained more individuals than did the single-locality populations. These results are comparable to those obtained by Carson (1961) with laboratory strains. Together, these studies indicate that the phenomenon is general to populations of fruit flies.

Laboratory experiments with *D. melanogaster* were designed to measure whether overdominance and overcompensation were associated with the superoxide dismutase (SOD) locus (Peng, Moya, and Ayala 1991). The data revealing overdominance associated with this locus were presented in chapter 4. Overcompensation is the enhanced efficiency of a mixture of genotypes relative to the performance of populations that contain just a single genotype. The performance of genotypes and mixtures of genotypes was estimated with production, measured as the number of flies emerging per female from a vial in which females had 24 hours to lay eggs. Replicated experiments were conducted at two temperatures, at two densities, and with SOD *FF*, *FS*, and *SS* genotypes, and with a mixtures of these genotypes. In 10 of 12 comparisons, the production of the mixtures of genotypes exceeded the production of pure genotypes. The production of a mixture of *SOD* genotypes exceeded by an average of 12% the production of populations containing single *SOD* genotypes. Thus, a mixture of genotypes was better able to utilize scant resources than was a group of flies bearing the same genotype.

The higher numbers of individuals and greater biomass in mixed populations suggest that genetic variability allows more efficient use of space and or nutrients. This tentative conclusion is consistent with studies of nutrient budgets of individuals exhibiting different levels of enzyme heterozygosity (chapter 7).

Population Size: Surveys of Natural Populations

In lizards, fishes, mammals, marine invertebrates, and *Drosophila*, there is a general pattern of increasing heterozygosity with increasing population size (Soulé 1976), a pattern this persists when all the data are pooled (figure 1.4). The correlation between heterozygosity and population size is $r = .70$. Population size, as used by Soulé (1976), does not refer to a representative population but, rather, is an eclectic measure based on density, geographic range, and vagility. Although theory based on neutral mutations predicts increasing variability with greater population size, there is a poor fit between these data and the neutral expectations, for small populations have more variation than expected and large populations have much less variation than expected. Soulé did not present protein variability as strictly neutral but mentioned life cycle and tissue heterogeneity as phenomena that would produce overdominance at structural loci. He referred to the influence of protein variabiltiy as "icing on the cake" (Soulé 1973).

The relationship between average heterozygosity and "species size" was examined in a survey of 1,111 species of plants and animals (Nevo et al. 1984). *Species size* is the total number of individuals in the species. The mean heterozygosities were 0.053, 0.066, 0.077, and 0.090 in species with small, medium, large, and very large numbers of individuals, respectively ($P < .01$).

Fecundity

All species produce an excess of offspring. To attain the average fitness in a population that is neither growing nor shrinking, a female of an obligately sexual, dioecious species needs to produce only two offspring that will live through the reproductive age. All species are more fecund than this necessary minimum, and the range in fecundity among species is great. *Fecundity* here refers to the total number of offspring that a lucky and healthy female could attempt to bring into the world. For example, a human female is capable of reproducing for 25 years and of giving birth approximately once a year. Thus the fecundity of humans is approximately 25. Although the prospect of having 25 children is inconceivable to most of us, it is important to appreciate here that this value of fecundity—shared by whales, elephant seals, and moose—is at the very bottom of the range of fecundity. At the other end of the scale we have eels, oysters, sturgeons, and pines, with fecundities in the dozens or hundreds of millions. This range of fecundity exceeds six orders of magnitude, and unlike the elusive "effective population size," it is easily estimated.

Regardless of her fecundity, the number of an obligately sexual female's offspring that survives to reproduce will be, on the average, two (when the population size is stable). Therefore, approximately 20 offspring of a female elephant seal will die, and 2 will live to reproduce. Contrast the elephant seal with a female American eel, who breeds once, spilling an excess of 2 million eggs into the sea. Again, 2 offspring will survive to reproduce, but more than 2 million will die. The reproductive excess of eels is thus five orders of magnitude greater than that of elephant seals.

What are the consequences of these different magnitudes of reproductive excess, or fecundity? Very simply, the opportunity for selection increases with fecundity (Williams 1966, 1975). This perspective does not assert that all mortality is selective—clearly it is not. For example, the larvae of mussels and the seeds of pines are often carried to substrates that are inimical to normal development and growth. But even if we were to assume that 90% of all mortality was a matter of luck, completely independent of genotype, the conclusion would be the same.

For simplicity, assume that random mortality precedes selective mortality, and consider the consequences of imposing random mortality that removes 90% of the populations of young elephant seals (fecundity = 20) and eels (fecundity = 2 million). Accordingly, random mortality will claim 18 of 20 seal pups and 1.8 million elvers. The 2 elephant seals that survive to reproductive maturity will be selected from the remaining family of 2, and the 2 eels will be selected from the remaining family of 200,000. Note that there is no room left for viability selection in the elephant seals but that there is an opportunity for very strong selection in eels. It is inconceivable that the intensity of selection typical of high fecundity and low fecundity species could be equal. Even if all the mortality of low fecundity species were selective and only 1% of the mortality of high fecundity species were selective, the opportunity for selection in highly fecund species would vastly exceed that in species with low fecundity.

Natural selection can be categorized into three general groups: directional selection, diversifying selection, and balancing selection (Dobzhansky et al. 1977). It is a commonly held opinion, weakly buttressed by empirical data (Endler 1986), that a major proportion of the selection in populations is balancing selection. If this is so, highly fecund species should be able to maintain more genetic variation in populations than species with low fecundity can.

The relationship of heterozygosity to fecundity was tested first among a large sample of plants for which estimates of gene diversity had been published (Hamrick, Linhart, and Mitton 1979) and then among a large set of animals. The sample of animals and their heterozygosities were taken from a review by Powell (1975). The fecundity of an animal species was estimated as the number of eggs that might be produced by a female enjoying both good health and a long life. Each plant species was placed into one of four classes of fecundity. Species in classes 1, 2, 3, and 4 produced $<10^2$, 10^2–10^3, 10^3–10^4, and $>10^4$ seeds, respectively. The values for heterozygosity and fecundity are listed for each species in appendix 3. The correlation between gene diversity and log fecundity among species of plants was .35 ($P < .01$, figure 9.4) and among species of animals was .66 ($P < .001$, figure 9.5).

The relationship of heterozygosity to fecundity was also tested with the data on an-

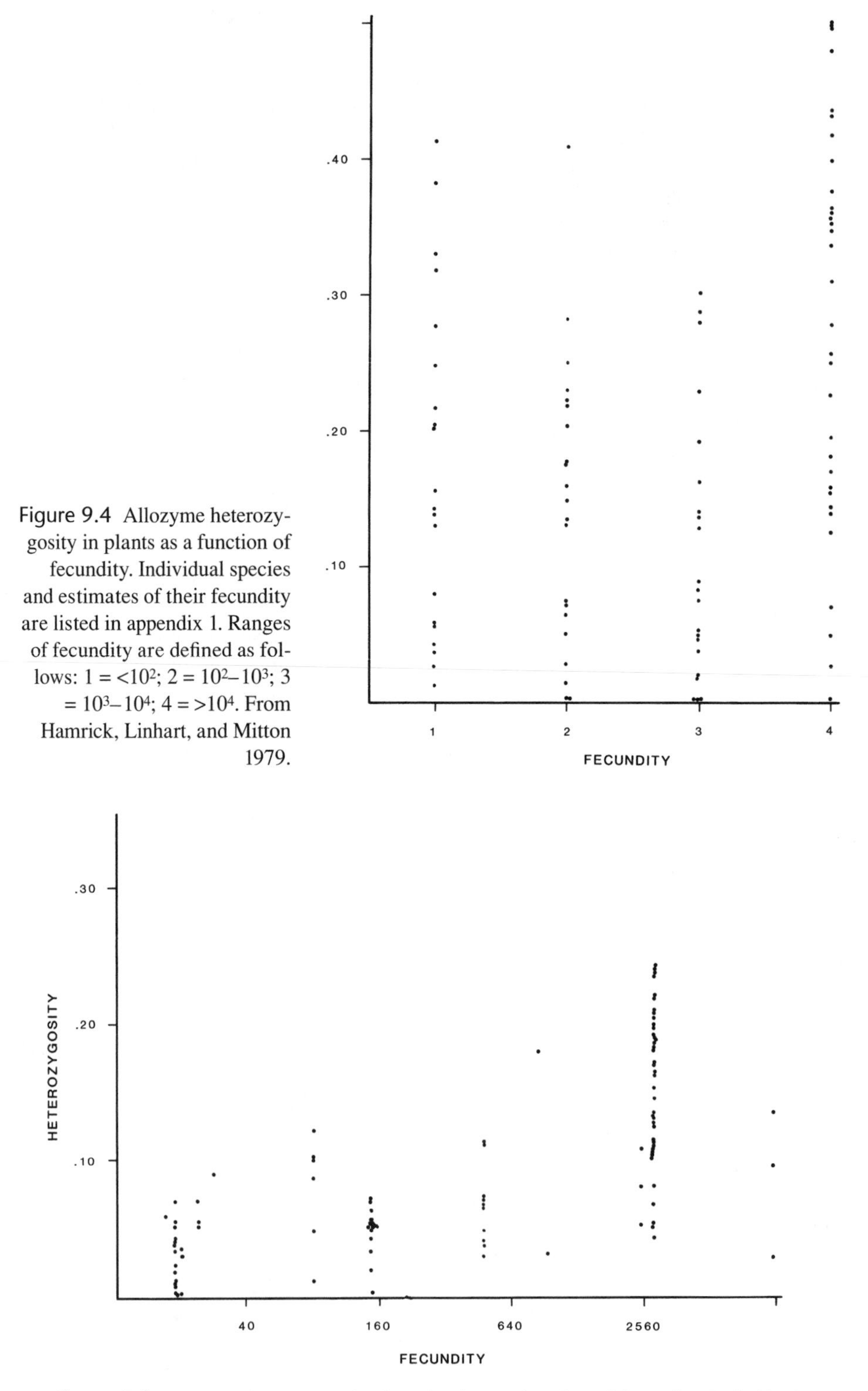

Figure 9.4 Allozyme heterozygosity in plants as a function of fecundity. Individual species and estimates of their fecundity are listed in appendix 1. Ranges of fecundity are defined as follows: 1 = $<10^2$; 2 = 10^2–10^3; 3 = 10^3–10^4; 4 = $>10^4$. From Hamrick, Linhart, and Mitton 1979.

Figure 9.5 Allozyme heterozygosity in animals as a function of fecundity. Individual species and estimates of their fecundity are listed in appendix 1. From Powell 1975.

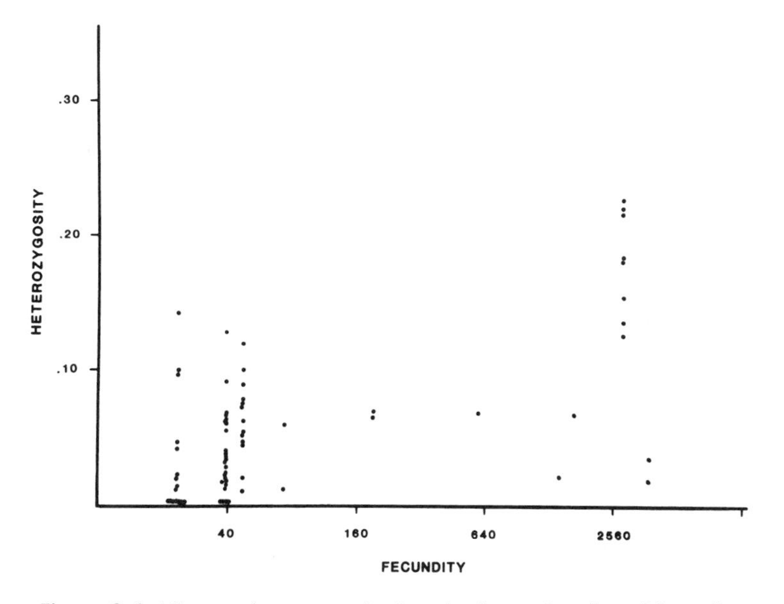

Figure 9.6 Allozyme heterozygosity in animals as a function of fecundity. Individual species and estimates of their fecundity are listed in appendix 1. From Nei and Graur 1984.

imals compiled by Nei and Graur (1984) after estimates of fecundity were added for each species (appendix 3). Just as in the earlier studies (figures 9.4 and 9.5), the level of genetic variation exhibited by a species increased with fecundity ($r = .65$, $P < .001$, figure 9.6).

Each of these analyses examines a heterogeneous set of species. Fruit flies and deer mice are included in one data set, and Engelmann spruce, orange trees, and annual ryegrass are included in another. If genetic variation is associated with life-history variation in general, and fecundity in particular, the relationships might be more apparent if the comparisons were restricted to a related group of species. To maximize the variation in genetic variation and in life-history traits within a group, we chose bony fishes for this analysis (Mitton and Lewis 1989). Bony fishes are the most numerous and the most diverse group of vertebrates. For example, in the 80 species we examined, average heterozygosity varied from 0.00 to 0.18, maximum weight varied from 4 to 73,900 grams, and lifetime fecundity varied from 69 to 30,000,000. Fecundity was estimated in three ways. F_t, the total lifetime fecuundity, was defined as the number of eggs produced by a healthy individual in a lifetime. F_w, the fecundity per gram, was defined as the F_t per unit body mass. F_a is F_t adjusted for mortality to estimate fecundity during an average lifetime rather than the maximal life span. All the correlations between fecundity and average heterozygosity were positive ($F_t = 0.15$, $F_a = 0.12$, $F_w = 0.25$), but only the correlation with F_w reached statistical significance ($P < .05$). The 12 species with the highest values of F_w ($F_w = 720$ eggs per gram, $H = 0.056$) had an average heterozygosity 50% higher than did the 12 species with the lowest values of F_w ($F_w = 13.2$ eggs per gram, $H = 0.037$). Similarly, the 3 species with F_t greater than 10 million (*Anguilla anguilla*, *Gadus morhua*, *Ictiobus cyprinellus*) had an average heterozygosity of $H =$

0.121, whereas the 3 species with fecundities lower than 1,000 (*Amblyopsis rosae*, *Amblyopsis spelaea*, *Chologaster agassizi*) had an average heterozygosity of $H = 0.011$ ($P < .05$). The 12 species with F_t greater than 2 million had an average heterozygosity of $H = 0.068$, and the 7 species with F_t less than 1,800 had an average heterozygosity of $H = 0.024$ ($P < .05$). The relationships between heterozygosity and fecundity were in the expected direction, but they were weak. Clearly, other life-history variables are expected to have an impact on the amount of genetic variation within populations.

Other relationships between genetic variation and life-history variables discovered in these data are summarized in figure 9.7. Heterozygosity was negatively correlated with maximum body size, length at maturity, and egg size. F_t increased with maximum body size and inversely with egg size. The strongest correlations were among life-history variables. Egg size and length at maturity were strongly correlated ($R_s = .84$, $P < .001$). This pattern of associations indicates that fishes that mature quickly and release small eggs have the highest levels of genetic variation. Fecundity increased with maximum body size and length at maturity.

The following generalizations can be made from these relationships: Larger fish tend to mature later, to produce larger eggs, and to be more fecund. Fecundity increases with decreasing egg diameter and increasing body size, but it is not related to length at maturity. Fishes that mature early and produce small eggs maintain the highest levels of genetic variation; these fishes have the highest capacity for increase. Thus, in bony fishes, high levels of genetic variation are found in species with life-history characteristics that conform to the general notion of *r*-selected species.

The comparative study of genetic variation and life-history variation in bony fishes (Mitton and Lewis 1989) was criticized for not having a random sample of bony fishes and for using relatively weak correlations to draw conclusions about evolutionary forces acting on genetic variation (Waples 1991). The available data (Nevo et al. 1984) were simply not a random sample of the approximately 21,000 species of bony fishes. Comparative studies rarely use a random sample of the taxa to which they are considered to apply. However, the 80 species contained a wide variety of taxa, and, perhaps more important, they provided an extraordinary range of easily quantifiable life-history characteristics. Associations between genetic variation and life-history variation have been overlooked, or have not been found, because they tend to be weak. Both life-history variables and estimates of genetic variation are made with substantial error, and therefore, comparative studies with small numbers of species, small numbers of loci, or narrow ranges in either life-history variables or genetic variability will probably not reveal statistically significant associations (Archie 1985; Schnell and Selander 1981).

Comparative studies with large numbers of species and wide ranges of variability have the capability of revealing statistically weak but biologically relevant associations of general evolutionary importance. If the study of bony fishes were the only study to discover an association between genetic variability and fecundity, this relationship could be justifiably overlooked in favor of stronger correlations (figure 9.7). But this analysis was initiated to further test the increases in heterozygosity with fecundity found in plants (Hamrick et al. 1979, figure 9.4) and in two heterogeneous sets of animals (figures 9.5, 9.6). The correlation between heterozygosity and fecundity was weak, but it was consistent with previous studies.

Thus, in both plants and animals, the level of genetic variation increases with fecundity, which may serve as a gauge of the opportunity for balancing selection. This hy-

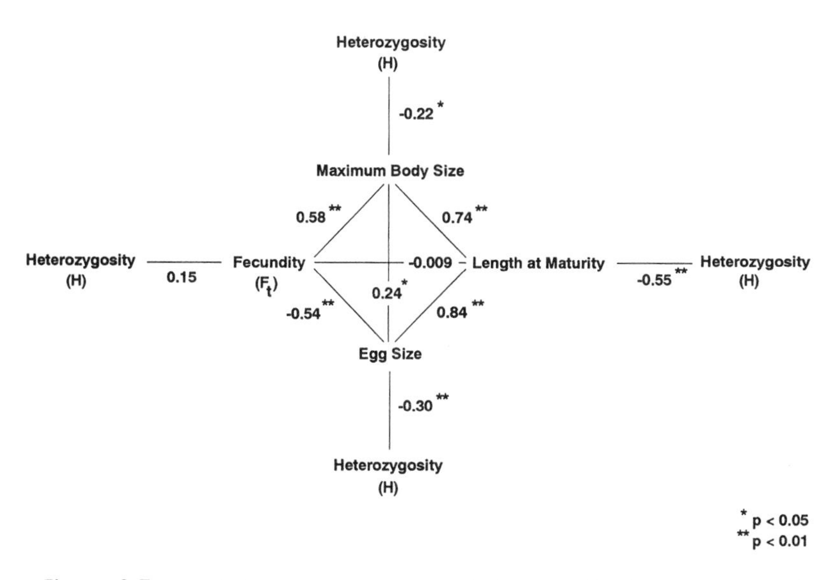

Figure 9.7 Associations among genetic variation and life-history characters in bony fishes. From Mitton and Lewis 1989.

pothesis, at least for species of animals, has approximately the same descriptive power as does the hypothesis based on population size (compare figures 1.4 and 9.5). Whereas the hypothesis based on population size relies on consequences of neutral mutations, the hypothesis based on fecundity relies solely on natural selection.

From Levins's work on fitness sets (1968) came the hypothesis that genetic variability would vary with an animal's size (Selander and Kaufmann 1973). Larger animals, by virtue of both their size and their mobility, might perceive their environments as fine grained and therefore converge on a strategy of genetic monomorphism. Small animals, owing to both their size and limited mobility, might see their world as coarse grained, necessitating polymorphisms to exploit the alternative patches. Much of the variation in heterozygosity among species is consistent with this hypothesis. Small poikilotherms, such as marine bivalves and fruit flies, have high levels of genetic variation, whereas large homeotherms, such as elephant seals, polar bears, walruses, moose, and elk, have little genetic variation. It is perfectly reasonable to propose that poikilotherms would perceive their world as more patchy than would homeotherms, and the disparity in size should compound the difference.

Environmental heterogeneity certainly has an important impact on the level of a population's genetic variability, but I suspect that size per se has very little to do with it. All the large animals such as bears, seals, moose, and elk have very low fecundities. Their life histories simply provide little opportunity for balancing selection to maintain genetic variation. Furthermore, very large animals have relatively small populations—consider the relative population sizes of bears and barnacles. The correlation between population size and fecundity in the set of animals studied by Nei and Graur (1984) is $r = .34$ ($P < .01$; appendix 3). Thus, very large vertebrates have low fecundities and tend to occur in small populations, and both of these characteristics are associated with low levels of genetic variation.

All the very large mammals have fecundities of approximately 25. Small mammals, such as house mice and deer mice, have fecundities that are approximately twice as great as those in large mammals, providing a greater potential for the occurrence of balancing selection. In mammals, as in all other groups of species, the opportunity for balancing selection increases with fecundity, but in mammals, size and fecundity are negatively correlated. Together, these patterns may explain the weak negative correlation between heterozygosity and size in mammals (Nevo et al. 1984; Wooten and Smith 1985). If someone will point to an immense homeotherm with high fecundity, I will predict that it will have lots of genetic variation.

Whereas the largest animals have little genetic variability, the largest plants have the highest level of heterozygosity of any group of species. Conifers have more genetic variation than do *Drososphila*, marine invertebrates, or amphibians (Hamrick, Mitton, and Linhart 1979; Mitton 1983, 1995b; Nevo, Beiles, and Ben-Shlomo 1984). Again, size is probably irrelevant. Multivariate analyses of life-history variables revealed that many aspects of ecology and life history covaried among the 113 taxa of plants examined (Hamrick, Linhart, and Mitton 1979). Principal-components analysis was used to examine the patterns of variation in the data, which included 15 variables for each species. The first principal axis explained 30% of the variation in the data, and this axis was most strongly associated with three measures of genetic variation. Other variables associated with this axis were generation length, mating system, pollination mechanism, fecundity, seed dispersal, and successional status. Species that were long lived and wind pollinated and had high fecundities tended to have the highest levels of genetic variation. This suite of characteristics is common to conifers.

Fecundity, geographic range, and the population size (Soulé 1976) covary among species; species with large population sizes (or populations with high densities) also tend to have large geographic ranges, and it is these species that most commonly have high levels of genetic variation. This coordinated set of axes can be used to predict rates of molecular evolution and rates of speciation. These predictions are presented in chapter 10).

Summary

Genetic variation varies dramatically among species of both plants and animals; fruit flies, marine mussels, and conifers have lots of genetic variation, whereas large vertebrates and weeds have much less. Widespread species tend to have more genetic variation than do species with small geographic ranges. Genetic variation increases with niche width in some studies but not in others. Neutral theory predicts that genetic variation will increase with population size, and empirical data are usually consistent with this prediction. Both the opportunity for selection and genetic variation increase with fecundity.

10

The Sisyphean Cycle

Hence it is the most flourishing, or, as they may be called, the dominant species—those which range widely over the world, are the most diffused in their own country, and are the most numerous in individuals—which often produce well marked varieties, or, as I consider them, incipient species.

C. Darwin, *On the Origin of Species by Means of Natural Selection* (1859)

Sisyphus repeatedly pushes a boulder up a steep slope until, on the verge of reaching the peak, it goes out of control and rolls to the bottom again. Analogously, an individual in the top end of the fitness distribution has achieved its near maximum of fitness by an only momentarily effective combination of genetics and individual history.

G. Williams, *Sex and Evolution* (1975)

At the species level, geographic range is the best predictor of levels of allozyme variation. Endemic species have the lowest genetic diversity whereas regionally distributed and widespread species maintain the most diversity.

J. Hamrick and M. J. Godt, "Allozyme Diversity in Plant Species" (1990)

In his book on the evolution of sexual reproduction (1975), George Williams coined the term *Sisyphean genotype* to refer to an optimal genotype selected by biotic and abiotic components of the environment. Williams proposed low heritability of fitness from generation to generation, for the fitness of a genotype is highly dependent on the environment in which it was tested, and environments are rarely constant. Because the optimal genotype varies with environmental conditions, natural selection, like Sisyphus, is never done.

In a sexually reproducing population, a highly heterozygous genotype is a Sisyphean

AaBbCc x AaBbCc

	ABC	ABc	AbC	aBC	Abc	aBc	abC	abc
ABC	0	1	1	1	2	2	2	3
ABc	1	0	2	2	1	1	3	2
AbC	1	2	0	2	1	3	1	2
aBC	1	2	2	0	3	1	1	2
Abc	2	1	1	3	0	2	2	1
aBc	2	1	3	1	2	0	2	1
abC	2	3	1	1	2	2	0	1
abc	3	2	2	2	1	1	1	0

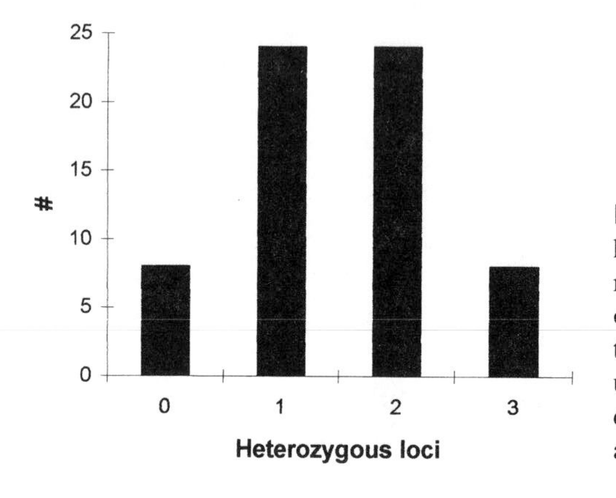

Figure 10.1 A cross between highly heterozygous individuals produces a range of individual heterozygosity in the offspring. In this example, a cross between individuals heterozygous at three unlinked loci yields a distribution of individual heterozygosity from 0 to 3, with a mean of 1.5.

genotype, for segregation and independent assortment produce many different genotypes in the progeny of a cross involving heterozygous parents. Figure 10.1 shows a simple example of this, a cross between two individuals heterozygous for the unlinked loci *A, B*, and *C*. Although both parents were heterozygous for all three loci and therefore had individual heterozygosities of 3, the mean heterozygosity of their offspring is only 1.5. In sexual populations, the work of truncation selection on the axis of individual heterozygosity is never done; it begins anew with each generation.

Here the *Sisyphean cycle* refers to the cycle of reproduction and truncation selection that is expected to be common to species that share the characteristics of high population densities, wide geographic range, high fecundity, and high levels of genetic variation. I propose that the Sisyphean cycle imposes on those species the common evolutionary characteristics of high rates of molecular evolution within lines, low rates of speciation, and stasis.

Correlated Axes

Among species of plants and species of animals, the axes of population density, population size, geographic range, fecundity, and the level of genetic variation are correlated. For example, in a study of genetic variation and life-history variation in 100 species of plants (Hamrick, Linhart, and Mitton 1979), fecundity was correlated with the level of genetic variation ($r = .36, P < .001$) and with geographic range ($r = .25, P = .01$). Approximately half the correlations among the 15 genetic and life-history variables were significant in that data set, necessitating multivariate analyses of the data. The first principal axis of a principal-components analysis of the data incorporated data from three genetic variables, generation length, mating system, pollination system, pollination mechanism, fecundity, seed dispersal, and successional status. This sort of result is representative of the results obtained from analyses within and among other groups of organisms (Armitage 1980; Bekoff, Diamond, and Mitton 1981). In this chapter I present first these axes of variation individually and then the correlations among them.

Population Density and Geographic Range

Species that tend to be abundant in any one locality tend also to have broad geographic ranges. Stated in another way, there generally is a positive correlation between the average density at a locality and the size of the geographic range. Although some negative results have been reported (Adams and Anderson 1982; Ricklefs 1972), positive associations between average local abundance and size of the geographic range were found for some vascular plants (Brown 1984; McNaughton and Wolf 1970; Raup 1975), insects (Hanski 1982a,b,c; Hanski and Koskela 1977, 1978; Muhlenberg et al. 1977; Price 1971), zooplankton (Brown 1984), rodents (Dueser and Shugart 1979), and birds (Able and Noon 1976; Bock 1984; Bock and Ricklefs 1983; Shugart and Patten 1972).

Geographic Range and Species Longevity

Numerous examples in the fossil record suggest that the longevity of a species increases with its geographic range. For example, according to the fossil record, the duration of species of bivalves and gastropods in the late Cretaceous increased with their geographic range ($r = .60, P < .001$) (Jablonski 1987). The extinction rate of bivalves in the late Cretaceous was not dependent on depth range or feeding mode but, instead, declined regularly with the number of marine provinces included in the geographic distribution (Jablonski 1995; Jablonski and Raup 1995).

The Gulf Coast neogastropods of the Lower Tertiary also exhibit a relationship between geographic distribution and species longevity. Neogastropods with pelagic larvae had larger geographic distributions and greater species longevities than did sympatric species without pelagic larvae (Hansen 1980). Hansen argued that the difference in species longevity is primarily attributable to the greater dispersal capability of pelagic larvae and the enhanced probability of their reaching appropriate environments when conditions change dramatically. Similar results and interpretation were presented by Scheltema (1977).

A study of 41 sympatric bivalve species living in the Upper Cretaceous revealed a consistent positive correlation between geographic range and species longevity (Koch 1980). Koch tabulated the presence, location, and abundance of 41 species of bivalves in the marine environments of western North America. After comparing several measures of geographic range, Koch settled on the linear distance between the most distant localities as the simplest and best measure. He analyzed data for infaunal species, epifaunal cemented species, and epifaunal uncemented species, and all species. Species longevity increased significantly in two of five analyses, and all the correlations between geographic distribution and population size were positive.

Jackson (1974) examined the relationship between geographic range and species longevity in 40 infaunal bivalves of the family Veneridae in the Caribbean Sea. The Caribbean was divided into 19 areas of approximately equal size, and the geographic range of a species was estimated as the number of areas from which the species had been collected. Species longevity was estimated from the duration of the species in the fossil record. The rank correlation between geographical range and species longevity was $r = .44$ ($P < .001$).

Population Size and Genetic Diversity

Soulé (1976) estimated the population sizes of a wide diversity of lizards, fish, mammals, marine invertebrates, and *Drosophila* and related the estimates of population size to the amount of genetic variation within populations (see figure 1.4). Clearly, the amount of genetic variation within a population increases with the estimated size of the population.

Nevo, Beiles, and Ben-Shlomo (1984) reported that, in a sample of 717 species of plants and animals, heterozygosity increased with the number of individuals in the species. For species categorized as small, medium, large, and very large, they reported heterozygosities of 0.039, 0.048, 0.077, and 0.090 ($P < .01$).

Geographic Range and Genetic Diversity

In both plants and animals, genetic variation increases with the size of the geographic range of the species (table 3.4). Figure 9.3 shows a specific example of this relationship presented for deer mice. Endemic species have the lowest levels of genetic variation, which is consistent with several interpretations. Endemic or relictual species often have small population sizes, and levels of genetic variation clearly increase with population size (figure 1.4). In addition, endemic species probably experience less environmental heterogeneity than do widespread species, and in many cases, genetic variation increases with environmental heterogeneity (tables 3.1, 3.2, 3.3). For example, in both vertebrates and invertebrates, genetic variation is higher in habitat generalists than in habitat specialists (table 3.4).

Fecundity and Genetic Diversity

In all sexual populations, population size will be stable if each female, on the average, leaves two offspring that reproduce. This leaves a vast potential for balancing selection in the mortality of highly fecund species (see chapter 4). Thus, we expect to see higher levels of genetic variation in species with high fecundity, a pattern seen in both plants and animals (figures 9.4 to 9.7).

Fecundity	**Low**	←——→	**High**
Population density or size	**Low**	←——→	**High**
Geographic range	**Restricted**	←——→	**Extensive**
Genetic variability	**Low**	←——→	**High**
Rate of molecular evolution	**Slow**	←——→	**Fast**
Rate of chromosomal evolution	**Fast**	←——→	**Slow**
Rate of morphological evolution	**Fast**	←——→	**Slow**

Figure 10.2 A set of correlate axes describes variation among species. Species with very high fecundity are predicted to have large populations or high population densities, large geographic ranges, high levels of genetic variation, a high rate of molecular evolution, and, in most (but not all) circumstances, low rates of chromosomal and morphological evolution. Species with low fecundities have the contrasting set of attributes.

Multiple Axes

Population densities, geographic ranges, lifetime fecundities, and levels of genetic variation tend to be positively correlated among species. Many species fit into the axis presented in figure 10.2, which incorporates the individual axes just discussed. Population density, population size, fecundity, geographic range, and genetic variability covary to define a single continuum. The species that, on average, are abundant in a representative locality (high population density, large population size) tend also to have high fecundities, large geographic ranges, and high genetic diversity. Representatives of this group are ponderosa pine and blue mussels. At the other end of the continuum are species limited to small populations in a few localities, which tend to have lower fecundities and genetic diversities. Cheetahs and Torrey pine are extreme representatives of this second group. I call this continuum the *fecundity axis*, for the fecundity of a species has a profound effect on its genetic diversity and is associated with other characters such as population size.

Clearly, some species do not fit neatly into the continuum described by the fecundity axis. This is approximately the first principal axis in the multivariate study of 100 species of plants (Hamrick, Linhart, and Mitton 1979), and the significance of this axis of life-history variation is little diminished by a minority of fascinating exceptions. If this axis describes a substantial proportion of the life-history variation among species, then it will also have important consequences for general patterns of rates of genetic change, rates of speciation, and stasis.

Chromosomal Evolution and Protein Diversity

There may indeed be a relationship between genetic diversity and the rate of biochemical evolution within genera (Table 5.1), but this does not necessarily imply that the rate of speciation increases with heterozygosity. The rates of anagenesis and cladogenesis may not be related. In fact, I argue that under most circumstances, those genera with higher genetic diversities exhibit lower rates of cladogenesis. My reasoning is based, once again, on the axis of fecundity (figure 10.2).

Evolutionary biologists typically estimate levels of genetic variation by surveying protein polymorphisms. The majority of the proteins employed in these surveys are metabolic enzymes. Estimates based on enzyme loci may or may not be representative of the genome; they appear to be associated with estimates of additive genetic variance (Beardmore 1983). Because estimates of genetic variation use genetic variation within metabolic pathways, enzyme polymorphisms may estimate metabolic flexibility (see chapter 2). Highly variable populations and species may be more capable of tolerating fluctuations in weather and also climate; this is suggested by their maintenance of higher levels of genetic diversity in variable environments (chapter 3) and the tendency for habitat generalists to have higher genetic diversities (chapter 9). Perhaps high levels of protein heterozygosity enable the population to track oscillations in climate and to adapt to shifts in the environment. If these sorts of perturbations are accommodated more readily in genetically diverse than in genetically depauperate populations, the highly variable populations will maintain larger population sizes and thus tend to avoid local extinctions and the bottlenecks in population size that may initiate the founder-flush mode of speciation (Carson 1971, 1975; Templeton 1980a).

Some evidence for a negative correlation between protein heterozygosity and the rate of cladogenesis was reported in a comparison of allozyme heterozygosity and the rate of chromosome evolution (Coyne 1984). Allozyme heterozygosities were negatively correlated with the rates of change of both chromosome number and chromosome arms in a broad survey of vertebrates and mollusks (figure 10.3). Chromosomal evolution is associated with speciation in some groups of plants and animals (White 1978). If the rate of molecular evolution increases with heterozygosity (figure 5.6) and the rate of chromosomal evolution is inversely associated with heterozygosity (figure 10.3), then the rate of molecular evolution may be inversely associated with the rate of cladistic evolution.

Under most circumstances, selection for high heterozygosity has a stabilizing influence on morphological characters (chapter 8), and thus cladistic evolution is inversely related to fecundity and heterozygosity. But there are conceivable circumstances under which high genetic variability might enhance the rate of cladistic evolution. These circumstances were modeled by the proliferation of inbred strains of the house mouse. Fitch and Atchley (1985) used genetic variation at 97 single-gene loci to measure differentiation among ten strains of laboratory mice. All strains were homozygous at these loci, as expected, but the degree of differentiation among strains was much greater than expected. The researchers considered, among other possibilities, that the extreme differentiation was a simple consequence of the husbandry in keeping and founding the stocks. Fearing the consequences of severe inbreeding, the largest, most vigorous animals were chosen as parents for the next generation and as founders of new strains. It is likely that this choice of animals inadvertently selected the most heterozygous of the available animals. Thus, new strains were founded with individuals from the upper end of the axis of individual heterozygosity, and the subsequent inbreeding and loss of genetic variability differentiated strains to a much greater extent than if average individuals had been chosen. The differentiation of laboratory strains of mice may provide a model for the differentiation of natural populations. If the founders of natural populations tend to be from the upper end of the axis of individual heterozygosity, new populations are initiated with atypically high levels of variability. This provides more opportunities for responding to selection in novel environments and for differentiating populations as variability is lost.

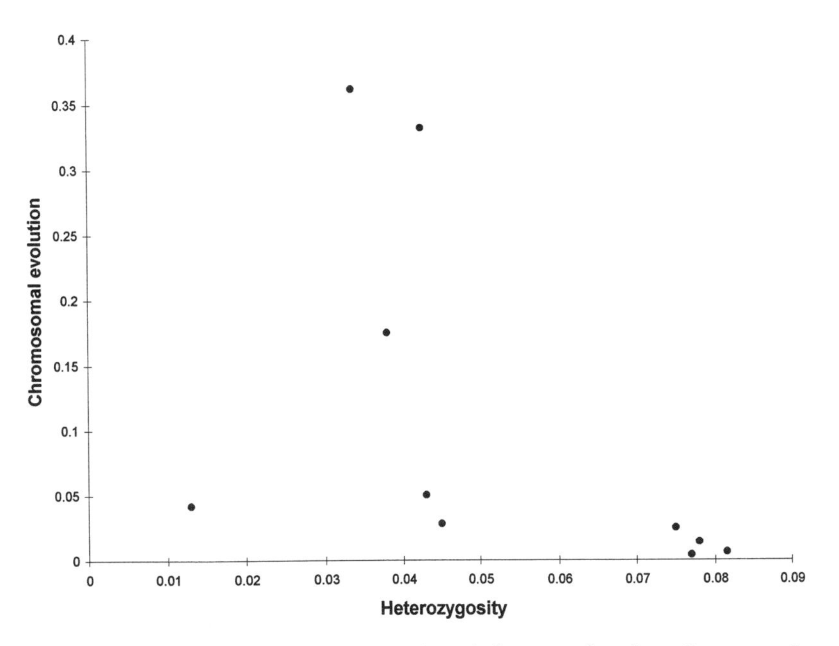

Figure 10.3 The rate of chromosomal evolution as a function of average allozyme heterozygosity in a survey of vertebrates and mollusks. Data from Coyne 1984.

The critical point here is the extent to which new populations are founded by individuals with unusually high levels of individual heterozygosity. Some data suggest that the dispersing individuals have higher levels of heterozygosity (Myers and Krebs 1971), and some data indicate that the dispersal distances of highly heterozygous individuals exceed those of other genotypes (Fleischer, Lowther, and Johnston 1984; Florence, Johnson, and Coster 1982). It also is interesting to note that for the taxa sampled by Avise and Aquadro in their analysis of genic heterozygosity and genetic distance (1982), the number of species per genus increased with heterozygosity across fishes, amphibians, reptiles, birds, and mammals (table 10.1). These reports are sufficient to identify an intriguing pattern, but they are not sufficient to support any generalizations.

Stasis and Diversity

Stasis can be defined as morphological stability through extended periods of time, and here the time is typically in millions of years (Eldredge and Gould 1972; Gould and Eldredge 1977). This subject has received ample attention and need not be discussed here beyond its relation to the axis of fecundity (figure 10.2). I predict stasis, or prolonged morphological stability, to be more common in species at the highly fecund end of this distribution and rare or absent in species at other end of the distribution. There are several reasons for making this prediction. First, stasis is identified as stability of morphological form. In order for stasis to be identified in a species, that species must be persistent in the fossil record. All other things being equal, or nearly so, the longevity of a

Table 10.1 Average heterozygosity and average number of species per genus in groups of vertebrates

Group	Heterozygosity	No. of Species/Genus
Fishes	.054	26.1
Amphibians	.058	95.8
Reptiles	.041	62.8
Birds	.037	17.5
Mammals	.039	18.2

Note: Data from Avise and Aquadro 1982. Heterozygosity and mean number of species/genus are not statistically significant by Pearson product moment correlation coefficient ($r = .61$, $P = .13$) but are related by both the Kendall ($r = .80$, $P < .05$) and the Spearman ($r = .90$, $P < .05$) rank-order correlation coefficients.

species increases with its geographic range and population size. A relatively large population size and an extensive geographic range may provide some insurance against extinctions caused by periodic events of severe weather and other geographically restricted challenges. Extensive geographic range, large population size, high fecundity, and high genetic diversity are traits found toward the high end of the fecundity axis. Species with these combinations of life-history traits should have a higher probability of persisting for extended periods of time and, therefore, of leaving extensive fossil records.

But persistence is not, by itself, sufficient to draw attention to a species as an example of stasis. Morphology also must be relatively constant for an extended period of time. Several factors, such as the persistence of a relatively stable environment, might enhance the probability of stasis, and among these factors is high variability at genes coding for metabolic enzymes. High genetic variability of metabolic enzymes enables populations to endure short-term fluctuations in weather and longer-term shifts in climate without the severe reductions in population size that might be experienced in genetically depauperate populations. Several empirical observations are consistent with this proposition. First, environmental heterogeneity in laboratory populations appears to maintain genetic variability, and habitat generalists seem to maintain more genetic variability than do habitat specialists (chapter 9). Second, all the associations between individual heterozygosity and viability, growth rate, developmental stability, and routine costs of metabolism are consistent with the hypothesis that high heterozygosity of metabolic enzymes provides high physiological efficiency to an individual. Individuals with high physiological efficiency should expend less energy in respiration and have relatively more energy left for growth or reproduction or homeostatic mechanisms during periods of severe stress.

These observations suggest that highly heterozygous populations—because of their enhanced physiological efficiency—may be able to tolerate environmental fluctuations and shifts that either push genetically depauperate populations to extinction or force the evolution of further adaptations that might be reflected in morphology. Highly fecund and heterozygous individuals, faced with a challenge of changing climate, have a high probability of producing offspring with the appropriate metabolic constitution to adapt to the new demands put on them. Species lacking this simple option must either evolve behavioral or morphological adaptations to the novel stimulus or suffer higher mortality and thus reduced population size. The increased genetic drift accompanying reduced population size might be recorded as morphological change. Paleontologists are less likely to offer the latter species as an example of stasis.

One additional characteristic of the Sisyphean cycle enhances the probability of finding stasis in highly fecund species. When selection favors highly heterozygous individuals, it simultaneously favors individuals with more average phenotypes, more symmetry, and greater developmental homeostasis (chapter 7). Imagine a population represented as a cloud of data points in multidimensional morphological space. Highly heterozygous individuals are buried in the center of the cloud, and highly homozygous individuals are sprinkled around the periphery (Mitton and Koehn 1985). This cloud of data points is like an onion—truncation selection peels away the outer shells of the onion, rejecting unusual phenotypes and stabilizing the morphology.

The Axis of Fecundity and the Sisyphean Cycle

The Sisyphean cycle is most common in species bearing those life-history traits that cluster at the high end of the fecundity axis. The most likely candidates are species like elms, oysters, mussels, cod, and pines. This cycle has little impact on the evolution of whales, walruses, elephant seals, or humans, as their fecundities are too low and do not provide sufficient opportunity for the strong balancing selection characteristic of the cycle.

The Sisyphean cycle operates within a life cycle, within a single generation. It begins with the prolific production of progeny by individuals with high individual heterozygosity for metabolic enzymes. The offspring are themselves highly variable, and they probably span a greater range of the axis of individual heterozygosity than did their parents (figures 6.1, 6.2, 10.1). Through the effects of individual heterozygosity on physiology, these offspring's large range in individual heterozygosity results in a large variance in their fitness potential. Differential survival, including normalizing selection on morphology, narrows the range of individual heterozygosity by culling individuals at the low end of the axis. This is the truncation selection considered in chapter 7.

Later in the life cycle, the fecundity of larger individuals and individuals with more favorable energy balances gives them a final advantage over predominantly homozygous individuals. At the end of the life cycle, highly heterozygous individuals have been favored by superior viability, developmental stability, growth rate, and fecundity, but this advance in individual heterozygosity during the life cycle is largely but not completely undone (Mitton et al. 1993) by the process of sexual reproduction (figure 10.1). When a locus segregates two alleles, a female heterozygous for that locus produces only 50% heterozygous offspring, regardless of the genotype of her mate or mates. The Sisyphean cycle begins anew with the production of the next set of progeny.

Although the Sisyphean cycle is expected to be a characteristic of populations of highly fecund species, the intensity of the selection need not be a constant—it probably varies from essentially nonexistent to intense, depending on such factors as the proportion of nonselective mortality and the quality of the environment that year. A population might experience intense recurrent truncation selection of the axis of individual heterozygosity in each generation, or the selection may be episodic and highly variable in its intensity.

The ideas presented here are compatible with some aspects of population genetic theory and are consistent with the substantial and growing corpus of empirical data. Admittedly, much of the scenario of the Sisyphean cycle is speculative, but I hope that this speculation will inspire empirical tests of hypotheses concerning the impact of protein polymorphisms on whole plant and animal physiology, the mechanism of fitness determination, the form of multilocus organization generated by truncation selection, and the intensity of natural selection. These tests must be conducted by population biologists and physiological ecologists, and they will have a profound impact on our perspectives on evolution.

Summary

A set of correlated axes describe some of the variation among species in fecundity, population density, geographic range, and genetic variation. At one end of this continuum are species with high fecundity, large populations, broad geographic ranges, and abundant genetic variation. At the other extreme are relatively rare, often endemic species with low fecundity and little genetic variation.

The opportunity for natural selection is highest at the highly fecund end of this set of correlated axes. In highly fecund species, selection is predicted to favor highly heterozygous individuals. Species at the high end of the fecundity axis have low rates of speciation, high rates of molecular evolution, long tenures in the fossil record, and stasis.

11

Comments on Natural Selection

It would be more correct to say that natural selection is mainly concerned with preventing evolution, not causing it. Much selection is concerned with the elimination of low-fitness genotypes produced by mutation or recombination. It can be at a generally intense level, but vary so in direction and strength at different times or places that little cumulative change takes place.

G. Williams, *Sex and Evolution* (1975)

I never intended this summary of evolutionary genetics to be comprehensive. I have focused primarily on enzyme polymorphisms and have gained some critical insights by comparing and contrasting the variation of enzyme polymorphisms with the variation of DNA sequences. Like all books, this one reflects the writer's personal experiences and favored hypotheses. This book is mainly a summary of empirical data, and from this I want to emphasize several points.

The Nature of Enzyme Polymorphisms

Enzymes—as well as functional proteins such as hemoglobin, transferrin, and haptoglobin—are catalysts that allow metabolic reactions to proceed at temperatures much lower than would otherwise be possible; metabolic rates would be imperceptibly slow without the aid of enzymes. Enzymes must effectively bind to substrates, change conformation to catalyze a reaction, and then release the products of the reaction. All this must be accomplished in a range of temperatures and a set of concentrations of the reacting ingredients determined by the ecology and physiology of the species. But more than allowing reactions to occur at physiological temperatures, enzymes control and regulate metabolism to enable honeybees to fly, cheetahs to sprint, and ponderosa pine to capture carbon from the air. These are not trivial side effects of randomly assembled neutral mutations but adaptations orchestrated by natural selection. Some fish are biochemically tuned for bursts of speed, others for endurance swimming, and still others for neither of these. The metabolic characteristics in these fishes are no more interchangeable than are the engines of dirt bikes, farm tractors, and diesel buses. Studies of closely related species in distinctly different environments have revealed the adaptive evolution of the enzymes' biochemical properties (chapter 2), allowing species to maintain physiological function and activity in their current environments.

Populations Are Rarely at Equilibrium

Populations are rarely at equilibrium because environmental conditions fluctuate with the climate, the weather, the seasons, and the activities of sympatric species. For example, the frequencies of inversions in *D. pseudoobscura* track environmental conditions each year, with inversion frequencies changing characteristically with the seasons (Dobzhansky 1948, 1970). Similarly, larvae of the blue mussel ride currents into Long Island Sound, obscuring steep clinal variation between salinity environments. But each fall, selection strikes the populations in and around the salinity gradient, reconstructing a steep cline in allelic frequencies (Koehn, Newell, and Immerman 1980). Each year the larvae of the American eel rise to the surface of the Sargasso Sea and are swept by the Gulf Stream first west and then north to bays and rivers in North America. Selection during these extended larval migrations creates similar clines in allelic frequencies each year (Koehn and Williams 1978; Williams, Koehn, and Mitton 1973). *Colias* butterflies of the Rocky Mountains live in environments that fluctuate, almost daily, beyond the upper and lower temperatures of the flight range. Some genotypes fare better during the morning chill, whereas others fare better during the occasional hot spells (chapter 4). In ponderosa pine, kinetic studies showed that the optimal temperatures of Per genotypes differ among genotypes, and this kinetic variation produces variation in genotypic frequencies among adjacent environments (Beckman 1977). South-facing slopes and low elevations have relatively high frequencies of the common heterozygote, which has a broad range of optimal temperatures. High elevations and north-facing slopes have relatively high frequencies of the common homozygote, which has a very low optimal temperature (Beckman and Mitton 1984). For plants and animals living in heterogeneous environments, a single genotype is rarely the optimal genotype in the full range of environments occupied by a species.

Natural Selection Varies with Environmental Conditions

Because environments are not perfectly stable, we should expect that natural selection would be heterogeneous in space and time, just as environments are. The enzyme kinetics of the LDH polymorphism in the killifish reveals that the genotypes' enzyme products differ at 15°C but not at 25°C (figure 4.1). Empirical studies show that measures of performance associated with the LDH polymorphism differ at 15°C but not at 25°C. Selection might differ between genotypes at 15°C, but because the genotypes are functionally equivalent and have similar physiological performances at 25°C, selection could not act on LDH at 25°C. Both the Lap polymorphism in blue mussels and the Gpt polymorphism in *Tigriopus* copepods play important roles in the maintenance of cell volume, and both mediate physiological responses to fluctuations of salinity. Genotypes may be functionally equivalent at a particular salinity but are critically different when salinity fluctuates.

Discordant Patterns of Population Structure Reveal Different Evolutionary Forces

Protein polymorphisms and DNA markers reveal discordant patterns of geographic variation in deer mice, cod, horseshoe crabs, oysters, lodgepole pine, the closed-cone pines of California, and limber pine (chapter 7). For example, discordant patterns of geographic variation in the marine copepod, *Tigriopus californicus*, were uncovered by surveys of allozyme variation and both nuclear and mitochondrial DNA markers (Burton 1986, 1987; Burton and Feldman 1981; Burton and Lee 1994). *Tigriopus* lives in splash pools above the intertidal zone along the California coast. The drying of pools during calm weather and their inundation during storms repeatedly cause local extinctions and colonizations, suggesting a high degree of gene flow among populations. Allozyme studies, however, found high frequencies of private alleles, indicating very low rates of gene flow among populations.

An analysis of genetic distances of allozymes uncovered two well-defined clusters of populations separated by Nei's genetic distance of 0.28. DNA sequences from the COI (cytochrome oxidase subunit I) locus in the mitochondrial genome and from the nuclear histone H1 locus also revealed two well-defined clusters of populations, marked by a minimum of 18% sequence divergence at COI and by fixed insertions and deletions at H1. Although both allozymes and DNA sequences revealed distinct geographic clusters of populations, the clusters differed. The border between the northern and southern clusters is south of Los Angeles in the allozyme analysis, but it is at Point Conception in the DNA analysis—a disparity of 150 miles. This difference cannot be attributed to differences between mitochondrial and nuclear markers, for the nuclear allozyme loci have a pattern unlike the pattern of the nuclear histone sequences.

Surveys of allozyme (Millar et al. 1988), cpDNA (Hong, Hipkins, and Strauss 1993), and mtDNA (Strauss, Hong, and Hipkins 1993) in Bishop pine, *Pinus muricata*, showed unexpected differences in the degree of differentiation among populations. In Bishop pine, allozymes are carried by both pollen and seed, cpDNA is carried only by pollen, and mtDNA is carried only by seed. Thus, from the potential for gene flow, we expect the G_{st}s of cpDNA and alloyzmes to be small and the G_{st} for mtDNA to be much greater. Allozymes revealed moderate differentiation among populations (G_{st} = 22%), but both the values of G_{st} estimated for cpDNA (87%) and mtDNA (96%) were significantly higher and were not different from each other. The disparity in values of G_{st} cannot be explained by the potential for gene flow in these markers.

In comparison to protein polymorphisms, DNA markers show more genetic variation and provide population biologists with choices among sets of loci with different patterns of inheritance. But DNA techniques have not entirely supplanted allozyme techniques because comparisons of these sets of genetic markers reveals much more than either does by itself. These discordant patterns are not attributable to alternative modes of inheritance but must be produced by differences in the evolutionary forces that influence the various compartments of the genome (Mitton 1994).

Natural Selection Discriminates Among the Genotypes at a Locus

The nature of natural selection and the possible targets of natural selection are the subjects of much debate (Brandon and Burian 1984; Dawkins 1976, 1982; Franklin and Lewontin 1970; Gould 1982; Gould and Lewontin 1979; Sober 1984; Sober and Lewontin 1982; Williams 1966, 1985). I do not want to summarize the entire debate but ask only, Can natural selection discriminate among the genotypes at a single locus, and can we measure the fitness consequences of variation at a single locus? Some biologists argue that genes are tiny segments of DNA embedded in immensely long strands, and, consequently, we cannot hope to discern the effects of a single gene apart from the influences of the numerous genes linked to it. Other biologists reason that if complex eukaryotes are composed of between 5,000 and 50,000 genes, then the physiological, phenotypic, and demographic consequences of variation at a single gene would be minuscule and probably below the limits of detection of empirical studies.

Genes are indeed embedded in chromosomes containing thousands of other genes. But it is not linkage that is relevant here but linkage disequilibrium. If linked genes segregate independently, then they may respond to selection as independently as if they were on separate chromosomes. If natural selection acts to truncate the axis of individual heterozygosity, favoring highly heterozygous individuals (chapters 7, 8), then we can make a specific prediction concerning the organization of the genome. Because the linkage disequilibrium between two loci is broken down only by recombination in doubly heterozygous individuals, linkage disequilibrium decays much faster than predicted by neutral theory when selection favors the reproduction of highly heterozygous individuals. Consequently, linkage disequilibrium is a rare and ephemeral phenomenon. Linkage disequilibrium is common and strong in selfing species, such as wild oats and barley (Allard 1975), but it is infrequent and weak in obligately sexual, outcrossing species such as fruit flies, ponderosa pine, cod, and marine mussels (Epperson and Allard 1987).

The strength and extent of linkage disequilibrium have been measured with sequence data, and these studies demonstrated that polymorphic sites in a locus may be in linkage equilibrium. Forty-four polymorphic sites in the Adh locus were scored in 1,533 *D. melanogaster* from 25 populations along the east coast of North America (Berry and Kreitman 1993). Linkage disequilibrium was generally high but was heterogeneous among polymorphic sites and did not show a simple decline with distance between sites. G_{st} values among sites were heterogeneous (less than 0.02 to greater than 0.08), and although the site responsible for the electrophoretic marker varied clinally with latitude, most other sites did not. Linkage disequilibrium also was measured among 359 polymorphic sites in the 3.5-kb region of the Adh locus in *D. pseudoobscura* (Schaeffer and Miller 1993). Only 127 of the 74,274 pairwise comparisons of polymorphic sites had significant levels of linkage disequilibrium, and these tended to be in two small clusters, conceivably associated with key features of mRNA secondary structure. If linkage disequilibrium does not extend the length of a gene's coding region, it is highly unlikely to tie together, in a statistical sense, a series of adjacent loci.

Kinetic variation among enzyme genotypes influences physiological variation and, consequently, components of fitness in both animals and plants (chapters 4, 7, 8). When kinetic variation leads to specific, testable hypotheses concerning physiological variation and when empirical data are consistent with specific predictions, the physiological variation is clearly attributable to the genetic variation at that locus. As clearly as we can see physiological variation attributable to the sickle cell polymorphism in humans (Allison 1955; Templeton 1982) and the hemoglobin polymorphism in deer mice (Snyder 1981), we can see the physiological consequences of the enzyme polymorphisms described in chapter 4. For example, the enzyme kinetics of PGI in *Colias* butterflies suggests a specific hypothesis (Watt, Cassin, and Swan 1983) concerning which genotypes would fly in specific ranges of temperatures. Field data were consistent with the hypothesis that heterozygotes fly over a greater range of temperatures than do homozygotes (chapter 4).

Under some ecological conditions, the physiological consequences of a single enzyme polymorphism can be great. For example, ADH heterozygotes in *Drosophila melanogaster* produce 14% to 64% more eggs than homozygotes do, depending on the environmental conditions and which homozygote is compared (Bijlsma-Meeles and Bijlsma 1988; Serradilla and Ayala 1983b). In *Colias* butterflies, the egg production of heterozygotes is 33% greater than of the common homozygotes and 100% greater than of the rare homozygotes (Watt 1992). In killifish, LDH homozygotes differed by 18 to 40% in the amount of oxygen delivered to tissues and by 20% in sustained swimming speeds (chapter 4).

Additional support for the hypothesis of selection acting on single loci comes in the conformance of some empirical data to the adaptive distance model. Estimates of fitness and allelic frequencies in cod, killifish, and conifers are consistent with the hypothesis of overdominance at the enzyme loci (chapter 7).

It is curious to me that evolutionary biologists embrace the notion that natural selection acts on the sickle cell hemoglobin polymorphism in humans (Allison 1955; Lewontin 1974; Templeton 1982) but that they usually regard as sloppy or uncritical science the notion that natural selection acts on other proteins, such as enzymes. The evidence that natural selection acts on Lap in mussels, LDH in killifish, and PGI in *Colias* butterflies matches or exceeds the data revealing selection on the sickle cell polymorphism. Both hemoglobins and enzymes are proteins with physiological functions, and we have no reason to believe that selection might act on one group but not the other.

Balancing Selection

For loci whose variation is maintained by balancing selection, fitness generally increases with the number of heterozygous loci (Ginzburg 1979; Turelli and Ginzburg 1983; chapter 7). This expectation from theoretical population genetics is consistent with many of the empirical data on protein polymorphisms. Numerous empirical studies report that viability, growth rate, fecundity, mating success, and developmental stability all increase with heterozygosity, suggesting that selection balances some of the variation at protein polymorphisms (chapter 8).

Although it is convenient to display fitnesses against an axis of heterozygosity, condensing multilocus genotypes into a single axis loses much information and hides much of the variation in fitness among genotypes. Overdominance is characterized by the superior fitness of heterozygotes, but it also predicts that common homozygotes will have relatively high fitness and that rare homozygotes will have relatively low fitness (chapter 7), a pattern that is also observed in the empirical data.

The majority of kinetic studies of enzyme polymorphisms show differences among the genotypes, usually with heterozygotes intermediate between homozygotes (Koehn, Zera, and Hall 1983; Zera, Koehn, and Hall 1983). However, a few studies found overdominance for enzyme kinetics (Carter 1997; Koehn 1969; Pogson 1991; Roth and Bergmann 1995; Watt 1977; Carter et al. 1997) and also reported measures of performance consistent with the molecular overdominance.

The Intensity of Natural Selection

The intensities of natural selection that are sometimes detected at enzyme polymorphisms will come as a surprise to theoretical population geneticists, who traditionally use selection coefficients of 0.001 to 0.01 in their models. Selection coefficients estimated with enzyme polymorphisms are often in the range of 0.2, and they occasionally exceed 0.5 (e.g., table 7.1), just as they do in morphological studies (figure 1.1). For example, the selection coefficients estimated at the Gly locus in pinyon pine in stressful environments are 0.57, 0.38, and 0.0, respectively, for the *FF*, *FS*, and *SS* genotypes, but no differences were detected among these genotypes on typical soils (Cobb, Mitton, and Whitham 1994). Selection coefficients for the survival of Soay sheep challenged by nematode infections were 0.41 and 0.17 for the *FF* and *SS* homozygotes, respectively (Albon et al. 1993; Gulland et al. 1993). In the desert fish *Poeciliopsis monacha*, selection coefficients for the +/+ and *v*/*v* homozygotes at the Ck-A locus were estimated to be approximately 0.45 and 0.60, respectively, during hypoxia (Vrijenhoek, Pfeiler, and Wetherington 1992). From my summary of the data on survival and fecundity of *Colias* butterflies (chapter 4), I estimated the selection coefficients against the PGI 33 and 44 genotypes to be 0.66 and 0.83, respectively.

Life-History Variables and Speciation

Some general patterns are revealed when species are grouped on the axis of fecundity (chapter 10). Species with high fecundity tend to have high population densities, large geographic ranges, and high levels of genetic variation. Through a complex nexus of interactions, these species also tend to have great longevities or persistence in fossil records. It is these species that are predicted to exhibit morphological stasis. Underlying the predicted morphological stasis of highly fecund species stasis are relatively high rates of molecular evolution and relatively low levels of karyotypic evolution. Thus, highly fecund species are usually characterized by morphological stasis and high species longevity but relatively low levels of speciation.

Although the rate of speciation of highly fecund species is low in large populations, I suspect that their potential for genetic change is greater when they are suddenly isolated in populations that experience great fluctuations in size. The founder-flush model (Carson 1987) and the genetic transilience (Templeton 1980a,b, 1981) models of speciation would have a greater number of possible outcomes, and therefore a greater probability of speciation, when the initial populations contain abundant genetic variation. Furthermore, if isolated populations tend to be at the periphery of geographic distributions and the degree of differentiation among populations increases with the amount of variability within populations, then isolated populations of highly fecund species might be highly differentiated, making it more probable that founder flush or transilience would produce a significantly differentiated population. Similarly, the choice of vigorous, conceivably highly heterozygous mice for the establishment of new lines may help explain the rapid differentiation among strains of laboratory mice (Fitch and Atcheley 1985; chapter 10).

Natural Populations Are Dynamic

Natural populations are frequently perturbed from equilibrium, and selection coefficients estimated with allozyme polymorphisms are often large. When these statements are joined together, images of dynamic populations emerge. For example, strong selection (table 7.1) was associated with the thinning of populations of annual ryegrass on the shore of Lake Berryessa, which, I suspect, happens in most years. Each fall, as the phytoplankton disappear from the waters of Long Island Sound, differential survival in the blue mussels rebuilds the strong cline in Lap frequencies at the boundary between the brackish sound water and the saltier ocean water (Koehn, Newell, and Immerman 1980).

Each summer, female *Colias* butterflies watch males hover in a nuptial dance, and by choosing those with high endurance, they favor heterozygous males (Watt 1992). Each spring and summer, as young eels ride the Gulf Stream from the Sargasso Sea to the rivers of North America, differential survival creates latitudinal clines in allelic frequencies (Koehn and Williams 1978; Williams, Koehn, and Mitton 1973). In the late fall and early spring, highly heterozygous brine shrimp produce the vast majority of overwintering cysts that start the population in the Great Salt Lake in the following spring (Gajardo and Beardmore 1989). As conifers age in the Rocky Mountains, differential survival favors highly heterozygous genotypes, producing excesses of heterozygotes in the big old trees that produce the greatest numbers of seeds (Mitton and Jeffers 1989).

Implications for Industry and Conservation

In addition to increasing our knowledge of the processes in natural populations, the studies summarized here have implications for both agriculture and conservation biology. For example, a collaborative study of performance in pigs was conducted with Tyson Foods, Inc., which selects its breeding stock by ranking them on a scale defined by a linear combination of growth to 100 days, feed efficiency between 100 and 180 days, and the shape of the animal. Our study showed that Tyson was effectively selecting heterozygotes at the 6-phosphogluconate dehydrogenase locus (Mitton, Zelenka, and Carter 1994). Subsequent studies of these lines revealed that two common forms of mtDNA are in the breeding lines and that the growth rates of these mtDNA haplotypes differ by 4%.

Numerous studies report that protein heterozygosity is related to physiological efficiency, viability, fecundity, mating success, and resistance to disease (chapters 2, 3, 4, 6, 7, 8). Given that the heterozygosities of parents and their offspring are correlated (Mitton et al. 1993), conservation biologists might consider using protein heterozygosity in their management programs (Hughes 1991; Vrijenhoek and Leberg 1991).

Appendices

Appendix 1. Average Heterozygosity and Genetic Distance Among Species

Species	H	D	Reference
1 Insects			
Aedes hendersoni	.165	.48	(Matthews and Munstermann 1983)
A. triseriatus			
Drosophila willistoni	.177	.809	(Ayala et al. 1974)
D. tropicalis			
D. equinoxialis			
D. paulistorum			
D. nebulosa			
Troglophilus cavicola	.123	.248	(Sbordoni et al. 1981)
T. andrenii			
2 Birds			
Ammodramus savannarum	.057	.223	(Avise, Patton, and Aquadro 1980c)
A. henslowi			
A. sandwichensis			
Camarhynchus pauper	.027	.003	(Yang and Patton 1981)
C. parvulus			
Catharus ustulatus	.042	.024	(Avise, Patton, and Aquadro 1980b)
C. fuscenscens			
C. guttatus			
C. minimus			
Contropus borealis	.085	.024	(Zink and Johnson 1984)
C. sordidulus			
C. virens			
Dendroica petechia	.028	.043	(Avise, Patton, and Aquadro 1980a)
D. pennsylvanica			
D. caerulescens			
D. pinus			
D. virens			
D. discolor			
D. tigrina			
D. fusca			
D. magnolia			

(*continued*)

Appendix 1. (*continued*)

Species	H	D	Reference
D. coronata			
D. palmarum			
D. castanea			
Empidomax flaviventris	.063	.126	(Zink and Johnson 1984)
E. virescens			
E. alnorum			
E. trailii			
E. euleri			
E. minimus			
E. hammondii			
E. wrightii			
E. oberholseri			
E. difficilis			
E. flavescens			
E. atriceps			
Geospiza forta	.055	.022	(Yang and Patton 1981)
G. scandens			
G. fuliginosa			
G. difficilis			
G. magnisostris			
G. conirostris			
Geothlypis trichas	.047	.024	(Avise, Patton, and Aquadro 1980a)
G. formosa			
Lophortyx gambelii	.026	.007	(Gutierrez, Zink, and Young 1983)
L. californicus			
Melospiza melodia	.046	.059	(Zink 1982)
M. lincolnii			
M. georgiana			
Seirus aurocapillus	.095	.248	(Avise, Patton, and Aquadro 1980a)
S. novaboracensis			
Sphyrapicus thyroideus	.033	.054	(Johnson and Zink 1983)
S. varius			
S. nuchalis			
S. ruber			
Spizella passerina	.074	.024	(Avise et al. 1980c)
S. pusilla			
Toxostoma rufum	.016	.084	(Avise, Aquadro, and Patton 1982)
T. curvirostre			
T. dorsale			
Vermivora varia	.056	.179	(Avise et al. 1980a)
V. chrysoptera			
V. pinus			
V. peregrina			
V. celata			
Vireo griseus	.036	.360	(Avise, Aquadro, and Patton 1982)
V. philadelphi			
V. flavifrons			
V. solitarius			
V. olivaceus			
Zonotrichia melodia	.050	.057	(Avise et al. 1980c)

(*continued*)

Appendix 1. (*continued*)

Species	H	D	Reference
Z. georgiana			
Z. albicollis			
Zonatrichia capensis	.034	.117	(Zink 1982)
Z. leucophrys			
Z. albicollis			
Z. atricapilla			
Z. querula			
3 Reptiles and Amphibians			
Anniella geronimensis	.054	.291	(Bezy et al. 1977)
A. pulchra			
Anolis angusticeps	.07	1.01	(Webster, Selander, and Yang 1972)
A. sangrei			
A. distichus			
A. carolinensis			
Bipes biporous	.011	.74	(Kim et al. 1976)
B. tridactylus			
B. canaliculatus			
Bolitoglossa subpalmata	.144	1.035	(Hanken and Wake 1982)
B. adspersa			
B. vallecula			
Bolitoglossa morio	.098	1.132	(Larson 1983)
B. flavimembris			
B. helmrichi			
B. cuchumatana			
B. rostrata			
B. lincolni			
B. rufescens			
B. occidentalis			
B. engelhardti			
B. dunni			
B. dunni			
B. dofleini			
Bolitoglossa meliana	.116	.670	(Wake and Lynch 1982)
B. franklini			
B. lincolni			
B. resplendens			
Bufo boreas	.264	.910	(Feder 1979)
B. punctatus			
Bufo americanus	.100	.120	(Green 1983, 1984)
B. fowlerii			
B. hemiophrys			
Chinemys kwangtungensis	.051	.490	(Sites et al. 1984)
C. reevesi			
Egernia modesta	.025	.668	(Milton, Hughes, and Mather 1983)
E. whitii			
Eumeces laticeps	.068	.299	(Murphy, Cooper, and Richardson 1983)
E. inexpectatus			
E. fasciatus			
Hyla regilla	.043	1.71	(Case, Haneline, and Smith 1975)

(*continued*)

Appendix 1. (*continued*)

Species	H	D	Reference
H. cadaverina			
H. eximia			
Lacerta melisellensis	.040	.75	(Gorman and Nevo 1974)
L. sicula			
L. oxycephala			
Litoria angiana	.073	.675	(Dessauer, Gartside, and Zweifel 1977)
L. arfakiana			
L. caerula			
L. congenita			
L. darlingtoni			
L. eucnemis			
L. iris			
L. micromembrana			
L. modica			
L. rothi			
L. thesaurensis			
L. wollastoni			
Pelobates syriacus	.018	.616	(Nevo 1976)
P. cultripes			
Pelobates varaldii	.030	.791	(Busack, Maxson, and Wilson 1985)
P. cultripes			
Plethodon vehiculum	.059	1.122	(Feder, Wurst, and Wake 1978)
P. elongatus			
P. dunni			
P. gordoni			
Plethodon kentucki	.063	.44	(Highton and MacGregor 1983)
P. glutinosus			
P. jordani			
P. yonahlosee			
Rana boylii	.033	.870	(Case et al. 1975)
R. muscosa			
R. cascadae			
R. aurora			
R. pretiosa			
R. tarahumarae			
R. catesbeiana			
Scaphiopus holbrooki	.049	.801	(Sattler 1980)
S. couchi			
S. bombifrons			
S. hammonds			
S. multiplicatus			
Taricha rivularis	.088	.466	(Hedgecock and Ayala 1974)
T. granulosa			
T. taricha			
Thamnophis sauritas	.085	.078	(Gartside, Rogers, and Dessauer 1977)
T. proximus			
Thorius pennatulus	.110	.975	(Hanken 1983)
T. pulmonaris			
T. narisovalis			
T. dubitus			
T. troglodytes			

(*continued*)

Appendix 1. (*continued*)

Species	H	D	Reference
T. minutissimus			
T. macdougalli			
T. schmidti			
T. maxillabrochus			
Typhlosaurus lineatus	.013	.036	(Kim, Gorman, and Huey 1978)
T. garienpensis			
Uma notata	.013	.18	(Adest 1977)
U. inornata			
U. scoparia			
U. exsul			
U. paraphygus			
4 Fish			
Albula neoguinaica	.014	1.16	(Shaklee and Tamaru 1981)
A. glossodonta			
Bathygobius ramosus	.051	.328	(Gorman, Kim, and Rubinott 1976)
B. andrie			
B. soporator			
Campostoma anomalum	.046	.17	(Buth and Burr 1978)
C. oligolepis			
C. ornatum			
Elassoma okefenokee	.007	.357	(Avise, Straney, and Smith 1977)
E. evergladei			
Hypentelium etowanum	.020	.183	(Buth 1980)
H. nigricans			
H. roanokense			
Lepomis humilis	.059	1.14	(Avise et al. 1977)
L. microlophus			
L. auritus			
L. gulosus			
L. megalotis			
L. cyanellus			
L. gibbosus			
L. marginatus			
L. macrochirus			
L. punctatus			
Lipochromis microdon	.027	.000	(Sage et al. 1984)
L. obesus			
Oncorhynchus gorbuscha	.015	.63	(Utter, Allendorf, and Hodgins 1973)
O. keta			
O. nerka			
O. kisutch			
O. tshawytscha			
Prognathochromis longirostris	.025	.006	(Sage et al. 1984)
P. macrognathus			
Puntius binotatus	.088	.760	(Kornfield and Carpenter 1984)
P. tumba			
P. lindog			
P. altus			
Scaphirynchus albus	.014	.000	(Phelps and Allendorf 1983)

(*continued*)

Appendix 1. (*continued*)

Species	H	D	Reference
S. platorynchus			
Sebastolobus altivelis	.048	.226	(Siebenaller 1978)
S. alascanus			
5 Mammals			
Arvicola sapidus	.025	.290	(Graf 1982)
A. terrestris			
Dipodomys panamintinus	.021	.490	(Johnson and Selander 1971)
D. elator			
D. microps			
D. spectabilis			
D. ordii			
D. deserti			
D. compactus D. agilis			
D. nitratoides			
D. heermanni			
D. merriami			
Geomys bursarius	.039	.50	(Penney and Zimmerman 1976)
G. personatus			
G. arenarius			
G. pinetis			
G. tropicalis			
Hylobates lar	.022	.130	(Bruce and Ayala 1979)
H. concolor			
Macaca mulatta	.059	.139	(Nozowa et al. 1977)
M. nemestrina			
M. fuscata			
M. cyclopis			
M. fasicularis			
M. speciosa			
M. radiata			
Macrotus californicus	.030	.256	(Greenbaum and Baker 1976)
M. waterhousii			
Microdipodops pallidus	.217	.256	(Hafner, Hafner, and Hafner 1979)
M. megacephalus			
Microtus arvalis	.055	.241	(Graf 1982)
M. agrestis			
M. multiplex			
M. mariae			
Mus musculus	.059	.346	(Britton and Thaler 1978)
M. spretus			
Neotomo albigula	.072	.150	(Zimmerman and Nejtek 1977)
N. micropus			
N. floridana			
Pan troglodytes	.014	.103	(Bruce and Ayala 1979)
P. paniscus			
Papio anubis	.024	.020	(Kawamoto, Shotake, and Nozowa 1982)
P. hamadryas			
Peromyscus attwateri	.044	.300	(Kilpatrick and Zimmerman 1975)
P. boylii			

(*continued*)

Appendix 1. (*continued*)

Species	H	D	Reference
P. pectoralis			
P. polius			
Pongo pygmaeus abelii	.033	.130	(Bruce and Ayala 1979)
P. p. pygmaeus			
Proechimys guaire	.089	.062	(Benado et al. 1979)
P. urichi			
Rattus bowersii	.013	1.238	(Chan, Dhaliwal, and Young 1979)
R. muelleri			
R. cremoriventer			
R. cameroni			
R. bukit			
R. sabanus			
R. edwardsi			
Sigmodon hispidus	.025	.27	(Johnson et al. 1972)
S. arizonae			
Spermophilus tridecemlineatus	.069	.140	(Cothran, Zimmerman, and Nadler 1977)
S. mexicanus			
s. spilosoma			
Zapus princeps	.019	.411	(Hafner, Petersen, and Yates 1981)
6 Invertebrates			
Glycera dibranchiata	.128	.759	(Nicklas and Hoffmann 1979)
G. americana			
Mytilus edulis	.214	.172	(Skibinski, Cross, and Ahmad 1980)
M. galloprovincialis			

References cited

Adest, G. A. 1977. Genetic relationships in the genus *Uma*. *Copeia* 1977:47–52.

Avise, J. C., C. F. Aquadro, and J. C. Patton. 1982. Evolutionary genetics of birds. V. Genetic distances within Mimidae (mimic thrushes) and Vireonidae (vireos). *Biochem. Genet.* 20:95–104.

Avise, J. C., J. C. Patton, and C. F. Aquadro. 1980a. Evolutionary genetics of birds: Comparative molecular evolution in New World warblers and rodents. *J. Hered.* 71:303–310.

Avise, J. C., J. C. Patton, and C. F. Aquadro. 1980b. Evolutionary genetics of birds I. Relationships among North American thrushes and allies. *Auk.* 97:135–147.

Avise, J. C., J. C. Patton, and C. F. Aquadro. 1980c. Evolutionary genetics of birds II. Conservative protein evolution in North American sparrows and relatives. *Syst. Zool.* 29:323–334.

Avise, J. C., D. O. Straney, and M. H. Smith. 1977. Biochemical genetics of sunfish IV. Relationships of centrarchid genera. *Copeia* 1977:250–258.

Ayala, F. J., M. L. Tracey, L. G. Barr, J. F. McDonald, and S. Perez-Salas. 1974. Genetic variation in natural populations of five *Drosophila* species and the hypothesis of the selective neutrality of polymorphisms. *Genetics* 77:343–384.

Benado, M., M. Aguilera, O. A. Reig, and F. J. Ayala. 1979. Biochemical genetics of chromosome forms of Venezuelan spiny rats of the *Proechimys guairae* and *Proechimys trinitatis* superspecies. *Genetica* 50:89–97.

Bezy, R. L., G. C. Gorman, Y. J. Kim, and J. W. Wright. 1977. Chromosomal and genetic divergences in the fossorial lizards of the family Anniellidae. *Syst. Zool.* 26:57.

Britton, J., and L. Thaler. 1978. Evidence for the presence of two sympatric species of mice (genus *Mus* L.) in southern France based on biochemical genetics. *Biochem. Genet.* 16:213–225.

Bruce, E. J., and F. J. Ayala. 1979. Phylogenetic relationships between man and the apes: Electrophoretic evidence. *Evolution* 33:1040–1056.

Busack, S. D., L. R. Maxson, and M. A. Wilson. 1985. *Pelobates varaldii* (Anura: Pelobatidae): A morphologically conservative species. *Copeia* 1985:107–112.

Buth, D. G. 1980. Evolutionary genetics and systematic relationships in the castomid genus *Hypentelium*. *Copeia* 1980:280–290.

Buth, D. G., and B. M. Burr. 1978. Isozyme variability in the cyprinid genus *Campostoma*. *Copeia* 1978:289–311.

Case, S. M. 1978a. Biochemical systematics of members of the genus *Rana* native to western North America. *Syst. Zool.* 27:299.

Case, S. M. 1978b. Electrophoretic variation in two species of ranid frogs, *Rana boylei* and *R. muscosa*. *Copeia* 1978:311–327.

Case, S. M., P. G. Haneline, and M. F. Smith. 1975. Protein variation in several species of *Hyla*. *Syst. Zool.* 24:281–295.

Chan, K. L., S. S. Dhaliwal, and H. S. Young. 1979. Protein variation and systematics of three subgenera of Malayan rats (Rodentia: Muridae, genus *Rattus* Fischer). *Comp. Biochem. Physiol.* 64:329–337.

Cothran, E. G., E. G. Zimmerman, and C. F. Nadler. 1977. Genic differentiation and evolution in the ground squirrels subgenus Ictiodomys (genus *Spermophilus*). *J. Mammol.* 58:610–622.

Dessauer, H. C., D. F. Gartside, and R. G. Zweifel. 1977. Protein electrophoresis and the systematics of some New Guinea hylid frogs (genus *Litoria*). *Syst. Zool.* 26:426–436.

Feder, J. H. 1979. Natural hybridization and genetic divergence between the toads *Bufo boreas* and *Bufo punctatus*. *Evolution* 33:1089–1097.

Feder, J. H., G. Z. Wurst, and D. B. Wake. 1978. Genetic variation in western salamanders of genus *Plethodon*, and the status of *Plethodon gordoni*. *Herpetologica* 34:64–69.

Gartside, D. F., J. S. Rogers, and M. C. Dessauer. 1977. Speciation with little genic and morphological differentiation in the ribbon snakes *Thamnophis proximus* and *T. sauritas*. *Copeia* 1977:697–705.

Gorman, G. C., Y. J. Kim, and R. Rubinoff. 1976. Genetic relationships of 3 species of *Bathygobius* from the Atlantic and Pacific sides of Panama. *Copeia* 1976:361–364.

Gorman, G. C., and E. Nevo. 1974. Evolutionary genetics of insular Adriatic lizards. *Evolution* 29:52–71.

Graf, J.-D. 1982. Genetique biochimique, zoogeographie et taxonomie des Arvicolidae (Mammalia: Rodentia). *Revue suisse zoologie* 89:749–787.

Green, D. M. 1983. Allozyme variation through a clinal hybrid zone between the toads *Bufo americanus* and *B. hemiophrys* in southeastern Manitoba. *Herpetologica* 39:28–40.

Green, D. M. 1984. Sympatric hybridization and allozyme variation in the toads *Bufo americanus* and *B. fowleri* in southern Ontario. *Copeia* 1984:18–26.

Greenbaum, I. F., and R. J. Baker. 1976. Evolutionary relationships in *Macrotus* (Mammalia: Chiroptera): Biochemical variation and karyology. *Syst. Zool.* 25:15–25.

Gutierrez, R. J., R. M. Zink, and S. Y. Young. 1983. Genetic variation, systematics and biographical relationships among some galliform birds. *Auk.* 100:33–47.

Hafner, D. J., J. C. Hafner, and M. S. Hafner. 1979. Systematic status of kangaroo mice, genus *Microdipodops*: Morphometric, chromosomal, and protein analyses. *J Mammol.* 60:1–10.

Hafner, D. J., K. E. Petersen, and T. L. Yates. 1981. Evolutionary relationships of jumping mice (genus *Zapus*) of the southwestern United States. *J. Mammol.* 62:501.

Hanken, J. 1983. Genetic variation in a dwarfed lineage, the Mexican salamander genus *Thorius*

(Amphibia: Plethodontidae): Taxonomic, ecologic, and evolutionary implications. *Copeia* 1983:1051–1073.

Hanken, J., and D. B. Wake. 1982. Genetic differentiation among plethodontid salamanders (genus *Bolitoglossa*) in Central and South America: Implications for the South American invasion. *Herpetologica* 38:272.

Hedgecock. D., and F. J. Ayala. 1974. Evolutionary divergence in the genus *Taricha*. *Copeia* 1974:738–747.

Highton, R., and J. R. MacGregor. 1983. *Plethodon kentucki* Mittleman: A valid species of Cumberland Plateau woodland salamander: *Herpetologica* 39:189–200.

Johnson, W. E., and R. K. Selander. 1971. Protein variation and systematics in kangaroo rats (genus *Dipodomys*). *Syst. Zool.* 20:377–405.

Johnson, W. E., R. K. Selander, M. H. Smith, and Y. J. Kim. 1972. Studies in genetics XIV. Genetics of sibling species of the cotton rat (*Sigmodon*). *Univ. Texas Publ.* #7213, pp. 297–305.

Johnson, N. K., and R. M. Zink. 1983. Speciation in sapsuckers. I Genetic differentiation. *Auk.* 100:871–884.

Kawamoto, Y., T. Shotake, and K. Nozowa. 1982. Genetic differentiation among three genera of family cercopithecidae. *Primates* 23:272–286.

Kilpatrick, C. W., and E. G. Zimmerman. 1975. Genetic variation and systematics of four species of mice of the *Peromyscus boylii* species group. *Syst. Zool.* 24:143–162.

Kim, Y. J., G. C. Gorman, and R. B. Huey. 1978. Genetic variation and differentiation in two species of fossorial African skink *Typhlosaurus* (Sauria: Scincidae). *Herpetologica* 34:192–194.

Kim, Y. J., G. C. Gorman, T. Papenfuss, and A. K. Roychoudhury. 1976. Genetic variation in the amphisbaenian genus *Bipes*. *Copeia* 1976:120–124.

Kornfield, I., and K. E. Carpenter. 1984. Cyprinids of Lake Lanao, Philippines: Taxonomic validity, evolutionary rates and speciation scenarios. Pp. 69–84. in A. A. Echelle and I. Kornfield (eds.) *Evolution of Fish Species Flocks*. University of Maine Press, Orono.

Larson, A. 1983. A molecular phylogenetic perspective on the origins of a lowland tropical salamander fauna. I. Phylogenetic inferences from protein comparisons. *Herpetologica* 39:85–99.

Matthews, T. C., and L. E. Munstermann. 1983. Genetic diversity and differentiation in a northern population of the tree-hole mosquito *Aedes hendersoni* (Diptera: Culicidae). *Ann. Entom. Soc. Amer.* 76:1005–1010.

Milton, D.A., J. M. Hughes, and P. B. Mather. 1983. Electrophoretic evidence for the specific distinctness of *Egernia modesta* and *E. whitii* (Lacertilla, Scincidae). *Herpetologica* 35:100–105.

Murphy, R. W., W. E. Cooper Jr., and W. S. Richardson. 1983 Phylogenetic relationships of the North American five-lined skinks, genus *Eumeces* (Sauria: Scincidae) *Herpetologica* 39:200–211.

Nevo, E. 1976. Genetic variation in constant environments. *Experientia* 32:858–859.

Nicklas, N. L., and R. J. Hoffmann. 1979 Genetic similarity between two morphologically similar species of polychaetes. *Mar. Biol.* 52:53–59.

Nozowa, K., T. Shotake, Y. Ohkura, and Y. Tanabe. 1977. Genetic variations within and between species of Asian Macaques. *Jose. J. Genet.* 52:15–30.

Patton, J. L. 1984. Biochemical genetics of the Galapagos giant tortoises. *Nat. Geog. Soc. Res. Rep.* 17:701–709.

Penney, D. F., and E. G. Zimmerman. 1976. Genic divergences and local population differentiation by random drift in the pocket gopher genus *Geomys*. *Evolution* 30:473–483.

Phelps, S. R., And F. W. Allendorf. 1983. Genetic identity of pallid and shovelnose sturgeon. *Copeia* 1983:696–700.

Sage, R. D., P. V. Loiselle, P. Basasibwaki, and A. C. Wilson. 1984. Molecular versus morphological change among cichlid fishes of Lake Victoria. Pp. 185–201 in A. A. Echelle and I. Kornfield (eds.) *Evolution of Fish Species Flocks*. University of Maine Press, Orono.

Sattler, P. W. 1980. Genetic relationships among selected species of North American *Scaphiopus*. *Copeia* 1980:605–610.

Sbordoni, V., G. Allegrucci, A. Caccone, D. Cesaroni, M. Cobolli Sbordoni, and E. DeMatthaeis. 1981. Genetic variability and divergence in cave populations of *Troglophilus cavicola* and *T. andrenii* (Orthoptera: Rhaphidophoridae). *Evolution* 35:226–233.

Shaklee, J. B., and C. S. Tamaru. 1981. Biochemical and morphological evolution of Hawaiian bonefishes (*Albula*). *Syst. Zool.* 30:125–146.

Siebenaller, J. F. 1978. Genetic variability in deep sea fishes of the genus *Sebastolobus*. Pp. 95–116 in B. Battaglia (ed.) *Marine Organisms: Genetics, Ecology, and Evolution*. New York, Plenum.

Sites, J. W., J. W. Bickham, B. A. Pytel, I. F. Greenbaum, and B. A. Bates. 1984. Biochemical characters and the reconstruction of turtle phylogenies: Relationships among batagurine genera. *Syst. Zool.* 33:137–158.

Skibinski, D. O. F., T. F. Cross, and M. Ahmad. 1980. Electrophoretic investigation of systematic relationships in the marine mussels *Modiolus modiolus* L., *Mytilus edulis* L., and *Mytilus galloprovincialis* Lmk. (Mytilidae: Mollusca). *Biol. J. Linn. Soc.* 13:65–73.

Utter, F. M., F. W. Allendorf, and H. O. Hodgins. 1973. Genetic variability and relationships in Pacific salmon and related trout based on protein variations. *Syst. Zool.* 22:257.

Wake, D. B., and J. F. Lynch. 1982. Evolutionary relationships among the Central American salamanders of the *Bolitoglossa franklini* group, with a description of a new species from Guatemala. *Herpetologica* 38:257–272.

Webster, T. P., R. K. Selander, and S. Y., Yang. 1972. Genetic variability and similarity in the Anolis lizards of Bimini. *Evolution* 26:523–535.

Yang, S., and J. L. Patton. 1981. Genetic variability and differentiation in the Galapagos finches. *Auk.* 98:230–242.

Zimmerman, E. G., and M. E. Nejtek. 1977. Genetics and speciation of three semispecies of *Neotoma*. *J. Mammol.* 58:391–402.

Zink, R. M. 1982. Patterns of genic and morphologic variation among sparrows in genera *Zonotrichia*, *Melospiza*, *Junco*, and *Passerella*. *Auk.* 99:632–649.

Zink, R. M., and N. K. Johnson. 1984. Evolutionary genetics of flycatchers I. Sibling species in the genera *Empidomax* and *Contropus*. *Syst. Zool.* 33:205–216.

Appendix 2. PASCAL Program for the Simulation of the Evolution of Female Choice

```
program short (input, output);
  { updated for Windows by Ken Mitton, 12/1/95 }
  { #ifdef windows }

uses
    wincrt;
  { #endif }

type
  int300x13 = array[1. .300, 1. .13] of integer;
  int400x13 = array[1. .400, 1. .13] of integer;
  int300x13ptr = ^int300x13;
  int400x13ptr = ^int400x13;

var
  gencount, dead, gen, top, mate, male, stud, i, j, ij, k, l, h:
    integer;
  ijk, inter: integer;
  loci, parents, zygotes: integer;
  g: array[1. .10] of real; { allelic frequencies }
  f: array[1. .10, 1. .3] of real; { fitnesses of genotypes }
  adult: int400x13ptr;
  zyg: int400x13ptr;
  fitness: array[1. .400] of real; { fitnesses in zyg }
  yellow: array[1. .400] of integer;
  hold, w, fit, choice, x, y, z, z1, z2, z3: real;
  lst: text;
  totalfixation: integer;

procedure report (adult: int300x13ptr; parents: integer; var
  choice: real);
{ reports allelic frequencies }

  var
    d, n: real;
    i: integer;
    count: array[1. .3] of integer;
  begin
    n :=parents;
    count[1] :=0;
    count[2] :=0;
    count[3] :=0;
    for i :=1 to parents do
       count [adult^[i, 12]] :=count[adult^[i, 12]] +1;
    choice :=(count[3] + 0.5 * count[2]) /n;
  end; {end of procedure }
```

```
procedure cross (adult: int300x13ptr; var zyg: int400x13ptr; l:
  integer; mate: intege integer);
  var
    g1, g2: integer;
    y: real;
  begin
{ adult is the array, l is the female, mate is the male, j=locus }
{ choose female gamete first }
    g1 :=0;
    if adult^[l, j] = 3 then
       g1 :=1;
    if adult^[l, j] = 2 then
      begin
        y :=random;
        if y >= 0.5 then
          g1 :=1;
      end;
{ now choose male gamete }
    g2 := 0;
    if adult^[mate, j] = 3 then
      g2 := 1;
    if adult^[mate, j] = 2 then
      begin
        y := random;
        if y >= 0.5 then
          g2 := 1;
    end;
  zyg^[k, j] := 1 + g1 + g2;
end;
zyg^[k, j] := 1 + g1 + g2;
end;                                    { end of procedure }

begin
  randomize;
  new(adult) ;
  new(zyg) ;
  assign (lst, 'c:\shortout.txt');
  rewrite(lst);
  writeln('begin') ;
  clrscr;
  gencount :=0;
  writeln('how many loci? (maximum of 10)') ;
  readln(loci) ;
  writeln(lst, 'loci: ', loci) ;
  writeln('how many parents? (maximum of 300) ' ) ;
  readln(parents) ;
  writeln(1st, 'parents: ' , parents) ;
  writeln( 'how many zygotes? (maximum of 400) ' ) ; { offspring
  before selection }
  readln(zygotes) ;
  writeln(lst, 'zygotes: ', zygotes);
```

```
{ now put in the allelic frequencies that you want }
  writeln( ' allelic frequencies for fitness loci ' );
  readln(z);
  for i := 1 to loci do
    begin
      g[i] := z
    end;
  write(1st, 'allelic freq.: ') ;
  for i :=1 to loci do
    write(lst, g[1]) ;
  writeln(lst, ' ') ;
{ now put in the distributions of fitnesses at each locus }
  writeln( ' fitnesses of three genotypes for the fitness loci ' ) ;
  readln(z1, z2, z3) ;
  for i :=1 to loci do
    begin
      f[i, 1] :=z1;
      f[i, 2] :=z2;
      f[i, 3] :=z3;
    end;
  writeln(lst, 'fitnesses: ', z1, z2, z3);
  writeln ('allelic frequency of the mate choice allele = ') ;
  readln- (choice);
  writeln(lst 'allelic freq. male choice: ', choice);
  z1 := 2 * choice * (1.0 - choice) ;
  writeln(' number of iterations = ') ;
  readln(inter) ;
  writeln(lst, 'iterations: ', inter) ;
  for ijk :=1 to inter do                    { do loop for iterations}
    begin
{ chapter 2 establish initial breeding population }
      ij := 0;
      for i :=1 to parents do
    begin
{ assigning sex—1 represents male, 0 represents female }
      z := random;
      if z > 0.5 then
        adult^[i, 11] := 1
      else
        adult^[i, 11] := 0;
      z := random;
      if z <= z1 then
        adult^[i, 12] := 2
      else
        adult^[i, 12] := 1;
      end;
    for j := 1 to parents do
      begin
        for i := 1 to parents do
          x := random;
```

```
        if x <= g[j] then
          k := 1
        else
          k :=0;
        y := random;
        if y <= g[j] then
          l := 1
        else
          l := 0;
        adult^[i, j] := 1 + k + l;
      end;
    end;
{ now assess heterozygosity of each individual in adult }
    for i := 1 to parents do
      begin
        h :=0;
        for j := 1 to loci do
          if adult^[i, j] = 2.0 then
            h := h + 1;
        adult^[i, 13] := h;
      end;
    report(adult, parents, choice) ;
    repeat        { generation loop }
      ij := ij + 1;
{ chapter 3 mate these individuals. 1's are male. column }
{ 11 is sex, col 12 is choice, with 2 and 3 being choosy, }
{ and 13 is heterozygosity }
{ get line numbers for most heterozygous males }
{ first identify most heterozygous }
      top := 0;
      for i := 1 to parents  do
        begin
          if adult^[i, 11] = 1 then { that is, if it is male }
            begin
              if adult^[i, 13] > top then
                top := adult^[i, 13];
            end;                               { if loop }
            end;                               { i loop }
{ now top is top 2 most het classes—find how many males }
{ are like this and place their line numbers in yellow }
      stud := 0
      for i :=1 to parents do
        begin
          if adult^[i, 11] = 1 then { that is, if it is male }
            begin
              if (adult^[i, 13] = top) or (adult^[i, 13] > top) then
                begin
                  stud := stud + 1;
                  yellow[stud] := i;
```

```
            end;                    { if loop }
          end;                  { if loop }
        end;                { i loop? }
    for k := 1 to zygotes do
      begin
                    { choose a random female }
        repeat
          l := random(parents) + 1; { randomly chosen female }
        until adult^[l, 11] = 0;
        if adult^[l, 12] > 1 then
          mate := yellow[random(stud) + 1]
        else
          repeat
            mate := random( parents) + 1]
          until adult^[mate, 11] = 1;

{ mate is the number of a male, randomly chosen or chosen by }
{ high heterozygosity, to be mated to l }
          for j := 1 to loci do
            cross(adult, zyg, l, mate, j) ;
          j := 12;
          cross(adult, zyg, l, mate, j) ;
  { choose sex of offspring }
          z := random:
          if z >= 0.5 then
            zyg^[k, 11] := 1
          else
            zyg^[k, 11] := 0;
        end;                        { end of the k loop }

{ lets have a look at the yellow pages—are they more heterozygous? }
{ writeln(lst, 'number of top males =' ,stud:8) ; }
{ for i:=1 to stud do writeln (lst,yellow[i]) ; }
{ writeln(lst, 'here come #, het of adults') ; }
{ for i:=1 to parents do writeln(lst,i:8, adult^[i,13]:8) ; }

{ now zygotes are piled up in the zyg matrix. calculate the }
{ fitness potential for each individual, then apply truncation
  selection, }
{ transferring only n=parents to the adult matrix. }

{ try reversing fitness every 4 generation }

    gencount := gencount + 1;
    if gencount > 4 then
      begin
        gencount := 0;
        for i := 1 to loci do
          begin
            hold := f[i, 1] ;
            f[i, 1] := f[i, 3] ;
```

```
            f[i, 3] := hold;
          end;                        { i loop }
        end;                                 { if loop }

    for i := 1 to zygotes do
      begin
        w := 1.0;
          for j := 1 to loci do
            w := w * f[j, zyg^[i, j]] ; { calculating fitness }
          fitness[i] := w;
        end;
{ now remove parents-zygotes from zyg, removing those with low
  fitness }

    for i := zygotes downto parents do
      begin { find individual with lowest fitness }
        fit := 1.0;
        for k :=1 to i do
          begin
            if fitness[k] < fit then
              begin
                dead := k;
                fit := fitness[k];
              end;       { if loop }
          end;         { k loop }
{ now dead is the number with the lowest fitness—rearrange }
{ both fitness and zyg }
        fitness[dead] := fitness[i] ;
        for k := 1 to  loci do
          zyg^[dead, k] := zyg^[i, k] ;
        for k := 11 to 12 do
          zyg^[dead, k] := zyg^[i, k] ;
          end;         { i loop }
{ now transfer survivors to adult matrix }
    for i := 1 to parents do
      begin
        for j := 1 to loci do
          adult^[i, j] := zyg^[i, j] ;
        for j := 11 to 12 do
          adult^[i, j] := zyg^[i, j] ;
  end                       { i loop }

{ now assess heterozygosity for each individual in adult }
    for i :=1 to parents do
      begin
        h :=0
        for j := 1 to loci do
          if adult^[i, j] = 2.0 then
            h := h + 1;
        adult^[i, 13] :=h;
```

```
        end;                          { i loop }
      report(adult, parents, choice);
    until (choice < 0.005) or (choice > 0.99);  { generation loop }
    if (choice > 0.9) then
      totalfixation := totalfixation + 1;
    writeln(' ') ;
    write( ' i=', ijk : 4, ' g=' , ij : 4, ' c=' , choice : 5 : 2) ;
    writeln(lst, ' iter=' , ijk : 6, ' generations=' , ij : 6,
      '  choice=' , choice : 8
3) ;
  end;                        { iteration loop }
  writeln(' ') ;
  writeln( ' total number at fixation: ' , totalfixation);
  writeln(lst, ' ') ;
  writeln(lst, ' total number at fixation: ' , totalfixation) ;
  close(lst) ;
  dispose(adult) ;
  dispose(zyg) ;
end.
```

Appendix 3. Heterozygosity and Maximum Lifetime Fecundity

Species[a]	Heterozygosity	Fecundity[a]
A. Species of animals, taken from Powell (1975)		
Aplonis metallica	.047	80
Aplonis cantoroides	.012	80
Zonatrichia capensis	.035	20
Mus musculus	.087	80
Mus musculus	.122	80
Mus musculus	.102	80
Mus musculus	.100	80
Peromyscus gossypinus	.051	25
Peromyscus leucopus	.070	25
Peromyscus floridanus	.055	25
Peromyscus polionotus	.059	15
Dipodomys merriami	.051	18
Dipodomys ordii	.018	18
Dipodomys panamintinus	.000	18
Dipodomys elator	.002	18
Dipodomys microps	.007	18
Dipodomys spectabilis	.008	18
Dipodomys deserti	.010	18
Dipodomys compactus	.023	18
Dipodomys agilis	.040	18
Dipodomys nitratoides	.040	18
Dipodomys humani	.042	18
Thomomys bottae	.070	18
Thomomys umbrinus	.033	18
Thomomys talpoides	.055	18
Mirounga angustirostris	.000	20
Mirounga leonina	.030	20
Drosophila busckii	.044	3000
Drosophila willistoni	.200	3000
Drosophila willistoni	.183	3000
Drosophila willistoni	.171	3000
Drosophila equinoxialis	.220	3000
Drosophila equinoxialis	.220	3000
Drosophila equinoxialis	.181	3000
Drosophila tropicalis	.198	3000
Drosophila paulistorum	.205	3000
Drosophila andean	.189	3000
Drosophila andean	.165	3000
Drosophila andean	.205	3000
Drosophila paulistorum	.134	3000
Drosophila paulistorum	.164	3000
Drosophila paulistorum	.188	3000
Drosophila paulistorum	.170	3000
Drosophila melanogaster	.154	3000
Drosophila simulans	.107	3000
Drosophila bipectinate	.241	3000
Drosophila parabipectinate	.134	3000
Drosophila pseudoananassae	.197	3000
Drosophila nigrens	.209	3000

(*continued*)

Appendix 3. (*continued*)

Species[a]	Heterozygosity	Fecundity[a]
Drosophila malerkotliana	.186	3000
Drosophila pallens	.235	3000
Drosophila anassae	.135	3000
Drosophila obscura	.109	3000
Drosophila subobscura	.105	3000
Drosophila bifasciata	.242	3000
Drosophila pseudoobscura	.125	3000
Drosophila pseudoobscura	.051	3000
Drosophila persimilis	.106	3000
Drosophila athabasca	.146	3000
Drosophila affinis	.238	3000
Drosophila mohavensis	.068	3000
Drosophila mohavensis	.054	3000
Drosophila mohavensis	.082	3000
Drosophila arizonesis	.128	3000
Drosophila mulleri	.113	3000
Drosophila aldrichi	.115	3000
Drosophila robusta	.110	3000
Drosophila pavani	.192	3000
Onchorhynchus	.032	1000
Fundulus heteroclitus	.180	900
Menidia menidia	.090	30
Menidia peninsulae	.054	150
Menidia beryllina	.055	150
Menidia audens	.070	150
Menidia extensa	.033	150
Bufo cognatus	.135	10000
Bufo speciousus	.097	10000
Bufo viridis	.134	10000
Bufo viridis	.029	10000
Taricha rivularis	.109	2500
Taricha granulosa	.081	2500
Taricha torosa	.053	2500
Uta stansburiana	.051	150
Uta stansburiana	.051	150
Anolis carolinensis	.049	150
Anolis distichus	.064	150
Anolis distichus	.043	150
Anolis angusticeps	.000	150
Anolis sagrei	.020	150
Sceloporous graciousus	.070	150
Lepomis humilis	.049	500
Lepomis microlophus	.037	500
Lepomis aurotus	.071	500
Lepomis gulogus	.030	500
Lepomis megalotis	.114	500
Lepomis cyanellus	.074	500
Lepomis punctatus	.113	500
Lepomis gibbosus	.067	500
Lepomis marginatus	.069	500
Lepomis macrochirus	.041	500

(*continued*)

Appendix 3. (*continued*)

Species[a]	Heterozygosity	Fecundity[a]
B. Species of animals taken from Nei and Graur (1984)		
Primates		
Homo sapiens	.143	25
Gorilla gorilla	.046	25
Pan troglodytes	.013	25
Macaca cyclopsis	.041	25
Macaca fuscata	.013	25
Macaca fascicularis	.096	25
Seals		
Pagophilus groenlandicus	.021	25
Mirounga angustirostris	.000	25
Ungulates		
Alces alces	.020	25
Odocoileus virginianus	.100	25
Cervus canadensis	.012	25
Lagomorphs		
Oryctolagus cuniculus	.059	75
Ochotona princeps	.011	75
Rodents		
Peromyscus guardia	.014	40
Peromyscus interparietalis	.000	40
Peromyscus dickeyi	.000	40
Peromyscus merriami	.016	40
Peromyscus stephani	.000	40
Peromyscus floridanus	.062	40
Peromyscus caniceps	.011	40
Peromyscus polionotus	.065	40
Peromyscus pectoralis	.022	40
Peromyscus maniculatus	.128	40
Peromyscus melanotis	.021	40
Peromyscus difficilis	.060	40
Peromyscus truei	.040	40
Eutamias panaminitus	.055	40
Sigmodon arizonae	.033	40
Sigmodon hispidus	.020	40
Spalax ehrenbergi(52)	.066	40
Spalax ehrenbergi(54)	.038	40
Spalax ehrenbergi(58)	.016	40
Spalax ehrenbergi(60)	.035	40
Thomomys umbrinus	.031	40
Thomomys bottae	.091	40
Geomys personatus	.027	40
Geomys tropicalis	.000	40
Geomys bursarius	.063	40

(*continued*)

Appendix 3. (*continued*)

Species[a]	Heterozygosity	Fecundity[a]
Carnivores		
Vulpes vulpes	.000	25
Mustela erminea	.000	25
Mustela putorius	.000	25
Martes foina	.000	25
Meles meles	.000	25
Acinomyx jubatus	.000	25
Lizards		
Anolis trinitatis	.061	50
Anolis carolinensis	.073	50
Anolis griseus	.020	50
Anolis segrei	.010	50
Anolis luciae	.089	50
Anolis critatellus	.120	50
Anolis wattsi	.046	50
Anolis disticus	.051	50
Anolis blanquillanus	.053	50
Anolis oculatus	.050	50
Anolis sabanus	.044	50
Anolis gingivinus	.100	50
Anolis grahami	.078	50
Anolis marmoratus	.051	50
Anolis roquet	.074	50
Anolis lividus	.053	50
Alligator mississippiensis	.021	1500
Newts		
Taricha rivularis	.068	600
Bony fishes		
Etheostoma spectabile	.069	200
Etheostoma caeruleum	.066	200
Oncorhynchus nerka	.018	4000
Salmo salar	.035	4000
Fruit flies		
Drosophila nebulosa	.218	1800
Drosophila tropicalis	.155	1800
Drosophila paulistorum	.228	1800
Drosophila willistoni	.183	1800
Drosophila equinoxialis	.185	1800
Drosophila pseudoobscura	.136	1800
Drosophila mimica	.222	1800
Drosophila engyochracea	.127	1800
Land snails		
Sphincterochilia aharoni	.067	1800

(*continued*)

Species[a]	Heterozygosity	Fecundity[a]
C. Species of plants, taken from Hamrick, Linhart, and Mitton (1979)		
Abies lasiocarpa	.399	4
Agrostis stolonifera	.181	4
Arabidopsis thaliana	.318	1
Avena barbata	.042	1
Avena fatua	.138	1
Avena hirtula	.012	1
Baptisia leucophaea	.287	3
Baptisia nuttaliana	.089	3
Baptisia sphaerocarpa	.229	3
Bromus mollis	.204	1
Bromus rubens	.202	1
Chenopodium fremontii	.019	3
Citrus aurantifolia	.250	4
Citrus aurantium	.154	4
Citrus grandis	.049	4
Citrus jambhiri	.500	4
Citrus limon	.499	4
Citrus medica	.070	4
Citrus paradisi	.125	4
Citrus reticulata	.361	4
Citrus sinensis	.310	4
Clarkia amoena	.071	2
Clarkia biloba	.203	2
Clarkia dudleyana	.250	2
Clarkia franciscana	.000	2
Clarkia lingulata	.175	2
Clarkia rubicunda	.177	2
Curcurbita foetidissima	.256	4
Elymus canadensis	.026	4
Eucalyptus obliqua	.351	4
Eucalyptus pauciflora	.278	4
Ficas carica	.530	4
Gaura demareei	.050	2
Gaura longifolia	.074	2
Helianthus annuus	.162	3
Hordeum distichum	.135	2
Hordeum jubatum	.192	3
Hordeum spontaneum	.282	2
Hordeum vulgare	.148	2
Hymenopappus artemisifolius	.222	2
Humenopappus scabiosaeus	.218	2
Larix decidua	.347	4
Liatris cylindracea	.158	4
Limnanthes alba	.159	2
Limnanthes floccosa	.130	1
Lolium multiflorum	.331	1
Lolium perenne	.280	3
Lupinus nanus	.248	1
Lupinus subcarnosus	.142	1

(*continued*)

Appendix 3. (*continued*)

Species[a]	Heterozygosity	Fecundity[a]
Lupinus succulentus	.080	1
Lupinus texensis	.414	1
Lycopersicon cheesmanii	.000	3
Lycopersicon chmielewskii	.053	3
Lycopersicon esculentum	.037	3
Lycopersicon esculentum cerasiformae	.018	3
Lycopersicon parviflorum	.000	3
Lycopersicon pimpinellifolium	.128	3
Mimulus guttatus	.336	4
Oenothera argillicola	.075	3
Oenothera biennis	.083	3
Oenothera hookeri	.000	3
Oenothera parviflora	.136	3
Oenothera strigosa	.046	3
Opuntia basilaris	.479	4
Origanum vulgare	.410	2
Oryza sativa	.383	1
Panicum maximum (sexual)	.377	4
Panicum maximum (asexual)	.139	4
Persea americana	.195	4
Phaseolus coccineus	.049	3
Phaseolus vulgaris	.064	2
Phlox cuspidata	.026	1
Phlox drummondii	.058	1
Phlox drummondii (cultivated)	.036	1
Phlox roemariana	.055	1
Picea abies	.418	4
Picea engelmannii	.432	4
Pinus longaeva	.364	4
Pinus pungens	.144	4
Pinus ponderosa	.226	4
Pinus resinosa	.000	4
Pinus rigida	.170	4
Pinus sylvestris	.359	4
Poncirus trifoliata	.500	4
Pseudotsuga menziesii	.436	4
Secale cereale	.216	1
Silene maritima	.140	3
Stephanomeria exigua carotifera	.156	1
Stephanomeria exigua coronaria	.277	1
Tragopogon dubius	.028	2
Tragopogon porrifolius	.014	2
Tragopogon pratensis	.000	2
Zea mays	.301	3
Zea mexicana (annual)	.230	2

Note: For fecundity in plants: 1 = $<10^2$; 2 = 10^2–10^3; 3 = 10^3–10^4; 4 = $>10^4$

a. My estimate of fecundity, the maximum number of offspring that a healthy and lucky female could produce.

Bibliography

Able, K. P., and B. R. Noon. 1976. Avian community structure along elevational gradients in the northeastern United States. *Oecologia* 26:275–294.

Adams, D. E., and R. C. Anderson. 1982. An inverse relationship between dominance and habitat breadth in Illinois forests. *Amer. Midl. Nat.* 107:192–195.

Albon, S. D., T. H. Clutton-Brock, O. F. Price, B. T. Genfell, J. M. Pemberton, and F. M. D. Gulland. 1993. Towards a more exact population demography. In D. McCullough (ed.) *Wildlife 2001: Populations*. American Elsevier, New York.

Allard, R. W. 1975. The mating system and microevolution. *Genetics* 79:115–126.

Allard, R. W., S. K. Jain, and P. L. Workman. 1968. The genetics of inbreeding populations. *Adv. Genet.* 14:55–131.

Allard, R. W., A. L. Kahler, and M. T. Clegg. 1977. Estimation of mating cycle components of selection in plants. Pp. 1–19 in F. B. Christiansen and T. M. Fenchel (eds.) *Measuring Selection in Natural Populations*. Springer-Verlag, New York.

Allendorf, F. W., F. B. Christiansen, T. Dobson, W. F. Eanes, and O. Frydenberg. 1979. Electrophoretic variation in large mammals. I. The polar bear, *Thalarctos maritimus*. *Hereditas* 91:19–22.

Allendorf, F. W., and R. F. Leary. 1986. Heterozygosity and fitness in natural populations of animals. Pp. 57–76. In M. E. Soulé (ed.) *The Science of Scarcity and Diversity*. Sinauer Associates, Sunderland, MA.

Allison, A. C. 1955. Aspects of polymorphism in man. *Cold Spring Harb. Symp. Quant. Biol.* 20:239–255.

Allison, A. C. 1964. Polymorphism and natural selection in human populations. *Cold Spring Harbor Symp. Quant. Biol.* 29:137–149.

Ananthakrishnan, R., and H. Walter. 1972. Some notes on the geographical distribution of the human red cell acid phosphatase phenotypes. *Humangenetik* 15:177–181.

Anderson, S. M., and J. F. McDonald. 1983. Biochemical and molecular analysis of naturally occurring ADH variants in *Drosophila melanogaster*. *Proc. Natl. Acad. Sci. USA* 80:4798–4802.

Anderson, S. M., M. Santos, and J. McDonald. 1980. Comparative study of the thermostability of crude and purified preparations of alcohol dehydrogenase (EC 1.1.1.1.) from *D. melanogaster*. *Drosophila Info. Serv.* 55:13–14.

Anderson, W. W., L. Levine, O. Olvera, J. R. Powell, M. E. De la Rosa, V. M. Salceda, M. I. Gaso, and J. Guzman. 1979. Evidence for selection by male mating success in natural populations of *Drosophila pseudoobscura*. *Proc. Natl. Acad. Sci. USA* 76:1519–1523.

Andersson, M. 1982a. Female choice selects for extreme tail length in a widowbird. *Nature* (London) 299:818–820.

Andersson, M. 1982b. Sexual selection, natural selection, and quality advertisement. *Biol. J. Linn. Soc.* 17:375–393.

Antonovics, J. 1971. The effects of a heterogeneous environment on the genetics of natural populations. *Amer. Scient.* 59:593–599.

Appels, R., and R. L. Honeycutt. 1986. RDNA: Evolution over a billion years. Pp. 81–135 in S. K. Dutta (ed.) *DNA Systematics*. CRC Press, Boca Raton, FL.

Archie, J. W. 1985. Statistical analysis of heterozygosity data: Independent sample comparisons. *Evolution* 39:623–637.

Armitage, K. B. 1981. Sociality as a life history tactic of ground squirrels. *Oecologia* 48:36–39.

Avise, J. C. 1991. Ten unorthodox perspectives on evolution prompted by comparative population genetic genetic findings on mitochondrial DNA. *Annu. Rev. Genet.* 25:45–69.

Avise, J. C. 1992. Molecular population structure and the biogeographic history of a regional fauna: A case history with lessons for conservation biology. *Oikos* 63:62–76.

Avise, J. C. 1994. *Molecular Markers, Natural History and Evolution*. Chapman and Hall, New York.

Avise, J. C., and C. F. Aquadro. 1982. A comparative summary of genetic distances in the vertebrates: Patterns and correlations. *Evol. Biol.* 15:151–185.

Avise, J. C., J. Arnold, R. M. Ball, E. Bermingham, T. Lamb, J. E. Neigel, C. A. Reeb, and N. C. Saunders. 1987. Intraspecific phylogeography: The mitochondrial DNA bridge between population genetics and systematics. *Annu. Rev. Ecol. Syst.* 18:489–522.

Avise, J. C., B. W. Bowen, and T. Lamb. 1990. DNA fingerprints from hypervariable mitochondrial genotypes. *Mol. Biol. Evol.* 6:258–269.

Avise, J. C., and R. A. Lansman. 1983. Polymorphism of mitochondrial DNA in populations of higher animals. Pp. 165–190 in M. Nei and R. K. Koehn (eds.) *Evolution of Genes and Proteins*. Sinauer Associates, Sunderland, MA.

Avise, J. C., R. A. Lansman, and R. O. Shade. 1979. The use of restriction endonucleases to measure mitochondrial DNA sequence relatedness in natuural populations. 1. Population structure and evolution in the genus *Peromyscus*. *Genetics* 92:279–295.

Avise, J. C., M. H. Smith, and R. K. Selander. 1979. Biochemical polymorphism and systematics in the genus *Peromyscus*. 7. Geographic differentiation in members of the *truei* and *maniculatus* species groups. *J. Mammal.* 60:177–192.

Ayala, F. 1965. Evolution of fitness in experimental populations of *Drosophila serrata*. *Science* 150:903–905.

Ayala, F. J. 1975. Genetic differentiation during the speciation process. *Evol. Biol.* 8:1–78.

Ayala, F. J., D. Hedgecock, G. S. Zumwalt, and J. Valentine. 1973. Genetic variation in *Tridacna maxima*, an ecological analog of some unsuccessful evolutionary lineages. *Evolution* 27:177–191.

Ayala, F. J., and J. A. Kiger. 1984. *Modern Genetics*. Benjamin/Cummings, Menlo Park, CA.

Ayala, F. J., and J. R. Powell. 1972. Enzyme variability in the *Drosophila willistoni* group: Levels of polymorphism and the physiological function of enzymes. *Biochem. Genet.* 7:331–345.

Ayala, F. J., and J. W. Valentine. 1974. Genetic variability in the cosmopolitan deep-water ophiuran *Ophiomusium lymani*. *Mar. Biol.* 27:51–57.

Ayala, F. J., and J. W. Valentine. 1979. Genetic variability in the pelagic environment: A paradox? *Ecology* 60:24–29.

Ayala, F. J., J. W. Valentine, L. G. Barr, and G. S. Zumwalt. 1974. Genetic variability in a temperate intertidal phoronid, *Phoronopsis viridis*. *Biochem. Genet.* 11:413–427.

Ayala, F. J., J. W. Valentine, T. E. DeLaca, and G. Zumwalt. 1975a. Genetic variability of the Antarctic brachiopod *Liothyrella notorcadensis* and its bearing on mass extinction hypotheses. *J. Paleont.* 49:1–9.

Ayala, F. J., J. W. Valentine, D. Hedgecock, and L. G. Barr. 1975b. Deep sea asteroids: High genetic variability in a stable environment. *Evolution* 29:203–212.

Ayala, F. J., J. W. Valentine, and G. S. Zumwalt. 1974. Genetic variability in Antarctic krill. *Antarctic J. U.S.* 9:300–301.

Ayala, F. J., J. W. Valentine, and G. S. Zumwalt. 1975. An electrophoretic study of the Antarctic zooplankter *Euphausia superba*. *Limnol. Oceanog.* 20:635–640.

Bachmair, A., D. Finley, and A. Varshavsky. 1986. In vivo half-life of a protein is a function of its amino-terminal residue. *Science* 234:179–186.

Batten, M. 1992. *Sexual Strategies*. Putnam, New York.

Bayne, B. L., and R. C. Newell. 1983. Physiological energetics in marine molluscs. Pp. 407–515 in A. S. M. Saleudden and K. M. Wilbur (eds.) *The Mollusca*. Vol. 4. Academic Press, New York.

Beardmore, J. A. 1983. Extinction, survival, and genetic variation. Pp. 125–151 in C. Schonewald-Cox, S. Chambers, B. MacBryde, and W. Thomas (eds.) *Genetics and Conservation: A Reference for Managing Wild Animal and Plant Populations*. Benjamin/Cummings, Menlo Park, CA.

Beardmore, J. A., T. Dobzhansky, and O. Pavlovsky. 1960. An attempt to compare the fitness of polymorphic and monomorphic experimental populations of *Drosophila pseudoobscura*. *Heredity* 14:19–33.

Beardmore, J. A., and L. Levine 1963. Fitness and environmental variation. 1. A study of some polymorphic populations of *Drosophila pseudoobscura*. *Evolution* 17:121–129.

Beardmore, J. A., and S. A. Shami. 1979. Heterozygosity and the optimum phenotype under stabilising selection. *Aquilo. Ser. Zool.* 20:100–110.

Beckman, J. S. 1977. Adaptive peroxidase differentiation in *Pinus ponderosa*. M.A. thesis, University of Colorado, Boulder.

Beckman, J. S., and J. B. Mitton. 1984. Peroxidase allozyme differentiation among successional stands of ponderosa pine. *Amer. Midl. Nat.* 112:43–49.

Bekoff, M., J. Diamond, and J. B. Mitton. 1981. Life history patterns and sociality in canids: Body size, reproduction, and behavior. *Oecologia* 50:386–390.

Bennett, A. F. 1987. Interindividual variability: An underutilized resource. Pp. 147–166 in M. E. Feder, A. F. Bennett, W. W. Burggren, and R. B. Huey (eds.) *New Directions in Ecological Physiology*. Cambridge University Press, Cambridge.

Berger, E. 1976. Heterosis and the maintenance of enzyme polymorphism. *Amer. Nat.* 110:823–839.

Bergmann, F., and H. R. Gregorius. 1993. Ecogeographical distribution and thermostability of isocitrate dehydrogenase (IDH) alloenzymes in European silver fir (*Abies alba*). *Biochem. System. Ecol.* 21:597–606.

Bergmann, F., and W. Ruetz. 1991. Isozyme genetic variation and heterozygosity in random tree samples and selected orchard clones from the same Norway spruce populations. *Forest Ecol. Manag.* 46:39–47.

Bergmann, F., and F. Scholz. 1984. Effects of selection pressure by SO_2 pollution on genetic structures of Norway spruce (*Picea abies*). Pp. 267–275 in H.-R. Gregorius (ed.) *Population Genetics in Forestry: Lecture Notes in Biomathematics*. Vol. 60. Springer-Verlag, Berlin.

Bergmann, F., and F. Scholz. 1987. The impact of air pollution on the genetic structure of Norway spruce. *Silvae Genet.* 36:80–83.

Bergmann, F., and F. Scholz. 1989. Selection effects of air pollution in Norway spruce (*Picea abies*) populations. Pp. 143–160 in F. Scholz, H.-R. Gregorius, and D. Rudin (eds.) *Genetic Effects of Air Pollutants in Forest Tree Populations*. Springer-Verlag, Berlin.

Bernardi, G., P. Sordino, and D. A. Powers. 1993. Concordant mitochondrial and nuclear DNA phylogenies for populations of the teleost *Fundulus heteroclitus*. *Proc. Natl. Acad. Sci. USA* 99:9271–9274.

Bernstein, S. C., L. H. Throckmorton, and J. L. Hubby. 1973. Still more genetic variability in natural populations. *Proc. Natl. Acad. Sci. USA* 70:3928–3931.

Berry, A., and M. Kreitman. 1993. Molecular analysis of an allzoyme cline: Alcohol dehydrogenase in *Drosophila melanogaster* on the east coast of North America. *Genetics* 134:869–893.

Berry, R. J., M. E. Jakson, and J. Peters. 1987. Inherited differences within an island population of the house mouse (*Mus domesticus*). *J. Zool.* (London) 211:605–618.

Bigelow, H. B., and W. C. Schroeder. 1953. *Fishes of the Gulf of Maine*. U.S. Government Printing Office, Washington, DC.

Bijlsma, R. 1978. Polymorphism at the glucose-6-phosphate dehydrogenase and 6-phosphogluconate dehydrogenase loci in *Drosophila melanogaster*. Part 2. Evidence for interaction in fitness. *Genet. Res.* 31:227–238.

Bijlsma-Meeles, E., and R. Bijlsma. 1988. The alcohol dehydrogenase polymorphism in *Drosophila melanogaster*: Fitness measurements and predictions under conditions with no alcohol stress. *Genetics* 120:743–753.

Birky, C. W., Jr. 1978. Transmission genetics of mitochondria and chloroplasts. *Annu. Rev. Genet.* 12:471–512.

Birky, C. W., Jr., T. Maruyama, and P. A. Fuerst. 1983. An approach to population and evolutionary genetic theory for genes in mitochondria and chloroplasts and some results. *Genetics* 103:513–527.

Black, F. L., and F. M. Salzano. 1981. Evidence for heterosis in the HLA system. *Amer. J. Hum. Genet.* 33:894–899.

Bock, C. E. 1984. Geographical correlates of abundance vs. rarity in some North American winter landbirds. *Auk* 101:266–273.

Bock, C. E., and R. E. Ricklefs. 1983. Range size and local abundance of some North American songbirds: A positive correlation. *Amer. Nat.* 122:295–299.

Bongarten, B. C., N. C. Wheeler, and K. S. Jech. 1985. Isozyme heterozygosity as a selection criterion for yield improvement in Douglas-fir. *Proc., Canadian Tree Impr. Assoc.*: 121–128.

Bonhomme, F., and R. K. Selander. 1978. The extent of allelic diversity underlying electrophoretic protein variation in the house mouse. Pp. 569–589 in H. C. Morse III (ed.) *Origins of Inbred Mice*. Academic Press, New York.

Bonnell, M. L., and R. K. Selander. 1974. Elephant seals: Genetic variation and near extinction. *Science* 184:908–909.

Booth, C. L., D. S. Woodruff, and S. J. Gould. 1990. Lack of significant associations between allozyme heterozygosity and phenotypic traits in the land snail *Cerion*. *Evolution* 44:210–213.

Borgia, G. 1979. Sexual selection and the evolution of mating systems. Pp. 19–80 in M. S. Blum and N. A. Blum (eds.) *Sexual Selection and Reproductive Competition in Insects*. Academic Press, New York.

Borsa, P., Y. Jousselin, and B. Delay. 1992. Relationships between allozymic heterozygosity, body size, and survival to natural anoxic stress in the palourde, *Ruditapes decussatus* L. (Bivalvia: Veneridae). *J. Exp. Mar. Biol. Ecol.* 155:169–181.

Bottini, E., F. Gloria-Bottini, P. Lucarelli, A. Polzonetti, F. Santoro, and A. Varveri. 1979. Genetic polymorphisms and intrauterine development: Evidence of decreased heterozygosity in light for dates human newborn babies. *Experientia* 35:1565–1567.

Boyce, T. M., M. E. Zwick, and C. F. Aquadro. 1989. Mitochondrial DNA in the bark weevils: Size, structure, and heteroplasmy. *Genetics* 123:825–836.

Boyer, S. H. 1961. Alkaline phosphatase in human sera and placentae. *Science* 134:1002–1004.

Boyle, T. J. B., and E. K. Morgenstern. 1986. Estimates of outcrossing rates in six populations of black spruce in central New Brunswick. *Silvae Genet.* 35:102–106.

Bradbury, J. W., and M. B. Andersson. 1987. *Sexual Selection: Testing the Alternatives*. Wiley, New York.

Brandon, R. N., and R. M. Burian. 1984. *Genes, Organisms, Populations: Controversies over the Units of Selection*. MIT Press, Cambridge, MA.

Brittnacher, J. G. 1981. Genetic variation and genetic load due to the male reproductive component of fitness in *Drosophila*. *Genetics* 97:719–730.

Brown, A. H. D. 1979. Enzyme polymorphism in plant populations. *Theor. Pop. Biol.* 15:1–42.

Brown, A. H. D., and R. W. Allard. 1970. Estimation of mating systems in open-pollinated maize populations using isozyme polymorphisms. *Genetics* 66:133–145.

Brown, A. H. D., D. R. Marshall, and L. Albrecht. 1974. The maintenance of alcohol dehydrogenase polymorphism in *Bromus mollis* L. *Aust. J. Biol. Sci.* 27:545–549.

Brown, A. H. D., D. R. Marshall, and L. Albrecht. 1975. Profiles of electrophoretic alleles in natural populations. *Genet. Res. Camb.* 25:137–143.

Brown, A. H. D., D. R. Marshall, and B. S. Weir. 1981. Current status of the charge state model for protein polyjorphism. Pp.15–43 in J. B. Gibson and J. G. Oakeshott (eds.) *Genetic Studies of* Drosophila *Populations: Proceedings of the 1979 Kioloa Conference*. Australian National University Press, Canberra.

Brown, A. J. L., and C. H. Langley. 1979. Reevaluation of level of genetic heterozygosity in natural population of *Drosophila melanogaster* by two-dimensional electrophoresis. *Proc. Natl. Acad. Sci. USA* 76:2381–2384.

Brown, J. H. 1984. On the relationship between abundance and distribution of species. *Amer. Nat.* 124:255–279.

Brown, J. L. 1997. A theory of mate choice based on heterozygosity. *Behav. Ecol.* 8:60–65.

Brown, W. M. 1983. Evolution of animal mitochondrial DNA. Pp. 62–88 in M. Nei and R. K. Koehn (eds.) *Evolution of Genes and Proteins*. Sinauer Associates, Sunderland, MA.

Brown, W. M. 1986. Nuclear and mitochondrial DNA comparisons reveal extreme variation in the molecular clock. *Science* 234:194–195.

Brown, W. M., M. George Jr., and A. C. Wilson. 1979. Rapid evolution of animal mitochondrial DNA. *Proc. Natl. Acad. Sci. USA* 76:1967–1971.

Bull, J. J. 1983. *Evolution of Sex Determining Mechanisms*. Benjamin/Cummings, Menlo Park, CA.

Bulmer, M. G. 1971. Protein polymorphism. *Nature* 234:410–411.

Burdon, J. J., D. R. Marshall, and A. H. D. Brown. 1983. Demographic and genetic changes in populations of *Echium plantagineum*. *J. Ecol.* 71:667–669.

Burke, T., and M. W. Bruford. 1987. DNA fingerprinting in birds. *Nature* 327:149–152.

Burke, T., N. B. Davies, M. W. Bruford, and B. J. Hatchwell. 1989. Parental care and mating behavior of polyandrous dunnocks, *Prunella modularis*, related to paternity by DNA fingerprinting. *Nature* 338:249–251.

Buroker, N. E. 1983. Population genetics of the American oyster *Crassostrea virginica* along the Atlantic coast and the Gulf of Mexico. *Mar. Biol.* 75:99–112.

Burton, R. S. 1986. Evolutionary consequences of restricted gene flow among natural populations of the copepod, *Tigriopus californicus*. *Bull. Mar. Sci.* 39:526–535.

Burton, R. S. 1987. Differentiation and integration of the genome in populations of the marine copepod *Tigriopus californicus*. *Evolution* 41:504–513.

Burton, R. S., and M. W. Feldman. 1981. Population genetics of *Tigriopus californicus*. 2. Differentiation among neighboring populations. *Evolution* 35:1192–1205.

Burton, R. S., and M. W. Feldman. 1982. Changes in free amino acid concentrations during osmotic response in the intertidal copepod *Tigriopus californicus*. *Comp. Biochem. Physiol.* 73A:441–445.

Burton, R. S., and M. W. Feldman. 1983. Physiological effects of an allozyme polymorphism: Glutamate–pyruvate transaminase and response to hyperosmotic stress in the copepod *Tigriopus californicus*. *Biochem. Genet.* 21:239–251.

Burton, R. S., and B.-N. Lee. 1994. Nuclear and mitochondrial gene genealogies and allozyme polymorphism across a major phylogeographic break in the copepod *Tigriopus californicus*. *Proc. Natl. Acad. Sci. USA* 91:5197–5201.

Burton, R. S., and A. R. Place. 1986. Evolution of selective neutrality: Further considerations. *Genetics* 114:1033–1039.

Bush, R. M., and P. E. Smouse. 1992. Evidence for the adaptive significance of allozymes in forest trees. *New Forests* 6:179–196. Also pp. 179–196 in W. T. Adams, S. H. Strauss, D. L. Copes, and A. R. Griffin (eds.) *Population Genetics of Forest Trees*. Kluwer Academic, Dordrecht, 1992.

Bush, R. M., P. E. Smouse, and F. T. Ledig. 1987. The fitness consequences of multiple-locus heterozygosity: The relationship between heterozygosity and growth rate in pitch pine (*Pinus rigida* Mill.). *Evolution* 41:787–798.

Byers, B. A., and J. B. Mitton. 1981. Habitat choice in a marine snail. *Mar. Biol.* 65:149–154.

Cameron, D. G., and E. R. Vyse. 1976. Genetic studies of Yellowstone elk (*Cervus canadensis*). *Genetics* 83:s12.

Cann, R. L., M. Stoneking, and A. C. Wilson. 1987. Mitochondrial DNA and human evolution. *Nature* 325:31–36.

Carson, H. L. 1959. Genetic conditions which promote or retard the formation of species. *Cold Spring Harb. Symp. Quant. Biol.* 24:87–105.

Carson, H. L. 1961. Heterosis and fitness in experimental populations of *Drosophila melanogaster*. *Evolution* 15:496–509.

Carson, H. L. 1965. Chromosomal morphism in geographically widespread species of *Drosophila*. Pp. 508–531 in H. B. Baker and G. L. Stebbins (eds.) *The Genetics of Colonizing Species*. Academic Press, New York.

Carson, H. L. 1971. Speciation and the founder principle. *Stadler Symp.* 3:51–70.

Carson, H. L. 1975. The genetics of speciation at the diploid level. *Amer. Nat.* 109:83–92.

Carson, H. L. 1987. The genetic system, the deme, and the origin of species. *Annu. Rev. Genet.* 21:405–423.

Carson, H. L., and W. B. Heed 1964. Structural homozygosity in marginal populations of neoarctic and neotropical species of *Drosophila* in Florida. *Proc. Natl. Acad. Sci. USA* 52:427–436.

Carter, P. A. 1992. Evolutionary genetic of Adh in tiger salamanders. Ph. D. thesis, University of Colorado, Boulder, CO.

Carter, P. A. 1997. Maintenance of the Adh polymorphism in *Ambystoma tigrinum nebulosum* (tiger salamanders). 1. Genotypic differences in time to metamorphosis in extreme oxygen environments. *Heredity* 78:101–109.

Carter, P. A., J. B. Mitton, T. D. Kocher, and J. Coelho. 1997. Enzyme kinetics and physiological consequences of the ADH polymorphism in the tiger salamander, *Ambystoma tigrinum*. Unpublished manuscript.

Carter, P. A., and W. B. Watt. 1988. Adaptation at specific loci. V. Metabolically adjacent enzyme loci may have very distinct experiences of selective pressures. *Genetics* 119:913–924.

Cavalli-Sforza, L. L. 1969. Human diversity. *Proc. 12th Internat. Congr. Genet.* 3:405–416.

Cavener, D. R., and M. T. Clegg. 1978. Dynamics of correlated genetic systems. IV. Multilocus effects of ethanol stress environments. *Genetics* 90:629–644.

Cavener, D. R., and M. T. Clegg. 1981. Evidence for biochemical and physiological differences between enzyme genotypes in *Drosophila melanogaster*. *Proc. Natl. Acad. Sci. USA* 78:4444–4447.

Chaisurisri, K., and Y. El-Kassaby. 1994. Genic diversity in a seed production population versus natural populations. *Biodiver. Conserv.* 3:512–523.

Chakraborty, R. 1981. The distribution of the number of heterozygous loci in an individual in natural populations. *Genetics* 98:461–466.

Chakraborty, R. 1987. Biochemical heterozygosity and phenotypic variability of polygenic traits. *Heredity* 59:19–28.

Chakraborty, R., P. A. Fuerst, and M. Nei. 1978. Statistical studies on protein polymorphism in natural populations. 2. Gene differentiation between populations. *Genetics* 88:367–390.

Chakraborty, R., and N. Ryman.1983. Relationship of mean and variance of genotypic values with heterozygosity per individual in a natural population. *Genetics* 103:149–152.

Chappell, M. A., and L. R. G. Snyder 1984. Biochemical and physiological correlates of deer mouse alpha-chain hemoglobin polymorphisms. *Publ. Natl. Acad. Sci. USA* 81:5484–5488.

Charlesworth, B. 1988. The evolution of mate choice in a fluctuating environment. *J. Theor. Biol.* 130:191–204.

Charlesworth, B., and D. Charlesworth. 1976. An experiment on recombinational load in *Drosophila melanogaster*. *Genet. Res. Camb.* 25:267–274.

Charlesworth, D., and B. Charlesworth. 1987. Inbreeding depression and its evolutionary consequences. *Annu. Rev. Ecol. Syst.* 18:237–268.

Charnov, E. R. 1982. *The Theory of Sex Allocation*. Princeton University Press, Princeton, NJ.

Cheliak, W. M., and B. P. Dancik. 1982. Genetic diversity of natural populations of a clone-forming tree *Populus tremuloides*. *Can. J. Genet. Cytol.* 24:611–616.

Cheliak, W. M., B. P. Dancik, K. Morgan, F. C. H. Yeh, and C. Strobeck. 1985. Temporal variation and the mating system in a natural population of jack pine. *Genetics* 109:569–584.

Chesser, R. K., and M. H. Smith. 1987. Relationship of genetic variation to growth and reproduction in the white-tailed deer. Pp. 168–177 in C. M. Wemmer (ed.) *Biology and Management of the Cervidae*. Smithsonian Institution Press, Washington, DC.

Ciandri, R., S. Maini, and L. Bullini. 1980 Genetic distance between pheromone strains of the European corn borer, *Ostrinia nubilalis*: Different contribution of variable substrate, regulatory and non regulatory enzymes. *Heredity* 45:383–388.

Clark, A. G., and R. K. Koehn. 1992. Enzymes and adaptation. Pp. 193–228 in R. J. Berry, T. J. Crawford, and G. M. Hewitt (eds.) *Genes in Ecology*. Blackwell Scientific, Oxford.

Clarke, B. 1975. The contribution of ecological genetics to evolutionary theory: Detecting the direct effects of natural selection on particular polymorphic loci. *Genetics* 79:101–108.

Clarke, B. C. 1979. The evolution of genetic diversity. *Proc. Roy. Soc. Lond. (Biol.)* 205:453–474.

Clegg, M. T. 1978. Dynamics of correlated genetic systems. 2. Simulation studies of chromosomal segments under selection. *Theor. Pop. Biol.* 13:1–23.

Clegg, M. T. 1980. Measuring plant mating systems. *Bioscience* 30:814–818.

Clegg, M. T., and R. W. Allard. 1973. Viability versus fecundity selection in the slender wild oat, *Avena barbata* L. *Science* 181:667–668.

Clegg, M. T., R. W. Allard, and A. L. Kahler. 1972. Is the gene the unit of selection? Evidence from two experimental plant populations. *Proc. Natl. Acad. Sci. USA* 69:2474–2478.

Clegg, M. T., A. L. Kahler, and R. W. Allard. 1978. Estimation of life cycle components of selection in an experimental plant garden. *Genetics* 89:765–792.

Clegg, M. T., J. F. Kidwell, and C. R. Horch. 1980. Dynamics of correlated genetic systems. 5. Rates of decay of linkage disequilibria in experimental populations of *Drosophila melanogaster*. *Genetics* 94:217–234.

Clutton-Brock, T. H. 1988. *Reproductive Success*. University of Chicago Press, Chicago.

Cobb, N., J. B. Mitton, and T. G. Whitham. 1994. Genetic variation associated with chronic water and nutrient stress in pinyon pine. *Amer. J. Bot.* 81:936–940.

Coelho, J. R., and J. B. Mitton. 1988. Oxygen consumption during hovering is associated with genetic variation of enzymes in honeybees. *Funct. Ecol.* 2:141–146.

Colwell, R. K., and D. Futuyma. 1971. On the measurement of niche breadth and overlap. *Ecology* 52:567–576.

Cothran, E. G., R. Chesser, M. H. Smith, and P. E. Johns. 1983. Influences of genetic variability and maternal factors on fetal growth in white-tailed deer. *Evolution* 37:282–291.

Coyne, J. A. 1976. Lack of genetic similarity between two sibling species of *Drosophila* as revealed by varied techniques. *Genetics* 84:593–607.

Coyne, J. A. 1982. Gel electrophoresis and cryptic protein variation. Pp. 1–32 in M. C. Rattazzi, J. G. Scandalios, and G. S. Whitt (eds.) *Isozymes: Current Topics in Biological and Medical Research*. Vol. 5. Alan R. Liss, New York.

Coyne, J. A. 1984. Correlation between heterozygosity and rate of chromosome evolution in animals. *Amer. Nat.* 123:725–729.

Coyne, J. A., A. A. Felton, and R. C. Lewontin. 1978. Extent of genetic variation at a highly polymorphic esterase locus in *Drosophila pseudoobscura*. *Proc. Natl. Acad. Sci. USA* 75:5090–5093.

Crawford, D. L., and D. A. Powers. 1989. Molecular basis of evolutionary adaptation at the lactate dehydrogenase-B locus in the fish *Fundulus heteroclitus*. *Proc. Natl. Acad. Sci. USA* 86:9365–9369.

Cronin, H. 1991. *The Ant and the Peacock*. Cambridge University Press, Cambridge.

Crow, J. F., and M. Kimura. 1970. *An Introduction to Population Genetics Theory*. Harper and Row, New York.

Crow, J. F., and M. Kimura. 1979. Efficiency of truncation selection. *Proc. Natl. Acad. Sci. USA* 76:396–399.

da Cunha, A. B., T. Dobzhansky, O. Pavlovsky, and B. Spassky. 1959. Genetics of natural populations. 28. Supplementary data on the chromosomal polymorphism in *Drosophila willistoni* in its relation to its environment. *Evolution* 13:389–404.

Danzmann, R. G., M. M. Ferguson, and F. W. Allendorf. 1986. Does enzyme heterozygosity influence developmental rate in rainbow trout? *Heredity* 56:417–425.

Danzmann, R. G., M. M. Ferguson, and F. W. Allendorf. 1987. Heterozygosity and oxygen-consumption rate as predictors of growth and developmental rate in rainbow trout. *Physiol. Zool.* 602:210–220.

Darwin, C. 1859. *On the Origin of Species by Means of Natural Selection, or the Preservation of Favoured Races in the Struggle for Survival*. John Murray, London.

Darwin, C. 1871. *The Descent of Man, and Selection in Relation to Sex*. John Murray, London.

Dawid, I. B., and A. W. Blackler. 1972. Maternal and cytoplasmic inheritance of mitochondrial DNA in *Xenopus*. *Devel. Biol.* 29:152–161.

Dawkins, R. 1976. *The Selfish Gene*. Oxford University Press, Oxford.

Dawkins, R. 1982. *The Extended Phenotype*. Freeman, New York.

Day, D., O. C. De Vos, D. Wilson, and H. Lambers. 1985. Regulation of respiration in the leaves and roots of two *Lolium perenne* populations with two contrasting mature leaf respiration rates and crop yields. *Plant Physiol.* 78:678–683.

Day, T. H., P. C. Hillier, and B. Clarke. 1974. The relative quantities and catalytic activities of enzymes produced by alleles at the alcohol dehydrogenase locus in *Drosophila melanogaster*. *Biochem. Genet.* 11:155–165.

Dempster, E. 1955. Maintenance of genetic heterogeneity. *Cold Spring Harb. Symp. Quant. Biol.* 20:25–32.

Diamond, S. A., M. C. Newman, M. Mulvey, P. M. Dixon, and D. Martinson. 1989. Allozyme genotype and time to death of mosquitofish, *Gambusia affinis* (Baird and Girard), during acute exposure to inorganic mercury. *Environ. Toxicol. Chem.* 8:613–622.

Diehl, W. J. 1989. Genetics of carbohydrate metabolism and growth in *Eisemia foetida* (Oligochatea: Lumbricidae). *Heredity* 61:379–387.

Diehl, W. J., P. M. Gaffney, and R. K. Koehn. 1986. Physiological and genetic aspects of growth in the mussel *Mytilus edulis*. 1. Oxygen consumption, growth, and weight loss. *Physiol. Zool.* 59:201–211.

Diehl, W. J., P. M. Gaffney, J. H. McDonald, and R. K. Koehn. 1985. Relationship between weight standardized oxygen consumption and multiple-locus heterozygosity in the marine mussel *Mytilus edulis* L. (Mollusca). Pp. 531–536 in P. Gibbs (ed.) *Proceedings of the 19th European Marine Biology Symposium*. Cambridge University Press, Cambridge.

Diehl, W. J., and R. K. Koehn. 1985. Multiple-locus heterozygosity, mortality, and growth in a cohort of *Mytilus edulis*. *Mar. Biol.* 88:265–271.

DiMichele, L., K. Paynter, and D. A. Powers. 1991. Lactate dehydrogenase-B allozymes directly affect development of *Fundulus heteroclitus*. *Science* 23:898–900.

DiMichele, L., and D. A. Powers 1982a. LDH-B genotype–specific hatching times of *Fundulus heteroclitus* embryos. *Nature* 296:563–565.

DiMichele, L., and D. A. Powers. 1982b. Physiological basis for swimming endurance differences between LDH-B genotypes of *Fundulus heteroclitus*. *Science* 216:1014–1016.

DiMichele, L., and D. A. Powers. 1984. Developmental and oxygen consumption rate differences between Ldh-B genotypes of *Fundulus heteroclitus* and their hatching time. *Physiol. Zool.* 57:52–56.

DiMichele, L., and M. H. Taylor. 1980. The environmental control of hatching in *Fundulus heteroclitus*. *J. Exp. Zool.* 214:181–187.

Dingle, H., and J. P. Hegmann. 1982. *Evolution and Genetics of Life Histories*. Springer-Verlag, New York.

Dobzhansky, Th. 1948. Genetics of natural populations. 16. Altitudinal and seasonal changes produced by natural selection in certain populations of *Drosophila pseudoobscura* and *Drosophila persimilis*. *Genetics* 33:158–176.

Dobzhansky, Th. 1970. *Genetics of the Evolutionary Process*. Columbia University Press, New York.

Dobzhansky, Th., F. J. Ayala, G. L. Stebbins, and J. W. Valentine. 1977. *Evolution*. Freeman, New York.

Dobzhansky, Th. and B. Spassky. 1960. Release of genetic variability through recombination. 5. Breakup of synthetic lethals by crossing over in *Drosophila pseudoobscura*. *Zool. Jahrb. Abt. Syst.* 88:57–66.

Dobzhansky, Th., and B. Spassky. 1963. Genetics of natural populations. 34. Adaptive norm, genetic load, and genetic elite in *D. pseudoobscura*. *Genetics* 48:1467–1486.

Dobzhansky. Th. B. Spassky, and T. Tidwell. 1963. Genetics of natural populations. 32. Inbreeding and the mutational and balanced genetic loads in natural populations of *Drosophila pseudoobscura*. *Genetics* 48:361–373.

Dobzhansky, Th., and B. Wallace. 1953. The genetics of homeostasis in *Drosophila*. *Proc. Natl. Acad. Sci. USA* 39:162–171.

Dong J., and D. B. Wagner. 1993. Taxonomic and population differentiation of mitochondrial DNA diversity in *Pinus banksiana* and *Pinus contorta*. *Theor. Appl. Genet.* 86:573–578.

Dong J., and D. B. Wagner. 1994. Paternally inherited chloroplast polymorphism in *Pinus*: Estimation of diversity and population subdivision, and tests of disequilibrium with a maternally inherited mitochondrial polymorphism. *Genetics* 136:1187–1194.

Dueser, R. D., and H. J. Shugart. 1979. Niche pattern in a forest floor small mammal fauna. *Ecology* 60:108–118.

Dykhuizen, D. E., J. DeFramond, and D. L. Hartl. 1984 Selective neutrality of glucose-6-phosphate dehydrogenase allozymes in *Escherichia coli*. *Mol. Biol. Evol.* 1:162–170.

Eanes, W. F. 1978. Morphological variance and enzyme heterozygosity in the monarch butterfly. *Nature* 276:263–264.

Eanes, W. F. 1981. Enzyme heterozygosity and morphological variance. *Nature* 290:609–610.

Eanes, W. F. 1984. Viability interactions, in vivo activity, and the G6PD polymorphism in *Drosophila melanogaster*. *Genetics* 106:95–107.

Eanes, W. F., P. M. Gaffney, R. K. Koehn, and C. M. Simon. 1977. A study of sexual selection in natural populations of the milkweed beetle, *Tetraopes tetraopthalmus*. Pp. 49–64 in F. B. Christiansen and T. M. Fenchel (eds.) *Measuring Selection in Natural Populations*. Springer-Verlag, Berlin.

Eanes, W. F., and R. K. Koehn. 1978. Subunit size and the number of electrophoretic variants in human enzymes. *Biochem. Genet.* 16:971–985.

Edwards, Y., and D. A. Hopkinson. 1980. Are abundant proteins less variable? *Nature* 284:511–512.

Eldredge, N., and S. J. Gould. 1972. Punctuated equilibria: An alternative to phyletic gradualism. Pp. 82–115 in T. J. J. Schopf (ed.) *Models in Paleobiology*. Freeman-Cooper, San Francisco.

El-Kassaby, Y. A. 1991. Genetic variation within and among conifer populations: Review and evaluation of methods. Pp. 61–76 in S. Fineschi, M. E. Malvolti, F. Cannata, and H. H. Hattemer (eds.) *Biochemical Markers in the Population Genetics of Forest Trees*. SPB Academic, The Hague.

El-Kassaby, Y., and K. Ritland. 1996. Impact of selection and breeding on the genetic diversity of Douglas-fir. *Biodiver. and Conserv.* 5:795–813.

Emlen, S. T., and L. W. Oring. 1977. Ecology, sexual selection, and the evolution of mating systems. *Science* 197:215–223.

Endler, J. A. 1986. *Natural Selection in the Wild*. Princeton University Press, Princeton, NJ.

Epperson, B. K., and R. W. Allard. 1987. Linkage disequilibrium between allozymes in natural populations of lodgepole pine. *Genetics* 15:341–352.

Eshel, I., and W. D. Hamilton. 1984. Parent offspring correlation in fitness under fluctuating selection. *Proc. Roy. Soc.* (Lond.) ser. B. 222:1–14.

Esteal, S. 1985. Generation time and the rate of molecular evolution. *Mol. Biol. Evol.* 2:450–453.

Falconer, D. S. 1989. *Introduction to Quantitative Genetics*. 3rd ed. Wiley, New York.

Farris, M. A., and J. B. Mitton. 1984. Population density, outcrossing rate, and heterozygote superiority in ponderosa pine. *Evolution* 38:1151–1154.

Feder, M. E., A. F. Bennett, W. A. Burggren, and R. B. Huey. 1987. *New Directions in Ecological Physiology*. Cambridge University Press, Cambridge.

Feder, M. E., and W. B. Watt. 1992. Functional biology of adaptation. Pp. 365–392 in R. J. Berry, T. J. Crawford, and G. M. Hewitt (eds.) *Genes in Ecology*. Blackwell Scientific, Oxford.

Felsenstein, J. 1975. The genetic basis of evolutionary change. *Evolution* 29:587–590.

Ferguson, M. M., and L. R. Drahushchak. 1990. Disease resistance and enzyme heterozygosity in rainbow trout. *Heredity* 64:413–417.

Field, K. G., G. J. Olsen, D. J. Lane, S. J. Giovannoni, M. T. Ghiselin, E. C. Raff, N. R. Pace, and R. A. Raff. 1988. Molecular phylogeny of the animal kingdom. *Science* 239:748–753.

Fields, R. D., K. R. Johnson, and G. H. Thorgaard. 1989. DNA fingerprints in rainbow trout detected by hybridization with DNA of bacteriophage M13. *Trans. Amer. Fish. Soc.* 118:78–81.

Fildes, R. A., and C. W. Parr. 1963. Human red cell phosphogluconate dehydrogenase. *Nature* 200:890–891.

Fincham, J. R. S. 1972. Heterozygous advantage as a likely general basis for enzyme polymorphisms. *Heredity* 28:387–391.

Fitch, W. M., and W. R. Atchley. 1985. Evolution in inbred strains of mice appears rapid. *Science* 228:1169–1175.

Fitch, W. M., and E. Margoliash. 1967. Construction of phylogenetic trees. *Science* 155:279–284.

Fitch, W. M., and E. Markowitz. 1970. An improved method for determining codon variability in a gene and its application to the rate of fixation of mutations in evolution. *Biochem. Genet.* 4:579–593.

Fitzsimmons, N. N., S. W. Buskirk, and M. H. Smith. 1995. Population history, genetic variability, and horn growth in bighorn sheep. *Conserv. Biol.* 9:314–323.

Fleischer, R. C., R. F. Johnston, and W. J. Klitz. 1983. Allozymic heterozygosity and morphological variation in house sparrows. *Nature* 304:628–630.

Fleischer, R. C., P. E. Lowther, and R. F. Johnston. 1984. Natal dispersal in house sparrows: Possible causes and consequences. *J. Field Ornithol.* 55:444–456.

Fleischer, R. C., and M. T. Murphy. 1992. Relationships among allozyme heterozygosity, morphology and lipid levels in house sparrows during winter. *J. Zool.* (London) 226:409–419.

Fletcher, T. S., F. J. Ayala, D. R. Thatcher, and G. K. Chambers. 1978. Structural analysis of the ADH^S electromorph of *Drosophila melanogaster*. *Proc. Natl. Acad. Sci. USA* 75:5609–5612.

Florence, L. Z., P. C. Johnson, and J. E. Coster. 1982. Behavioral and genic diversity during dispersal—analysis of a polymorphic esterase locus in southern pine beetle, *Dendroctonus frontalis* (Coleoptera, Scolytidae). *Environ. Entomol.* 11:1014–1018.

Franklin, I. R., and R. C. Lewontin. 1970. Is the gene the unit of selection? *Genetics* 65:707–734.

Frelinger, J. A. 1971. Maternally derived transferrin in pigeon squabs. *Science* 171:1260–1261.

Frelinger, J. A. 1972. The maintenance of transferrin polymorphisms in pigeons. *Proc. Natl. Acad. Sci. USA* 69:326–329.

Freriksen, A., D. Seykens, W. Scharloo, and P. W. H. Heinstra. 1991. Alcohol dehydrogenase controls the flux from ethanol into lipids in *Drosophila* larvae. *J. Biol. Chem.* 266:21399–21403.

Frydenberg, O., D. Moller, G. Naevdal, and K. Sick. 1965. Haemoglobin polymorphisms in Norwegian cod populations. *Hereditas* 53:257–271.

Fuerst, P. A., R. Chakraborty, and M. Nei. 1977. Statistical studies on protein polymorphism in natural populations. *Genetics* 76:837–848.

Fujio, Y. 1982. A correlation of heterozygosity with growth rate in the Pacific oyster, *Crassostrea gigas*. *Tohoku J. Agric. Res.* 33:66–75.

Fujio. Y. Y. Nakamura, and M. Sugita. 1979. Selective advantage of heterozygotes at catalase locus in the Pacific oyster. *Crassostrea gigas*. *Japanese J. Genet.* 54:359–366.

Futuyma, D. J. 1986. *Evolutionary Biology*. Sinauer Associates, Sunderland, MA.

Gaffney, P. M. 1990. Enzyme heterozygosity, growth rate, and viability in *Mytilus edulis*: Another look. *Evolution* 44:204–210.

Gaffney, P. M., and T. M. Scott. 1984. Genetic heterozygosity and production traits in natural and hatchery populations of bivalves. *Aquaculture* 42:289–302.

Gaffney, P. M., T. M. Scott, R. K. Koehn, and W. J. Diehl. 1990. Interrelationships of heterozygosity, growth rate and heterozygote deficiencies in the coot clam, *Mulinia lateralis*. *Genetics* 124:687–699.

Gajardo, G. M., and J. A. Beardmore. 1989. Ability to switch reproductive mode in *Artemia* is related to maternal heterozygosity. *Mar. Ecol. Prog. Ser.* 55:191–195.

Garton, D. W. 1984. Relationahip between multiple locus heterozygosity and physiological energetics of growth in the estuarine gastropod *Thais haemastoma*. *Physiol. Zool.* 57:530–543.

Garton, D. W., and W. R. Haag. 1991. Heterozygosity, shell length and metabolism in the European mussel, *Dreissena polymorphha* from a recently established population in Lake Erie. *Comp. Biochem. Physiol.* 99A:45–48.

Garton, D. W., R. K. Koehn, and T. M. Scott. 1984. Multiple-locus heterozygosity and the physiological energetics of growth in the coot clam, *Mulinia lateralis*, from a natural population. *Genetics* 108:445–455.

Gauldie, R. W. 1984. A reciprocal relationship between heterozygosities of the phosphoglucomutase and glucose phosphate isomerase loci. *Genetica* 63:93–103.

Geburek, T., F. Scholz, W. Knabe, and A. Vornweg. 1987. Genetic studies by isozyme loci on tolerance and sensitivity in an air polluted *Pinus sylvestris* field trial. *Silvae Genet.* 36:49–53.

Geist, V. 1966. The evolutionary significance of mountain sheep horns. *Evolution* 20:558–566.

Geist, V. 1971. *Mountain Sheep: A Study in Behavior and Evolution*. University of Chicago Press, Chicago.

Gentili, M. R., and A. R. Beaumont. 1988. Environmental stress, heterozygosity, and growth rate in *Mytilus edulis*. *J. Exp. Mar. Biol. Ecol.* 120:145–153.

Ghiselin, M. T. 1974. *The Economy of Nature and the Evolution of Sex*. University of California Press, Berkeley and Los Angeles.

Giannini, R. 1991. Effects of pollution on the genetic structure of forest tree populations. Proceedings of meeting, Rome, April 3, 1990. Consiglio nazionale delle ricerche (C.N.R.)

Gibson, R., and J. Bradbury. 1987. Lek organization in sage grouse: Variations on a territorial theme. *Auk* 104:77–84.

Gillespie, J. H. 1973. Natural selection with varying selection coefficients—A haploid model. *Genet. Res.* 21:115–120.

Gillespie, J. H. 1976. A general model to account for enzyme variation in natural populations. 2. Characterization of the fitness functions. *Amer. Nat.* 110:809–821.
Gillespie, J. H. 1977. A general model to account for enzyme variation in natural populations. 3. Multiple alleles. *Evolution* 31:85–90.
Gillespie, J. H. 1978a. A general model to account for enzyme variation in natural populations. 4. The quantitative genetics of viability mutants. Pp. 301–314 in F. B. Christiansen and T. M. Fenchel (eds.) *Measuring Selection in Natural Populations*. Springer-Verlag, New York.
Gillespie, J. H. 1978b. A general model to account for enzyme variation in natural populations. 5. The SAS–CFF model. *Theor. Popul. Biol.* 14:1–45.
Gillespie, J. H. 1991. *The Causes of Molecular Evolution*. Oxford University Press, New York.
Gillespie, J. H., and K. Kojima. 1968. The degree of polymorphism in enzymes involved in energy production compared to that in non-specific enzymes in two *Drosophila ananassae* populations. *Proc. Natl. Acad. Sci. USA* 61:582–585.
Gillespie, J. H., and C. H. Langley 1974. A general model to account for enzyme variation in natural populations. *Genetics* 76:837–848.
Gillespie, J. H., and C. H. Langley. 1979. Are evolutionary rates really variable? *J. Mol. Evol.* 13:27–34.
Gingerich, P. D. 1983. Rates of evolution: Effects of time and temporal scaling. *Science* 222:159–161.
Gingerich, P. D. 1984. Technical comment. *Science* 226:995–996.
Ginzburg, L. R. 1979. Why are heterozygotes often superior in fitness? *Theor. Pop. Biol.* 15:264–267.
Ginzburg, L. R. 1983. *Theory of Natural Selection and Population Growth*. Benjamin/Cummings, Menlo Park, CA.
Goldstein, D. B., A. Ruiz Linares, L. L. Cavalli-Sforza, and M. W. Feldman. 1995. Genetic absolute dating based on microsatellites and the origin of modern humans. *Proc. Natl. Acad. Sci. USA* 92:6723–6727.
Goolish, E. M., and R. S. Burton. 1989. Energetics of osmoregulation in an intertidal copepod: Effects of anoxia and lipid reserves on the pattern of free amino acid accumulation. *Func. Ecol.* 3:81–89.
Gould, S. J. 1982. Darwinism and the expansion of evolutionary theory. *Science* 216:380–387.
Gould, S. J. 1984. Smooth curve of evolutionary rate: A psychological and mathematical artifact. *Science* 226:994–995.
Gould, S. J., and N. Eldredge. 1977. Punctuated equilibria: On the tempo and mode of evolution reconsidered. *Paleobiology* 3:115– 151.
Gould, S. J., and R. C. Lewontin. 1979. The spandrels of San Marco and the Panglossian paradigm: A critique of the adaptationist program. *Proc. Roy. Soc.* (Lond.), ser. B 205:581–598.
Goulson, D. 1993. Allozyme variation in the butterfly, *Maniola jurtina* (Lepidoptera: Satyrinae) (L.): Evidence for selection. *Heredity* 71:386–393.
Graves, J., R. H. Rosenblatt, and G. N. Somero. 1983. Kinetic and electrophoretic differentiation of lactate dehydrogenases of teleost species-pairs from the Atlantic and Pacific coasts of Panama. *Evolution* 37:30–37.
Graves, J. E., and G. N Somero. 1982. Electrophoretic and functional enzymic evolution in four species of eastern Pacific barracudas from different thermal environments. *Evolution* 36:97–106.
Green, R. H., S. M. Singh, B. Hicks, and J. M. McCuaig. 1983. An arctic intertidal population of *Macoma balthica* (Mollusca, Pelecypoda): Genotypic and phenotypic components of population structure. *Can. J. Fish. Aquat. Sci.* 40:1360–1371.
Grell, E. H., K. B. Jacobson, and J. B. Murphy. 1965. Alcohol dehydrogenase in *Drosophila melanogaster*: Isozymes and genetic variants. *Science* 149:80–82.
Gulland, F. M. D., S. D. Albon, J. M. Pemberton, P. R. Moorcroft, and T. H. Clutton-Brock. 1993.

Parasite-associated polymorphism in a cyclic ungulate population. *Proc. R. Soc. Lond. B.* 254:7–13.

Guries R. P., and F. T. Ledig. 1982. Genetic diversity and population structure in pitch pine (*Pinus rigida* Mill.). *Evolution* 36:387–399.

Halanych, K. M., J. D. Bacheller, A. A. Aguinaldo, S. M. Liva, D. M. Hillis, and J. A. Lake. 1995. Evidence from 18S ribosomal DNA that the lophophorates are prototome animals. *Science* 267:1641–1643.

Haldane, J. B. S. 1954. *The Biochemistry of Genetics*. George Allen and Unwin, London.

Haldane, J. B. S. 1957. The cost of natural selection. *J. Genet.* 55:511–524.

Haldane, J. B. S., and S. D. Jayakar. 1963. Polymorphism due to selection varying direction. *J. Genet.* 58:237–242.

Haldorson, L., and J. L. King. 1976. Unimodality, symmetry, and step-state hypothesis of electrophoretic variation in natural populations. *J. Mol. Evol.* 8:351–356.

Hall, J. G. 1985. Temperature-related kinetic differentiation of glucosephosphate isomerase alleloenzymes isolated from the blue mussel, *Mytilus edulis*. *Biochem. Genet.* 23:705–728.

Hall, J. G., and R. K. Koehn. 1983. The evolution of enzyme catalytic efficiency and adaptive inference from steady-state kinetic data. *Evol. Biol.* 16:53–96.

Halliday, T. R. 1987. Physiological constraints on sexual selection. Pp. 247–264 in J. W. Bradbury and M. B. Andersson (eds.) *Sexual Selection: Testing the Alternatives*. Wiley, New York.

Hamilton, W. D. 1964. The genetical evolution of social behavior, 1. *J. Theor. Biol.* 7:1–16.

Hamilton, W. D. 1980. Sex versus non-sex versus parasite. *Oikos* 35:282–290.

Hamilton, W. D., and M. Zuk. 1982. Heritable true fitness and bright birds: A role for parasites. *Science* 218:384–387.

Hamrick, J. L. 1983. The distribution of genetic variation within and among natural plant populations. Pp. 335–348 in C. M. Schonewald-Cox, S. M. Chambers, B. MacBryde, and W. L. Thomas (eds.) *Genetics and Conservation: A Reference for Managing Wild Animals and Plant Populations*. Benjamin/Cummings, Menlo Park, CA.

Hamrick, J. L., and M. J. Godt. 1990. Allozyme diversity in plant species. Pp. 43–63 in A. H. D. Brown, M. T. Clegg, A. L. Kahler, and B. S. Weir (eds.) *Plant Population Genetics, Breeding and Genetic Resources*. Sinauer Associates, Sunderland, MA.

Hamrick, J. L., Y. B. Linhart, and J. B. Mitton. 1979. Relationships between life history characters and electrophoretically detectable genetic variation in plants. *Annu. Rev. Ecol. Syst.* 10:173–200.

Hamrick, J. L., J. B. Mitton, and Y. B. Linhart. 1979. Levels of genetic variation in trees: Influence of life history characteristics. Pp. 35–41 in M. T. Conkle (ed.) *Isozymes of North American Forest Trees and Forest Insects*. Pacific Southwest Forest and Range Experiment Station, Berkeley, CA.

Handford, P. 1980. Heterozygosity at enzyme loci and morphological variation. *Nature* 286:261–262.

Hanken, J. 1983. Genetic variation in a dwarfed lineage, the Mexican salamander genus *Thorius* (Amphibia: Plethodontidae): Taxonomic, ecologic, and evolutionary implications. *Copeia* 1983:1051–1073.

Hansen, T. A. 1980. Influence of larval dispersal and geographic distribution on species longevity in neogastropods. *Paleobiology* 1980:193–207.

Hanski, I. 1982a. Communities of bumblebees: Testing the core-satellite species hypothesis. *Annu. Zool. Fenn.* 19:65–73.

Hanski, I. 1982b. Distributional ecology of anthropochorus plants in villages surrounded by forest. *Annu. Bot. Fenn.* 19:1–15.

Hanski, I. 1982c. Dynamics of regional distribution: The core and satellite species hypothesis. *Oikos* 38:210–221.

Hanski, I., and H. Koskela. 1977. Niche relationships among dung-inhabiting beetles. *Oecologia* 28:203–231.

Hanski, I., and H. Koskela. 1978. Stability, abundance, and niche width in the beetle community inhabiting cow dung. *Oikos* 31:290–298.

Harris, H. 1966. Enzyme polymorphisms in man. *Proc. Roy. Soc.* (Lond.), ser. B 164:298–310.

Harris, H., D. A. Hopkinson, and Y. G. Edwards. 1977. Polymorphism and the subunit structure of enzymes: A contribution to the neutralist–selectionist controversy. *Proc. Natl. Acad. Sci. USA* 74:698–701.

Harris, H., D. A. Hopkinson, and J. Luffman. 1968. Enzyme diversity in human populations. *Ann. N.Y. Acad. Sci.* 151:232–242.

Harrison, J. F., D. I. Nielsen, and R. E. Page Jr. 1996. Malate dehydrogenase phenotype, temperature and colony effects on flight metabolic rate in the honey-bee, *Apis mellifers*. *Funct. Ecol.* 10:81–88.

Hartl, G. B., G. Lang, F. Klein, and R. Willing. 1991. Relationships between allozymes, heterozygosity and morphological characters in red deer. (*Cervus elaphus*), and the influence of selective hunting on allele frequency distributions. *Heredity* 66:343–350.

Hartl, G. B., R. Willing, and K. Nadlinger. 1994. Allozymes in mammalian population genetics and systematics: Indicative function of a marker system reconsidered. Pp. 299–310 in B. Schierwater, S. Struit, G. P. Wagner, and R. DeSalle (eds.) *Molecular Ecology and Evolution: Approaches and Applications*. Springer-Verlag, Basel.

Hartl, G. B., F. Suchentrunk, R. Willing, and R. Petznek. 1995. Allozyme heterozygosity and fluctuating asymmetry in the brown hare (*Lepus europaeus*): A test of the developmental homeostasis hypothesis. *Proc. Roy. Soc.* (Lond.), ser. B 350:313–323.

Hawkins, A. J. S., B. L. Bayne, and A. J. Day. 1986. Protein turnover, physiological energetics and heterozygosity in the blue mussel, *Mytilus edulis*: The basis of variable age-specific growth. *Proc. Roy. Soc.* (Lond.), ser. B 229:161–176.

Hawkins, A. J. S., B. L. Bayne, A. J. Day, J. Rusin, and C. M. Worrall. 1989. Genotype-dependent interrelations between energy metabolism, protein metabolism and fitness. Pp. 283–292 in J. S. Ryland and P. A. Tyler (eds.) *Reproduction, Genetics and Distributions of Marine Organisms*. Olsen and Olsen, Fredensborg.

Heagler, M. G., M. C. Newman, M. Mulvey, and P. M. Dixon. 1993. Allozyme genotype in mosquitofish, *Gambusia holbrooki*, during mercury exposure: Temporal stability, concentration effects and field verification. *Environ. Toxicol. Chem.* 19:385–395.

Hedrick, P. W. 1974. Genetic variation in a heterogeneous environment. 1. Temporal heterogeneity and the absolute dominance model. *Genetics* 78:757–770.

Hedrick, P. W. 1976. Genetic variation in a heterogeneous environment. 2. Temporal heterogeneity and directional selection. *Genetics* 84:145–157.

Hedrick, P. W. 1986. Genetic polymorphism in heterogeneous environments: A decade later. *Annu. Rev. Ecol. Syst.* 17:535–566.

Hedrick, P. W., M. E. Ginevan, and E. P. Ewing. 1976. Genetic polymorphism in heterogeneous environments. *Annu. Rev. Ecol. Syst.* 7:1–32.

Hedrick, P., S. Jain, and L. Holden. 1978. Multilocus systems in evolution. *Evol. Biol.* 11:101–184.

Hiebert, R. D., and J. L. Hamrick. 1983. Patterns and level of genetic variation in Great Basin bristlecone pine, *Pinus longaeva*. *Evolution* 37:302–310.

Hilbish, T. J., and R. K. Koehn. 1985. Dominance in physiological phenotypes and fitness at an enzyme locus. *Science* 229:52–54.

Hilbish, T. J., and R. K. Koehn. 1985. The physiological basis of natural selection at the LAP locus. *Evolution* 39:1302–1317.

Hill, A. V. S. 1985. Use of minisatellite DNA probes for determination of twin zygosity at birth. *Lancet* 1985-I:1394.

Hines, S. A., D. P. Philipp, W. F. Childers, and G. S. Whitt. 1983. Thermal kinetic differences between allelic isozymes of malate dehydrogenase (Mdh-B locus) of largemouth bass, *Micropterus salmoides*. *Biochem. Genet.* 21:1143–1151.

Hochachka, P. W., and G. N. Somero. 1973. *Strategies of Biochemical Adaptation*. Saunders, Philadelphia.

Hochachka, P. W., and G. N. Somero. 1984. *Biochemical Adaptation*. Princeton Univeristy Press, Princeton, NJ.

Hoffmann, R. J. 1981a. Evolutionary genetics of *Metridium senile*. 1. Kinetic differences in phosphoglucose isomerase allozymes. *Biochem. Genet.* 19:129–144.

Hoffmann, R. J. 1981b. Evolutionary genetics of *Metridium senile*. 2. Geographic patterns of allozyme variation. *Biochem. Genet.* 19:145–154.

Hoffmann, R. J. 1983. Temperature modulation of the kinetics of phosphoglucose isomerase genetic variants from the sea anemone *Metridium senile*. *J. Exp. Zool.* 227:361–370.

Hoffmann, R. J., J. L. Boore, and W. M. Brown. 1992. A novel mitochondrial genome organization for the blue mussel, *Mytilus edulis*. *Genetics* 131:397–412.

Hogg, J. T. 1987. Intrasexual competition and mate choice in Rocky Mountain bighorn sheep. *Ethology* 75:119–144.

Hong, Y. P., V. D. Hipkins, and S. H. Strauss. 1993. Chloroplast DNA diversity among trees, populations and species in the California closed-cone pines (*Pinus radiata, Pinus muricata, and Pinus attenuata*). *Genetics* 135:1187–1196.

Hopkinson, D. A., N. Spencer, and H. Harris. 1964. Genetical studies on human red cell acid phosphatase. *Amer. J. Hum. Genet.* 16:141–154.

Jablonski, D. 1987. Heritability at the species level: Analysis of geographic ranges of Cretaceous mollusks. *Science* 238:360–362.

Jablonski, D. 1995. Extinctions in the fossil record. Pp. 25–44 in J. H. Lawton and R. M. May (eds.) *Extinction Rates*. Oxford University Press, Oxford.

Jablonski, D., and D. M. Raup. 1995. Selectivity of end-Cretaceous marine bivalve extinctions. *Science* 268:389–391.

Jackson, J. B. C. 1974. Biogeographic consequences of eurytopy and stenotypy among marine bivalves and their evolutionary significance. *Amer. Nat.* 198:541–560.

Jeffreys, A. J., and D. B. Morton. 1987. DNA fingerprints of dogs and cats. *Anim. Genet.* 18:1–15.

Jeffreys, A. J., N. J. Royle, V. Wilson, and Z. Wong. 1988. Spontaneous rates to new length alleles at tandem-repetitive hypervariable loci in human DNA. *Nature* 332:278–281.

Jeffreys, A. J., V. Wilson, R. Kelly, B. A. Taylor, and G. Bulfield. 1987. Mouse DNA "fingerprints": Analysis of chromosome localization and germline stability of hypervariable loci in recombinant inbred strains. *Nucleic Acids Res.* 15:2823–2836.

Jeffreys, A. J., V. Wilson, and S. L. Thein. 1985a. Hypervariable "minisatellite" regions in human DNA. *Nature* 314:67–73.

Jeffreys, A. J., V. Wilson, and S. L. Thein. 1985b. Individual-specific "fingerprints" of human DNA. *Nature* 316:76–79.

Jelinski, D. E. 1993. Associations between environmental heterogeneity, heterozygosity, and growth rates of *Populus tremuloides* in a cordilleran landscape. *Arctic and Alpine Res.* 255:183–188.

Jelinski, D. E., and W. M. Cheliak. 1992. Genetic diversity and spatial subdivision of *Populus tremuloides* (Salicaceae) in a heterogeneous landscape. *Amer. J. Bot.* 79:728–736.

Johnson, F. M., C. G. Kanapi, R. H. Richardson, M. R. Wheeler, and W. S. Stone. 1966a. An analysis of polymorphisms among isozyme loci in dark and light *Drosophila ananassae* strains from American and Western Samoa. *Proc. Natl. Acad. Sci. USA* 56:119–125.

Johnson, F. M., C. G. Kanapi, R. H. Richardson, M. R. Wheeler, and W. S. Stone. 1966b. An operational classification of *Drosophila* esterase for species comparison. *Univ. Texas Publ.* 6615:517–532.

Johnson, G. B. 1973a. Importance of substrate variability to enzyme polymorphisms. *Nature New Biol.* 243:151–153.

Johnson, G. B. 1973b. Relationship of enzyme polymorphism to species diversity. *Nature* 242:193–194.

Johnson, G. B. 1974. Enzyme polymorphism and metabolism. *Science* 184:28–37.

Johnson, G. B. 1976. Genetic polymorphism and enzyme function. Pp. 46–59 in F. J. Ayala (ed.) *Molecular Evolution.* Sinauer Associates, Sunderland, MA.

Johnson, G. B. 1977. Assessing electrophoretic similarity: The problem of hidden heterogeneity. *Annu. Rev. Ecol. Syst.* 8:309–328.

Johnson, G. B. 1979. Increasing the resolution of polyacrylamide gel electrophoresis by varying the degree of gel crosslinking. *Biochem. Genet.* 17:499–516.

Johnson, M. S., and M. F. Mickevich 1977. Variability and evolutionary rates of proteins. *Evolution* 31:642–648.

Jukes, T. H., R. Holmquist, and H. Moise. 1975. Amino acid composition of proteins: Selection against the code. *Science* 189:50–51.

Kahler, A. L., R. W. Allard, and R. D. Miller. 1984. Mutation rates for enzyme and morphological loci in barley (*Hordeum vulgare* L.). *Genetics* 106:729–734.

Karl, S. A., and J. C. Avise. 1992. Balancing selection at allozyme loci in oysters: Implications from nuclear RFLP's. *Science* 256:100–102.

Karlin, A. A., S. I. Guttman, and S. L. Rathbun. 1984. Spatial autocorrelation analysis of heterozygosity and geographic distribution in populations of *Desmognathus fuscus* (Amphibia: Plethodontidae). *Copeia* 1984:341–354.

Kat, P. W. 1982. The relationship between heterozygosity for enzyme loci and developmental homeostasis in peripheral populations of aquatic bivalves (Unionidae). *Amer. Nat.* 119:824–832.

Kimball, R. T. 1995. Sexual selection in house sparrows, *Passer domesticus*. Ph.D. thesis, University of New Mexico, Albuquerque.

Kimura, M. 1968. Evolutionary rate at the molecular level. *Nature* 217:624–626.

Kimura, M. 1979. The neutral theory of molecular evolution. *Scientif. Amer.* 241:94–104.

Kimura, M. 1983. *The Neutral Theory of Molecular Evolution.* Cambridge University Press, Cambridge.

Kimura, M., and J. F. Crow. 1964. The number of alleles that can be maintained in a finite population. *Genetics* 49:725–738.

Kimura, M., and J. F. Crow. 1978. Effect of overall phenotypic selection on genetic change at individual loci. *Proc. Natl. Acad. Sci. USA* 75:6168–6171.

Kimura, M., and T. Ohta. 1971. *Theoretical Aspects of Population Genetics.* Princeton University Press, Princeton, NJ.

King, J. L. 1967. Continuously distributed factors affecting fitness. *Genetics* 55:483–492.

King, J. L. 1972. Genetic polymorphism and environment. *Science* 176:545.

King, J. L. 1973. The probability of electrophoretic identity of proteins as a function of amino acid divergence. *J. Mol. Evol.* 2:317–322.

King, J. L., and T. H. Jukes. 1969. Non-Darwinian evolution: Random fixation for selectively neutral alleles. *Science* 164:788–798.

King, M. P., and G. Attardi. 1989. Human cells lacking mtDNA: Repopulation with exogenous mitochondria by complementation. *Science* 246:500–503.

Kirkpatrick, M. 1982. Sexual selection and the evolution of female choice. *Evolution* 36:1–12.

Kirkpatrick, M. 1986. Sexual selection and cycling parasites: a simulation study of Hamilton's hypothesis. *J. Theor. Biol.* 119:263.

Klitz, W., G. Thomson, N. Borot, and A. Cambon-Thomsen. 1992. Evolutionary and population perspectives of the human HLA complex. *Evol. Biol.* 26:35–72.

Kluge, A. G., and W. C. Kerfoot. 1973. The predictability and regularity of character divergence. *Amer. Nat.* 107:426–442.

Kobyliansky, E., and G. Livshits. 1983. Relationship between levels of biochemical heterozygosity and morphological variability in human populations. *Ann. Hum. Genet.* 47:215–223.

Kobyliansky, E., and G. Livshits. 1985. Differential fertility and morphological constitution of spouses. *J. Morph. Anthrop.* 76:95–105.

Koch, C. F. 1980. Bivalve species duration, areal extent and population size in a Cretaceous sea. *Paleobiology* 6:184–192.

Kochan, I., C. A. Golden, and J. A. Bukovic. 1968. Mechanisms of tuberculostasis in mammalian serum. 2. Induction of serum tuberculostatis in Guinea pigs. *J. Bacteriol.* 100:64–69.

Kodric-Brown, A., and J. H. Brown. 1984. Truth in advertising: The kinds of traits favored by sexual selection. *Amer. Nat.* 124:309–323.

Koehn, R. K. 1969. Esterase heterogeneity: Dynamics of a polymorphism. *Science* 163:943–944.

Koehn, R. K. 1978. Physiology and biochemistry of enzyme variation: The interface of ecology and population genetics. Pp. 51–71 in P. Brussard (ed.) *Ecological Genetics: The Interface.* Springer-Verlag, New York.

Koehn, R. K. 1987. The importance of genetics to physiological ecology. Pp. 170–188 in M. E. Feder, A. F. Bennett, W. W. Burggren, and R. B. Huey (eds.) *New Directions in Ecological Physiology*. Cambridge University Press, Cambridge.

Koehn, R. K. 1991. The cost of enzyme synthesis in the genetics of energy balance and physiological performance. *Biol. J. Linn. Soc.* 44:231–247.

Koehn, R. K., and B. L. Bayne. 1988. Towards a physiological and genetical understanding of the energetics of the stress response. *Biol. J. Linn. Soc.* 37:157.

Koehn, R. K., B. L. Bayne, M. N. Moore, and J. F. Siebenaller. 1980. Salinity related physiological and genetic differences between populations of *Mytilus edulis*. *Biol. J. Linn. Soc.* 14:319–334.

Koehn, R. K., W. J. Diehl, and T. M. Scott. 1988. The differential contribution by individual enzymes of glycolysis and protein catabolism to the relationship between heterozygosity and growth rate in the coot clam, *Mulinia lateralis*. *Genetics* 118:121–130.

Koehn, R. K., and W. F. Eanes. 1977. Subunit size and genetic variation of enzymes in natural populations of *Drosophila*. *Theor. Pop. Biol.* 11:330–341.

Koehn, R. K., and W. F. Eanes. 1978. Molecular structure and protein variation within and among populations. *Evol. Biol.* 11:39–100.

Koehn, R. K., and P. M. Gaffney. 1984. Genetic heterozygosity and growth rate in *Mytilus edulis*. *Mar. Biol.* 82:1–7.

Koehn, R. K., J. G. Hall, and A. J. Zera. 1980. Parallel variation of genotype-dependent aminopeptidase-I activity between *Mytilus edulis* and *Mercenaria mercenaria*. *Mar. Biol. Letters* 1:245–263.

Koehn, R. K., and T. J. Hilbish. 1987. The adaptive importance of genetic variation. *Amer. Scient.* 75:134–140.

Koehn, R. K., and F. W. Immerman. 1981. Biochemical studies of aminopeptidase polymorphism in *Mytilus edulis*. 1. Dependence of enzyme activity on season, tissue, and genotype. *Biochem. Genet.* 19:1115–1142.

Koehn, R. K., R. Milkman, and J. B. Mitton. 1976. Population genetics of marine pelecypods. 4. Selection, migration and genetic differentiation in the blue mussel *Mytilus edulis*. *Evolution* 30:2–32.

Koehn, R. K., and J. B. Mitton. 1972. Population genetics of marine pelecypods. 1. Ecological heterogeneity and evolutionary strategy at an enzyme locus. *Amer. Nat.* 106:47–56.

Koehn, R. K., R. J. E. Newell, and F. Immerman. 1980. Maintenance of an aminopeptidase allele frequency cline by natural selection. *Proc. Natl. Acad. Sci. USA* 77:5385–5389.

Koehn, R. K., and D. I. Rasmussen. 1967. Polymorphic and monomorphic serum esterase heterogeneity in catostomid fish populations. *Biochem. Genet.* 1:131–144.

Koehn, R. K., and S. E. Shumway. 1982. A genetic/physiological explanation for differential growth rate among individuals of the American oyster, *Crassostrea virginica* (Gmelin). *Mar. Biol. Ltrs* 3:35–42.

Koehn, R. K., and J. F. Siebenaller. 1981. Biochemical studies of aminopeptidase polymorphism in *Mytilus edulis*. 2. Dependence of reaction rate on physical factors and enzyme concentration. *Biochem. Genet.* 19:1143–1162.

Koehn, R. K., F. J. Turano, and J. B. Mitton. 1973. Population genetics of marine pelecypods. 2. Genetic differences in microhabitats of *Modiolus demissus*. *Evolution* 27:100–105.

Koehn, R. K., and G. C. Williams. 1978. Genetic differentiation without isolation in the American eel, *Anguilla rostrata*. 2. Temporal stability of geographic patterns. *Evolution* 32:624–637.

Koehn, R. K., A. J. Zera, and J. G. Hall. 1983. Enzyme polymorphism and natural selection. Pp. 115–136 in M. Nei and R. K. Koehn (eds.) *Evolution of Genes and Proteins*. Sinauer Associates, Sunderland, MA.

Kojima, K., J. Gillespie, and Y. N. Tobari. 1970. A profile of *Drosophila* spp. enzymes assayed by electrophoresis. 1. Number of alleles, heterozygosities, and linkage disequilibrium in glucose-metabolizing systems and some other enzymes. *Biochem. Genet.* 4:627–637.

Kojima, K., and R. C. Lewontin. 1970. Evolutionary significance of linkage and epistasis. Pp. 367–388 in K. Kojima (ed.) *Mathematical Topics in Population Genetics*. Springer-Verlag, New York.

Kreitman, M., B. Shorrocks, and C. Dytham. 1992. Genes and ecology: Two alternative perspectives using *Drosophila*. Pp. 281–312 in R. J. Berry, T. J. Crawford, and G. M. Hewitt (eds.) *Genes in Ecology*. Blackwell Scientific, Oxford.

Lande, R. 1981. Models of speciation by sexual selection on polygenic traits. *Proc. Natl. Acad. Sci. USA* 78:3721–3725.

Langley, C. H. 1977. Nonrandom associations between allozymes in natural populations of *Drosophila melanogaster*. Pp. 265–273 in F. B. Christiansen and T. M. Fenchel (eds.) *Measuring Selection in Natural Populations*. Springer-Verlag, Berlin.

Langley, C. H., and W. M. Fitch. 1973. The constancy of evolution. A statistical analysis of the a and b hemoglobins, cytochrome c, and fibrinopeptide A. Pp. 246–262 in N. E. Morton (ed.) *Genetic Structure of Populations*. University of Hawaii Press, Honolulu.

Langley, C. H., and W. M. Fitch. 1974. An examination of the constancy of the rate of molecular evolution. *J. Mol. Evol.* 3:161–177.

Lansman, R. A., J. C. Avise, C. F. Aquadro, J. F. Shapira, and S. W. Daniel. 1983. Extensive genetic variation in mitochondrial DNA's among geographic populations of the deer mouse, *Peromyscus maniculatus*. *Evolution* 37:1–16.

Lansman, R. A., R. O. Shade, J. F. Shapira, and J. C. Avise. 1981. The use of restriction endonucleases to measure mitochondrial DNA sequence relatedness in natural populations. 3. Techniques and potential applications. *J. Mol. Evol.* 17:214–226.

Lassen, H. H., and F. J. Turano. 1978. Clinal variation in heterozygote deficit at the Lap-locus in *Mytilus edulis*. *Mar. Biol.* 49:245–254.

Latta, R., and J. B. Mitton. 1997. A comparison of population differentiation across four classes of gene marker in limber pine, *Pinus flexilis*. *Genetics* 146:1153–1163.

Leary, R. F., F. W. Allendorf, and K. L. Knudsen. 1983. Developmental stability and enzyme heterozygosity in rainbow trout. *Nature* 301:71–72.

Leary, R. F., F. W. Allendorf, and K. L. Knudsen. 1984a. Major morphological effects of a regulatory gene: Pgm1-t in rainbow trout. *Mol. Biol. Evol.* 1:183–194.

Leary, R. F., F. W. Allendorf, and K. L. Knudsen. 1984b. Superior developmental stability of heterozygotes at enzyme loci in salmonid fishes. *Amer. Nat.* 124:540–551.

Leary, R. F., F. W. Allendorf, and K. L. Knudsen. 1985. Inheritance of meristic variation and the evolution of developmental stability in rainbow trout. *Evolution* 39:308–314.

Leary, R. F., F. W. Allendorf, and K. L. Knudsen. 1994. Null alleles at two lactate dehydrogenase loci in rainbow trout are associated with decreased developmental stability. Pp. 5–15 in T. A. Markow (ed.) *Developmental Instability: Its Origins and Evolutionary Implications*. Kluwer Academic, Dordrecht.

Leberg, P., M. H. Smith, and I. L. Brisbin Jr. 1992. Influence of sex, habitat, and genotype on the growth patterns of white-tailed deer. Pp. 343–350 in R. D. Brown (ed.) *The Biology of Deer*. Springer-Verlag, New York.

Leberg, P. L., M. H. Smith, and O. E. Rhodes Jr. 1990. The association between heterozygosity and the growth of deer fetuses is not explained by effects of the loci examined. *Evolution* 44:454–459.

Ledig, F. T., R. P. Guries, and B. A. Bonefield. 1983. The relation of growth to heterozygosity in pitch pine. *Evolution* 37:1227–1238.

Lee, Y. M., H. P. Misra, and F. J. Ayala. 1981. Superoxide dismutase in *Drosophila melanogaster*: Biochemical and structural characterization of allozyme variants. *Proc. Natl. Acad. Sci. USA* 78:7052–7055.

Lee, Y. M., D. J. Friedman, and F. J. Ayala. 1985. Superoxide dismutase: An evolutionary puzzle. *Proc. Natl. Acad. Sci. USA* 82:824–828.

Leigh Brown, A. J. 1977. Physiological correlates of an enzyme polymorphism. *Nature* 269:803–804.

Lerner, I. M. 1954. *Genetic Homeostasis*. Oliver and Boyd, Edinburgh.

Lesica, P., and F. W. Allendorf. 1992. Are small populations of plants worth preserving? *Conserv. Biol.* 6:135–139.

Levene, H. 1953. Genetic equilibrium when more than one ecological niche is available. *Amer. Nat.* 87:331–333.

Levins, R. 1968. *Evolution in Changing Environments*. Princeton University Press, Princeton, NJ.

Levinton, J. 1973. Genetic variation in a gradient of environmental variability. *Science* 180:75–76.

Levinton, J. S. 1975. Levels of genetic polymorphism at two enzyme encoding loci in eight species of the genus *Macoma* (Mollusca: Bivalvia). *Mar. Biol.* 33:41–47.

Levinton, J. S., and R. K. Koehn. 1976. Population genetics of mussels. Pp. 357–384 in B. L. Bayne (ed.) *Marine Mussels: Their Ecology and Physiology*. Cambridge University Press, Cambridge.

Lewandowski, A., J. Burczyk, and L. Meinartowicz. 1991. Genetic structure and the mating system in an old stand of Polish larch. *Silvae Genet.* 40:75–79.

Lewis, N., and J. Gibson. 1978. Variation in amount of enzyme protein in natural populations. *Biochem. Genet.* 16:159–170.

Lewontin, R. C. 1964. The interaction of selection and linkage. 1. General considerations: Heterotic models. *Genetics* 49:49–67.

Lewontin, R. C. 1974. *The Genetic Basis of Evolutionary Change*. Columbia University Press, New York.

Lewontin, R. C. 1988. On measures of gametic disequilibrium. *Genetics* 120:849–852.

Lewontin, R. C. 1991. Twenty-five years ago in GENETICS. Electrophoresis in the development of evolutionary genetics: Milestone or millstone? *Genetics* 128:657–662.

Lewontin, R. C., and J. L. Hubby. 1966. A molecular approach to the study of genic heterozygosity in natural populations. 2. Amount of variation and degree of heterozygosity in natural populations of *Drosophila pseudoobscura*. *Genetics* 54:595–609.

Linhart, Y. B., and M. C. Grant. 1996. Evolutionary significance of local genetic differentiation in plants. *Annu. Rev. Ecol. Syst.* 27:237–277

Linhart, Y. B., and J. B. Mitton. 1985. Relationships among reproduction, growth rates, and protein heterozygosity in ponderosa pine. *Amer. J. Bot.* 72:181–184.

Linhart, Y. B., J. B. Mitton, K. B. Sturgeon, and M. L. Davis. 1981. Genetic variation in space and time in a population of ponderosa pine. *Heredity* 46:407–426.

Lively, C. M., C. Craddock, and R. C. Vrijenhoek. 1990. The Red Queen hypothesis supported by parasitism in sexual and clonal fish. *Nature* 344:864–866.

Livshits, G., and E. Kobyliansky. 1984. Biochemical heterozygosity as a predictor of developmental homeostasis in man. *Ann. Hum. Genet.* 48:173–184.

Livshits, G., and E. Kobyliansky. 1985. Lerner's concept of developmental homeostasis and the problem of heterozygosity level in natural populations. *Heredity* 55:341–353.

Livshits, G., and P. E. Smouse. 1994. Relationship between fluctuating asymmetry, morphological modality and heterozygosity in an elderly Israeli population. Pp. 157–168 in T. A. Markow (ed.) *Developmental Instability: Its Origins and Evolutionary Implications*. Kluwer Academic, Dordrecht.

Long, T. 1970. Genetic effects of fluctuating temperature in populations of *Drosophila melanogaster*. *Genetics* 66:401–416.

Lotrich, V. A. 1975. Summer home range and movements of *Fundulus heteroclitus* (Pisces: Cyprinodontidae) in a tidal creek. *Ecology* 56:191–198.

Loudenslager, E. J. 1978. Variation in genetic structure of *Peromyscus* populations. 1. Genetic heterozygosity—Its relation to adaptive divergence. *Biochem. Genet.* 16:1165–1179.

Loveless, L. D., and J. L. Hamrick. 1984. Ecological determinants of genetic structure in plant populations. *Annu. Rev. Ecol. Syst.* 15:65–95.

Lynch, M. 1984. The selective value of alleles underlying polygenic traits. *Genetics* 108:1021–1033.

Mackay, T. F. C. 1981. Genetic variation in varying environments. *Genet. Res.* 37:79–93.

MacRae, A. F., and W. W. Anderson. 1988. Evidence for non-neutrality of mitochondrial DNA haplotypes in *Drosophila pseudoobscura*. *Genetics* 120:485–494.

MacRae, A. F., and W. W. Anderson. 1990. Can mating preferences explain changes in mtDNA haplotype frequencies? *Genetics* 124:999–1001.

Mane, S. D., L. Tompkin, and R. C. Richmond. 1983. Male esterase-6 catalyzes the synthesis of a sex pheromone in *Drosophila melanogaster* females. *Science* 222:419–421.

Manning, J. T., and M. A. Hartley. 1991. Symmetry and ornamentation are correlated in the peacock's train. *Anim. Behav.* 42:1020–1021.

Manning, J. T., and L. Ockenden. 1994. Fluctuating asymmetry in racehorses. *Nature* 370:185–186.

Markow, T., P. W. Hedrick, K. Zuerlein, J. Danilovs, J. Martin, T. Vyvial, and C. Armstrong. 1993. HLA polymorphism in the Havasupai: Evidence for balancing selection. *Amer. J. Hum. Genet.* 53:943–952.

Markowski, J., P. Osmulski, W. Duda, E. Dyner, and A. Swiatecki. 1990a. Could haptotypes be an indicator of health status of brown hare populations? *Lagomorph Nltr* 11:14–16.

Markowski, J., P. Osmulski, W. Duda, E. Dyner, A. Swiatecki, M. Ulanska, and T. Janiszewski. 1990b. Relation between haptoglobin polymorphism and the health status of brown hare populations in Poland. *Acta Theriol.* 35:215–224.

Maynard-Smith, J. 1956. Fertility, mating behavior and sexual selection in *Drosophila subobscura*. *J. Genet.* 54:261–279.

Maynard-Smith, J. 1978. *The Evolution of Sex*. Cambridge University Press, Cambridge.

Mayr, E. 1963. *Animal Species and Evolution*. Harvard University Press, Cambridge, MA.

McAlpine, S. 1993. Genetic heterozygosity and reproductive success in the green treefrog, *Hyla cinera*. *Heredity* 70:553–558.

McAndrew, J. J., R. D. Ward, and J. A. Beardmore. 1986. Growth rate and heterozygosity in the plaice, *Pleuronectes platessa*. *Heredity* 57:171–180.

McConkey, E. H. 1982. Molecular evolution, intracellular organization, and the quinary structure of proteins. *Proc. Natl. Acad. Sci. USA* 79:3236–3240.

McConkey, E. H., B. J. Taylor, and D. Phan. 1979. Human heterozygosity: A new estimate. *Proc. Natl. Acad. Sci. USA* 76:6500–6504.

McDonald, J. F. 1983. The molecular basis of adaptation: A critical review of relevant ideas and observations. *Annu. Rev. Ecol. Syst.* 14:77–102.

McDonald, J. F., S. M. Anderson, and M. Santos. 1980. Biochemical differences between products of the Adh locus in *Drosophila*. *Genetics* 95:1013–1022.

McDonald, J. F., and F. J. Ayala. 1974. Genetic response to environmental heterogeneity. *Nature* 250:572–574.

McNaughton, S. J., and L. L. Wolf. 1970. Dominance and the niche in ecological systems. *Science* 167:131–139.

Messier, S., and J. B. Mitton. 1996. Heterozygosity at the malate dehydrogenase locus and developmental homeostasis in *Apis mellifera*. *Heredity* 76:612–622.

Michod, R. E. 1982. The theory of kin selection. *Annu. Rev. Ecol. Syst.* 13:23–55.

Mickevich, M. F., and M. S. Johnson. 1976. Congruence between morphological and allozyme data in evolutionary inference and character evolution. *Syst. Zoos.* 25:260–270.

Miki, Y., S. I. Chigusa, and E. T. Matsuura. 1989. Complete replacement of mitochondrial DNA in *Drosophila*. *Nature* 341:551–552.

Milkman, R. D. 1967. Heterosis as a major cause of heterozygosity in nature. *Genetics* 55:493–495.

Milkman, R. 1978. Selection differentials and selection coefficients. *Genetics* 88:391–403.

Milkman, R. 1982. Toward a unified selection theory. Pp. 105–118 in R. Milkman (ed.) *Perspectives on Evolution*. Sinauer Associates, Sunderland, MA.

Millar C. I., S. H. Strauss, M. T. Conkle, and R. Westfall. 1988. Allozyme differentiation and biosystematics of the Californian closed-cone pines. *Syst. Bot.* 13:351–370.

Millar, C. I., and R. D. Westfall. 1992. Allozyme markers in forest genetic conservation. *New Forests* 6:347–371. Also pp. 347–371 in W. T. Adams, S. H. Strauss, D. L. Copes, and A. R. Griffin (eds.) *Population Genetics of Forest Trees*. Kluwer Academic, Dordrecht.

Minawa, A., and A. J. Birley. 1975. Genetic and environmental diversity in *Drosophila melanogaster*. *Nature* 255:702–704.

Mitton, J. B. 1978. Relationship between heterozygosity for enzyme loci and variation of morphological characters in natural populations. *Nature* 273:661–662.

Mitton, J. B. 1983. Conifers. Pp. 443–472 in S. D. Tanksley and T. J. Orton (eds.) *Isozymes in Plant Genetics and Breeding, Part B*. Elsevier, Amsterdam.

Mitton, J. B. 1989. Physiological and demographic variation associated with allozyme variation. Pp. 127–145 in D. Soltis and P. Soltis (eds.) *Isozmes in Plant Biology*. Dioscorides Press, Portland, OR.

Mitton, J. B. 1993a. Enzyme heterozygosity, metabolism, and developmental stability. *Genetica* 89:47–65. Also pp. 49–67 in T. A. Markow (ed.) *Developmental Instability: Its Origins and Evolutionary Implications*. Kluwer Academic, Dordrecht, 1994.

Mitton, J. B. 1993b. Theory and data pertinent to the relationship between heterozygosity and fitness. Pp. 17–41 in N. W. Thornhill (ed.) *The Natural History of Inbreeding and Outbreeding*. University of Chicago Press, Chicago.

Mitton, J. B. 1994. Molecular approaches to population biology. *Annu. Rev. Ecol. Syst.* 25:45–69.

Mitton, J. B. 1995a. Enzyme heterozygosity and developmental stability. *Acta Theriol.* (suppl.) 3:33–54.

Mitton, J. B. 1995b. Genetics and the physiological ecology of conifers. Pp. 1–36 in W. K. Smith and T. M. Hinckley (eds.) *Ecophysiology of Coniferous Forests*. Academic Press, New York.

Mitton, J. B., C. Carey, and T. D. Kocher. 1986. The relation of enzyme heterozygosity to standard and active oxygen consumption and body size of tiger salamanders, *Ambystoma tigrinum*. *Physiol. Zool.* 59:574–582.

Mitton, J. B., P. A. Carter, and A. S. DiGiacomo. 1997. Relationships between the lactate dehydrogenase polymorphism and respiration in the wood louse, *Porcellio scaber* Latr. *Proc. R. Soc. Land. B.*, in press.

Mitton, J. B., and M. C. Grant. 1980. Observations on the ecology and evolution of quaking aspen, *Populus tremuloides*, in the Colorado Front Range. *Amer. J. Bot.* 67:202–209.

Mitton, J. B., and M. C. Grant. 1984. Associations among protein heterozygosity, growth rate, and developmental homeostasis. *Annu. Rev. Ecol. Syst.* 15:479–499.

Mitton, J. B., and M. C. Grant. 1995. Genetics and the natural history of quaking aspen. *BioScience* 46:25–31.

Mitton, J. B., and R. M. Jeffers. 1989. The genetic consequences of mass selection for growth rate in Engelmann spruce. *Silvae Genet.* 38:6–12.

Mitton, J. B., and R. K. Koehn. 1975. Genetic organization and adaptive response of allozymes to ecological variables in *Fundulus heteroclitus*. *Genetics* 79:97–111.

Mitton, J. B., and R. K. Koehn. 1976. Morphological adaptation to thermal stress in a marine fish, *Fundulus heteroclitus*. *Biol. Bull.* 151:546–559.

Mitton, J. B., and R. K. Koehn. 1985. Shell shape variation in the blue mussel, *Mytilus edulis*, and its association with enzyme heterozygosity. *J. Exp. Mar. Biol. Ecol.* 90:73–80.

Mitton, J. B., and W. M. Lewis Jr. 1989. Relationships between genetic variability and life history features of bony fishes. *Evolution* 43:1712–1723.

Mitton, J. B., and W. M. Lewis Jr. 1992. Response to Waples's comment on heterozygosity and life-history variation in bony fishes. *Evolution* 46:576–577.

Mitton, J. B., and P. H. Odense. 1985. Muscle esterase variation and size variation in the sand launce, *Ammodytes dubius*. *Mar. Biol.* 87:279–283.

Mitton, J. B., and B. A. Pierce. 1980. The distribution of individual heterozygosity in natural populations. *Genetics* 95:1043–1054.

Mitton, J. B., W. S. F. Schuster, E. G. Cothran, and J. De Fries. 1993. The correlation in heterozygosity between parents and their offspring. *Heredity* 71:59–63.

Mitton, J. B., D. J. Zelenka, and P. A. Carter. 1994. Selection of breeding stock in pigs favours 6PGD heterozygotes. *Heredity* 73:177–184.

Møller, A. P. 1990. Fluctuating asymmetry in male sexual ornaments may reliably reveal male quality. *Anim. Behav.* 40:1185–1187.

Møller, A. P. 1992. Female swallow preference for symmetrical male sexual ornaments. *Nature* 357:238–240.

Møller, A. P. 1993. Female preference for apparently symmetrical male sexual ornaments in the barn swallow *Hirundo rustica*. *Behav. Ecol. Sociobiol.* 32:371–376.

Møller, A. P. 1994a. *Sexual Selection and the Barn Swallow*. Oxford University Press, Oxford.

Møller, A. P. 1994b. Sexual selection in the barn swallow (*Hirundo rustica*). 4. Patterns of fluctuating asymmetry and selection against asymmetry. *Evolution* 48:658–670.

Moore, M. N., R. K. Koehn, and B. L. Bayne. 1980. Leucine aminopeptidase (aminopeptidase-I), *N*-acetyl-B-hexosaminidase and lysosomes in the mussel, *Mytilus edulis* L., in response to salinity changes. *J. Exp. Zool.* 214:239–249.

Mopper, S., J. B. Mitton, T. G. Whitham, N. S. Cobb, and K. M. Christensen. 1991. Genetic differentiation and heterozygosity in pinyon pine associated with resistance to herbivory and environmental stress. *Evolution* 45:989–999.

Mopper, S., and T. G. Whitham. 1986. Natural bonsai of Sunset Crater. *Natl. Hist.* 95:42–47.

Morgante, M., G. G. Vendramin, P. Rossi, and A. M. Olivieri. 1993. Selection against inbreds in early life-cycle phases in *Pinus leucodermis*. *Ant. Hered.* 70:622–627.

Moritz, C., T. E. Dowling, and W. M. Brown. 1987. Evolution of animal mitochondrial DNA: Relevance for population biology and systematics. *Annu. Rev. Ecol. Syst.* 18:269–292.

Mork, J., N. Ryman, G. Stahl, F. Utter, and G. Sundnes. 1985. Genetic variation in Atlantic cod (*Gadus morhua*) throughout its range. *Can. J. Fish. Aquat. Sci.* 42:1580–1587.

Mork, J., and G. Sundnes. 1985. 0-group cod (*Gadus morhua*) in captivity: Differential survivial of certain genotypes. *Helgolander Meeresuntersuchungen* 39:63–70.

Mouches, C., M. Magnin, J. B. Gerge, M. De Silvestri, V. Beyssat, N. Pasteur, and G. P. Georghiou. 1987. Overproduction of detoxifying esterases in organophosphate-resistant *Culex* mosquitoes and their presence in other insects. *Proc. Natl. Acad. Sci. USA* 84:2113–2116.

Mouches, C., N. Pasteur, J. B. Berge, O. Hyrien, M. Raymond, B. Robert de Saint Vincent, M. De Silvestri, and G. P. Georghiou. 1986. Amplification of an esterase gene is responsible for insecticide resistance in a California *Culex* mosquito. *Science* 233:778.

Muhlenberg, M., D. Leipold, H. H. Mader, and B. Steinhauer. 1977. Island ecology of arthropods. 2. Niches and relative abundance of Seychelles ants (Formicidae) in different habitats. *Eocologia* 29:135–144.

Mukai, T. 1969. Maintenance of polygenic and isoallelic variation in populations. *Proc. 12th Intern. Congr. Genet.* 3:293–308.

Mukai, T., and T. Maruyama. 1971. The genetic structure of natural populations of *Drosophila melanogaster*. 9: A prediction of genetic equilibrium. *Genetics* 68:105–126.

Mukai, T., L. E. Mettler, and S. Chigusa. 1971. Linkage disequilibrium in a local population of *Drosophila melanogaster*. *Proc. Natl. Acad. Sci. USA* 68:1065–1069.

Mukai, T., and O. Yamaguchi. 1974. The genetic structure of natural populations of *Drosophila melanogaster*. 11. Genetic variability in a local population. *Genetics* 76:339–366.

Mukai, T., and T. Yamazaki. 1980. Test for selection on polymorphic isozyme genes using the population cage method. *Genetics* 96:537–542.

Muller-Stark, G. 1985. Genetic differences between "tolerant" and "sensitive" beeches (*Fagus sylvatica* L.) in an environmentally stressed adult forest stand. *Silvae Genet.* 34:241–247.

Mulvey, M. T. M. Goater, G. W. Esch, and A. E. Crews. 1987. Genotype frequency differences in *Halipegus occidualis*–infected and uninfected *Helisoma anceps*. *J. Parasit.* 73:757–761.

Muona, O., G. F. Moran, and J. C. Bell. 1991. Hierarchical patterns of correlated mating in *Acacia melanoxylon*. *Genetics* 127:619–626.

Myers, J. H., and C. J. Krebs. 1971. Genetic, behavioral, and reproductive attributes of dispersing field voles *Microtus pennsylvanicus* and *Microtus ochrogaster*. *Ecol. Monogr.* 41:53–78.

Namkoong G., H. C. Kang, and J. S. Brouard. 1988. *Tree Breeding: Principles and Strategies*. Springer-Verlag, Berlin.

Neale, D. B., and W. T. Adams. 1985. Allozyme and mating-system variation in balsam fir (*Abies balsamea*) across a continuous elevational transect. *Can. J. Bot.* 63:2448–2453.

Nei, M. 1972. Genetic distance between populations. *Amer. Nat.* 106:283–292.

Nei, M. 1975. *Molecular Population Genetics and Evolution*. North-Holland, Amsterdam.

Nei., M. 1995. Genetic support for the out-of-Africa theory of human evolution. *Proc. Natl. Acad. Sci. USA* 92:6720–6722.

Nei, M., P. A. Fuerst, and R. Chakraborty. 1978. Subunit molecular weight and genetic variability of proteins in natural populations. *Proc. Natl. Acad. Sci. USA* 75:3359–3362.

Nei, M., and D. Graur. 1984. Extent of protein polymorphism and the neutral mutation theory. Pp. 73–118 in M. K. Hecht, B. Wallace, and G. T. Prance (eds.) *Evolutionary Biology*. Vol. 17. Plenum, New York.

Nei, M., T. Maruyama, and R. Chakraborty. 1975. The bottleneck effect and genetic variability in populations. *Evolution* 29:1–10.

Nei, M., and F. Tajima. 1981. DNA polymorphism detectable by restriction endonucleases. *Genetics* 97:145–163.

Nelson, K., and D. Hedgecock 1980. Enzyme polymorphism and adaptive strategy in the decapod crustacea. *Amer. Nat.* 116:238–280.

Nevo, E. 1978 Genetic variation in natural populations: Patterns and theory. *Theor. Pop. Biol.* 13:121–177.

Nevo, E., and A. Beiles. 1988. Genetic parallelism of protein polymorphism in nature: Ecological test of the neutral theory of molecular evolution. *Biol. J. Linn Soc.* 35:229–245.

Nevo, E., A. Beiles, and R. Ben-Shlomo. 1984. The evolutionary significance of genetic diversity: Ecological, demographic, and life histroy correlates. Pp. 13–213 in G. S. Mani (ed.) *Lecture Notes in Biomathematics*. Springer-Verlag, Berlin.

Nevo, E., B. Lavie, and R. Noy. 1987. Mercury selection of allozymes in marine gastropods: Prediction and verification in nature revisited. *Envir. Monitor. Assess.* 9:233–238.

Nevo, E. R. Noy, B. Lavie, A. Beiles, and S. Muchtar. 1986. Genetic diversity and resistance to marine pollution. *Biol. J. Linn. Soc.* 29:139–144.

Nevo, E., T. Shimony, and. M. Libney. 1978. Pollution selection of allozyme polymorphisms in barnacles. *Experientia* 34:1562–1564.

Newman, M. C., S. A. Diamond, M. Mulvey, and P. Dixon. 1989. Allozyme genotype and time to death of mosquitofish, *Gambusia affinis* (Baird and Girard) during acute toxicant exposure: A comparison of arsenate and inorganic mercury. *Aquat. Toxicol.* 15:141–156.

Newton, M. F., and J. Peters. 1983. Physiological variation of mouse haemoglobin. *Proc. Roy. Soc.* (Lond.), ser. B 218:443–453.

Noltmann, E. A. 1972. Aldos-ketose isomerases. Pp. 271–354 in P. D. Boyer (ed.) *Enzymes*. Vol. 6. Academic Press, New York.

Norris, K. J. 1990. Female choice and the evolution of the conspicuous plumage coloration of monogamous male Great Tits. *Behav. Ecol.* 26:129–138.

Norris, K. 1993. Heritable variation in a plumage indicator of viability in male great tits *Parus major*. *Nature* 362:537–539.

Novak, J. M., O. E. Rhodes Jr., M. H. Smith, and R. K. Chesser. 1993. Morphological asymmetry in mammals: Genetics and homeostasis reconsidered. *Acta Theriol.* (suppl.) 38 2:7–18.

Oakeshott, J. G., J. B. Gibson, P. R. Anderson, W. R. Knibb, D. G. Anderson, and G. K. Chambers. 1982. Alcohol hehydrogenase and glycerol-3 phosphate dehydrogenase clines in *Drosophila melanogaster* on different continents. *Evolution* 36:86–96.

Oakeshott, J. G., S. W. McKechnie, and G. K. Chambers. 1984. Population genetics of the metabolically related Adh, Gpdh, and Tpi polymorphisms in *Drosophila melanogaster*. 1. Geographic variation in Gpdh and Tpi allele frequencies in different continents. *Genetica* 63:21–29.

O'Brien, S. J., M. H. Gail, and D. L. Levin 1980. Correlative genetic variation in natural populations of cats, mice, and men. *Nature* 288:580–583.

O'Brien, S. J., M. E. Roelke, L. Marker, A. Newman, C. A. Winkler, D. Meltzer, L. Colly, J. F. Evermann, M. Bush, and D. E. Wildt. 1985. Genetic basis for species vulnerability in the cheetah. *Science* 227:1428–1434.

O'Brien, S. J., D. E. Wildt, D. Goldman, D. Merril, and M. Bush. 1983. The cheetah is depauperate in genetic variation. *Science* 221:459–462.

O'Farrell, P. H. 1975. High resolution of two-dimensional electrophoresis of proteins. *J. Biol. Chem.* 250:4007–4021.

Ohta, T., and M. Kimura. 1971. Amino acid composition of proteins as a product of molecular evolution. *Science* 174:150–153.

Ohta, T., and M. Kimura. 1973. A model of mutation appropriate to estimate the number of electrophoretically detectable alleles in a genetic population. *Genet. Res.* 22:201–204.

O'Malley, D. M., F. W. Allendorf, and G. M. Blake. 1979. Inheritance of isozyme variation and heterozygosity in *Pinus ponderosa*. *Biochem. Genet.* 17:233–250.

Oppenoorth, F. J. 1985. Biochemistry and genetics of insecticide resistance. Pp. 731–773 in G. A. Kerkut and L. J. Gilbert (eds.) *Comprehensive Insect Physiology, Biochemistry and Pharmacology*. Vol. 12. Pergamon, New York.

Palmer, A. R., and C. Strobeck. 1986. Fluctuating asymmetry: Measurement, analysis, patterns. *Annu. Rev. Ecol. Syst.* 17:391–421.

Parsons, P. A. 1971. Extreme environment heterosis and genetic loads. *Heredity* 26:479–483.

Parsons, P. A. 1973. Genetics of resistance to environmental stresses in *Drosophila* populations. *Annu. Rev. Genet.* 7:239–265.

Parsons, P. A. 1987. Evolutionary rates under environmental stress. *Evol. Biol.* 21:311–347.

Parsons, P. A. 1990. Fluctuating asymmetry: An epigenetic measure of stress. *Biol. Rev.* 63:131–145.

Partridge, L. 1980. Mate choice increases a component of offspring fitness in fruit flies. *Nature* 283:290–291.

Partridge, L. 1983. Non-random mating and offspring fitness. Pp. 227–255 in P. Bateson (ed.) *Mate Choice*. Cambridge University Press, Cambridge.

Pauling, L., H. A. Itano, S. J. Singer, and I. C. Wells. 1949. Sickle cell anemia, a molecular disease. *Science* 110:543–548.

Paynter, K. T., L. DiMichele, S. C. Hand, and D. A. Powers. 1991. Metabolic implications of LDH-B genotype during early development in *Fundulus heteroclitus*. *J. Exp. Zool.* 257:24–33.

Peng, T. X., A. Moya, and F. J. Ayala. 1986. Irradiation-resistance conferred by superoxide dismutase: Possible adaptive role of a natural polymorphism in *Drosophila melanogaster. Proc. Natl. Acad. Sci. USA* 83:684–687.

Peng, T. X., A. Moya, and F. J. Ayala. 1991. Two modes of balancing selection in *Drosophila melanogaster*: Overcompensation and overdominance. *Genetics* 128:381–391.

Petras, M. L., and J. C. Topping. 1983. The maintenance of polymorphisms at two loci in house mouse (*Mus musculus*) populations. *Can. J. Genet. Cytol.* 25:190–201.

Petrie, M. 1994. Improved growth and survival of offspring of peacocks with more elaborate trains. *Nature* 371:598–599.

Petrie, M., T. R. Halliday, and C. Sanders. 1991. Peahens prefer peacocks with elaborate trains. *Anim. Behav.* 41:323–331.

Pierce, B. A., and J. B. Mitton. 1979. A relationship of genetic variation within and among populations: An extension of the Kluge–Kerfoot phenomenon. *Syst. Zool.* 28:63–70.

Pierce, B. A., and J. B. Mitton. 1980. Patterns of allozyme variation in *Ambystoma tigrinum mavortium* and *A. T. nebulosum*. *Copeia* 1980:938–941.

Pierce, B. A., and J. B. Mitton. 1982. Allozyme heterozygosity and growth in the tiger salamander, *Ambystoma tigrinum*. *J. Hered.* 73:250–253.

Place, A. R., and D. A. Powers. 1979. Genetic variation and relative catalytic efficiencies: Lactate dehydrogenase B allozymes of *Fundulus heteroclitus*. *Proc. Natl. Acad. Sci. USA* 76:2354–2358.

Place, A. R., and D. A. Powers. 1984a. The lactate dehydrogenase (LDH-B) allozymes of *Fundulus heteroclitus* (Lin). 1. Purification and characterization. *J. Biol. Chem.* 259:1299–1308.

Place, A. R., and D. A. Powers. 1984b. The LDH-B allozymes of *Fundulus heteroclitus*: 2. Kinetic analyses. *J. Biol. Chem.* 259:1309–1318.

Plessas, M. E., and S. H. Strauss. 1986. Allozyme differentiation among populations, stands, and cohorts in Monterey pine. *Can. J. For. Res.* 16:1155–1164.

Pogson, G. H. 1991. Expression of overdominance for specific activity at the phosphoglucomutase-2 locus in the Pacific oyster *Crassostrea gigas*. *Genetics* 128:133–141.

Pogson, G. H., K. A. Mesa, and R. G. Boutilier. 1995. Genetic population structure and gene flow in the Atlantic cod *Gadus morhua*: A comparison of allozyme and nuclear RFLP loci. *Genetics* 139:375–385.

Pogson, G. H., and E. Zouros. 1994. Allozyme and RFLP heterozygosities as correlates of growth in the scallop *Placopecten magellanicus*: A test of the associative overdominance hypothesis. *Genetics* 137:221–231.

Politov, D. V., and K. V. Krutovskii. 1994. Allozyme polymorphism, heterozygosity, and mating system of stone pines. Pp. 36–42 in W. C. Schmidt and F.-K. Holtmeier (eds.) *Proceedings—International Workshop on Subalpine Stone Pines and Their Environment: The Status of Our Knowledge*. USDA For. Ser. Gen. Tech. Rep. INT-GTR-309.

Pomiankowski, A. 1987. Sexual selection: The handicap principle does work sometimes. *Proc. Roy. Soc.* (Lond.) ser. B. 231:123–145.

Pomiankowski, A., and A. P. Møller. 1995. A resolution of the lek paradox. *Proc. Roy. Acad. Sci.* London B. 260:21–29.

Poulik, M. D. 1957. Starch gel electrophoresis in a discontinuous system of buffers. *Nature* 180:1477–1479.

Powell, J. R. 1971. Genetic polymorphisms in varied environments. *Science* 174:1035–1036.

Powell, J. R. 1975. Protein variation in natural populations of animals. *Evol. Biol.* 8:79–119.

Powell, J. R., A. Caccone, G. D. Amato, and C. Yoon. 1986. Rates of nucleotide substitution in *Drosophila* mitochondrial DNA and nuclear DNA are similar. *Proc. Natl. Acad. Sci. USA* 83:9090–9093.

Powell, J. R., and C. E. Taylor. 1979. Genetic variation in ecologically diverse environments. *Amer. Scient.* 67:590–596.

Powell, J. R., and H. Wistrand. 1978. The effect of heterogeneous environments and a competitor on genetic variation in *Drosophila*. *Amer. Nat.* 112:935–947.

Powers, D. A. 1987. A multidisciplinary approach to the study of genetic variation within species. Pp. 102–130 in M. E. Feder, A. F. Bennett, W. W. Burggren, and R. B. Huey (eds.) *New Directions in Ecological Physiology*. Cambridge University Press, Cambridge.

Powers, D. A., L. DiMichele, and A. R. Place. 1983. The use of enzyme kinetics to predict differences in cellular metabolism, developmental rate, and swimming performance between LDH-B genotypes of the fish, *Fundulus heteroclitus*. In G. Whitt and G. C. Markert (eds.) *Isozymes: Current Topics in Biological and Medical Research*. Vol. 10. Academic Press, New York.

Powers, D. A., G. S. Greaney, and A. R. Place. 1979. Physiological correlation between lactate dehydrogenase genotype and haegmoglobin function in killifish. *Nature* 277:240–241.

Powers, D. A., and A. R. Place. 1978. Biochemical genetics of *Fundulus heteroclitus* (L.). I. Temporal and spatial variation in gene frequencies of Ldh-B, Mdh-A, Gpi-B, and Pgm-A. *Biochem. Genet.* 16:593–607.

Powers, D. A., M. Smith, I. Gonzalez-Villasenor, L. DiMichele, D. Crawford, G. Bernardi, and T. Lauerman. 1994. A multidisciplinary approach to the selection/neutralist controversy using the model teleost *Fundulus heteroclitus*. Pp. 43–107 in D. Futuyma and J. Antonovics (eds.) *Oxford Surveys in Evolutionary Biology*. Vol. 9. Oxford University Press, Oxford.

Preleuthner, M., W. Pinsker, L. Kruckenhauser, W. J. Miller, and H. Prosl. 1995. Alpine marmots in Austria: The present population structure as a result of the postglacial distribution history. *Acta Theriol.* (suppl.) 3:87–100.

Price, P. W. 1971. Niche breadth and dominance of parasitic insects sharing the same host species. *Ecology* 52:587–596.

Prody, C. A., P. Dreyfus, R. Zamir, H. Zakut, and H. Soreq. 1989. De novo amplification within a silent human cholinesterase gene in a family subjected to prolonged exposure ot organophosphorous insecticides. *Proc. Natl. Acad. Sci. USA* 86:690.

Qiao, C.-L., and M. Raymond. 1995. The same esterase B1 haplotype is amplified in insecticide-resistant mosquitoes of the *Culex pipiens* complex from the Americas and China. *Heredity* 74:339–345.

Quattro, J. M., and R. C. Vrijenhoek. 1989. Fitness differences among remnant populations of the endangered Sonoran topminnow. *Science* 245:976–978.

Racine, R. R., and C. H. Langley. 1980. Genetic heterozygosity in a natural population of *Mus musculus* assessed using two-dimensional electrophoresis. *Nature* 283:855–857.

Rainey, D. Y., J. B. Mitton, and R. K. Monson. 1987. Associations between enzyme genotypes and dark respiration in perennial reygrass, *Lolium perenne* L. *Oecologia* 74:335–338.

Rainey, D. Y., J. B. Mitton, R. K. Monson, and D. Wilson. 1990. Effects of selection for dark respiration rate on enzyme genotypes in *Lolium perenne* L. *Ann. Bot.* 66:649–654.

Rainey-Foreman, D., and J. B. Mitton. 1995. Glucose utilization by 6Pgd genotypes in *Lolium perenne*. *Mol. Ecol.* 4:231–237.

Ramshaw, J. A. M., J. A. Coyne, and R. C. Lewontin. 1979. The sensitivity of gel electrophoresis as a detector of genetic variation. *Genetics* 93:1019–1037.

Rand, D. M., and R. G. Harrison. 1989. Molecular population genetics of mtDNA size variation in crickets. *Genetics* 121:551–569.

Raup, H. M. 1975. Species versatility in shore habitats. *J. Arnold Arboretum* 56:126–163.

Raymond, M., V. Beyssat-Arnaouty, N. Sivasubramanian, C. Mouches, G. P. Georghiou, and N. Pasteur. 1989. Amplification of various esterase B's responsible for organophosphate resistance in *Culex* mosquitoes. *Biochem. Genet.* 27:417–243.

Raymond, M., E. Poulin, V. Boiroux, E. Dupont, and N. Pasteur. 1993. Stability of insecticide resistance due to amplification of esterase genes in *Culex pipiens*. *Heredity* 70:301–307.

Reeb, C. A., and J. C. Avise. 1990. A genetic discontinuity in a continuously distributed species: Mitochondrial DNA in the American oyster, *Crassostrea virginica*. *Genetics* 124:397–406.

Reeve, E. C. R. 1960. Some genetic tests on asymmetry of sternopleural chaeta number in *Drosophila*. *Genet. Res.* 1:151–172.

Rehfeldt, G. E. 1982. Differentiation of *Larix occidentalis* populations from the northern Rocky Mountains. *Silvae Genet.* 31:13–19.

Rehfeldt, G. E. 1988. Ecological genetics of *Pinus contorta* from the Rocky Mountains (USA): A synthesis. *Silvae Genet.* 37:131–135.

Rehfeldt, G. E. 1989. Ecological adaptations in Douglas-fir (*Pseudotsuga menziesii* var. *glauca*): A synthesis. *Forest Ecol. Manage.* 28: 203–215.

Rehfeldt, G. E. 1993. Genetic variation in the *ponderosae* of the southwest. *Amer. J. Bot.* 80:330–343.

Richardson, R. H., M. E. Richardson, and P. E. Smouse. 1975. Evolution of electrophoretic mobility in the *Drosophila mulleri* complex. Pp. 533–535 in C. L. Markert (ed.) *Isozymes 4. Genetics and Evolution*. Academic Press, New York.

Richmond, R. C., D. G. Gilbert, K. B. Sheehan, M. H. Gromko, and F. M. Butterworth. 1980. Esterase-6 and reproduction in *Drosophila melanogaster*. *Science* 297:1483–1485.

Ricklefs, R. E. 1972. Dominance and the niche in bird communities. *Amer. Nat.* 106:538–545.

Riska, B. 1979. Character variability and evolutionary rate in *Menidia*. *Evolution* 33:1001–1004.

Robertson, F. W., and E. C. R. Reeve. 1952. Heterozygosity, environmental variation and heterosis. *Nature* 170:286–287.

Robson, E. B., and H. Harris. 1965. Genetics of the alkaline phosphatase polymorphism of the human placenta. *Nature* 207:1257–1259.

Rodhouse, P. G., and P. M. Gaffney. 1984. Effect of heterozygosity on metabolism during starvation in the American oyster *Crassostrea virginica*. *Mar. Biol.* 80:179–188.

Rodhouse, P. G., J. H. McDonald, R. I. E. Newell, and R. K. Koehn. 1986. Gamete production, somatic growth and multiple locus heterozygosity in *Mytilus edulis* L. *Mar. Biol.* 90:209–214.

Rogers, S., R. Wells, and M. Rechsteiner. 1986. Amino acid sequences common to rapidly degraded proteins: The PEST hypothesis. *Science* 234:364–368.

Rohlf, F. J., A. J. Gilmartin, and G. Hart. 1983. The Kluge–Kerfoot phenomenon—A statistical artifact. *Evolution* 37:180–202.

Rolán-Alvarez, E., C. Zapata, and G. Alvarez. 1995. Multilocus heterozygosity and sexual selection in a natural population of the marine snail *Littorina mariae* (Gastropoda: Prosobranchia). *Heredity* 75:17–25.

Rothe, G. M., and F. Bergmann. 1995. Increased efficiency of Norway spruce heterozyogus phosphoenolpyruvate carboxylase phenotype in response to heavy air pollution. *Angew. Bot.* 69:27–30.

Royle, N. J., R. Clarkson, Z. Wong, and A. J. Jeffreys. 1987. Preference localization of hypervariable minisatellies near human telomeres. *Cytogenet. Cell Genet.* 46:685–686.

Ryman, N., G. Beckman, G. Brun-Petersen, and C. Reuterwall. 1977. Variability of red cell enzymes and genetic implications of management policies in Scandinavian moose (*Alces alces*). *Hereditas* 85:157–165.

Sabath, M. D. 1974. Niche breadth and genetic variability in sympatric natural populations of *Drosophilid* flies. *Amer. Nat.* 108:533–540.

Sage, R. D., P. V. Loiselle, P. Basasibwaki, and A. C. Wilson. 1984. Molecular versus morphological change among cichlid fishes of Lake Victoria. Pp. 185–197 in A. A. Echelle and I. Kornfield (eds.) *Evolution of Fish Species Flocks*. University of Maine Press, Orono.

Sampsell, B., and S. Simms. 1982. Interaction of Adh genotype and heat stress on alcohol tolerance in *Drosophila melanogaster*. *Nature* 296:853–855.

Sassaman, C. 1978. Dynamics of a lactate dehydrogenase polymorphism in the wood louse *Porcellio scaber* Latr.: Evidence for partial assortative mating and heterosis in natural populations. *Genetics* 88:591–609.

Saunders, N. C., L. G. Kessler, and J. C. Avise. 1986. Genetic variation and geographic differentiation in mitochondrial DNA of the horseshoe crab, *Limulus polyphemus*. *Genetics* 112:613–627.

Savolainen, O., and P. Hedrick. 1995. Heterozygosity and fitness: No association in scots pine. *Genetics* 140:755–766.

Schaal, B. A., and D. A. Levin. 1976. The demographic genetics of *Liatris cylindracea* Michx. *Amer. Nat.* 110:191–206.

Schaeffer, S. W., and E. L. Miller. 1993. Estimates of linkage disequilibrium and the recombination parameter determined from segregating nucleotide sites in the alcohol dehydrogenase region of *Drosophila pseudoobscura*. *Genetics* 135:541–552.

Scheltema, R. S. 1977. Dispersal of marine invertebrate organisms: Paleobiogeographic and biostratigraphic implications. Pp. 73–108 in E. G. Kauffman and J. E. Hazel (eds.) *Concepts and Methods of Biostratigraphy*. Dowden, Hutchinson and Ross, Stroudsburg, PA.

Scherer, S. 1990. The protein molecular clock: Time for a reevaluation. *Evol. Biol.* 24:83–148.

Schnell, D. G., and R. K. Selander. 1981. Environmental and morphological correlates of genetic variation in mammals. Pp. 60–99 in M. H. Smith and J. Joule (eds.) *Mammalian Population Genetics*. University of Georgia Press, Athens.

Scholz, F., and F. Bergmann. 1984. Selection pressure by air pollution as studied by isozyme-gene-systems in Norway spruce exposed to sulphur dioxide. *Silvae Genet.* 33:238–241.

Scholz, F., H.-R. Gregorius, and D. Rudin (eds.). 1989. *Genetic Effects of Air Pollutants in Forest Tree Populations*. Springer-Verlag, Berlin.

Schuster, W. S., D. L. Alles, and J. B. Mitton. 1989. Gene flow in limber pine: evidence from pollination phenology and genetic differentiation along an elevational transect. *Amer. J. Bot.* 76:1395–1403.

Scott, T. M., and R. K. Koehn. 1990. The effect of environmental stress on the relationship of heterozygosity to growth rate in the coot clam *Mulinia lateralis* (Say). *J. Exp. Mar. Biol. Ecol.* 135:109–116.

Scribner, K. T., and M. H. Smith. 1990. Genetic variability and antler development. Pp. 460–473 in G. A. Bubenik and A. B. Bubenik (eds.) *Horns, Pronghorns, and Antlers*. Springer-Verlag, New York.

Scribner, K. T., M. H. Smith, and P. E. Johns. 1984. Age, condition, and genetic effects on incidence of spiked bucks. *Proc. Annu. Conf. Southeast. Fish Wildl. Agencies* 38:23–32.

Scribner, K. T., M. H. Smith, and P. E. Johns. 1989. Environmental and genetic components of antler growth in white-tailed deer. *J. Mamm.* 70:284–291.

Seager, R. D., and F. J. Ayala. 1982. Chromosome interactions in *Drosophila melanogaster*. 1. Viability studies. *Genetics* 102:467–483.

Seager, R. D., F. J. Ayala, and R. W. Marks. 1982. Chromosome interactions in *Drosophila melanogaster*. 2. Total fitness. *Genetics* 102:485–502.

Selander, R. K. 1976. Genetic variation in natural populations. Pp. 21–45 in F. J. Ayala (ed.) *Molecular Evolution*. Sinauer Associates, Sunderland, MA.

Selander, R. K., and D. W. Kaufmann. 1973. Genetic variability and strategies of adaptation in animals. *Proc. Natl. Acad. Sci. USA* 70:1875–1877.

Selander, R. K., S. Y. Yang, R. C. Lewontin, and W. E. Johnson. 1970. Genetic variation in the horseshoe crab (*Limulus polyphemus*), a phylogenetic "relic." *Evolution* 24:402–414.

Serradilla, J. M., and F. J. Ayala. 1983a. Alloprocoptic selection: A mode of natural selection promoting polymorphism. *Proc. Natl. Acad. Sci. USA* 80:2022–2025.

Serradilla, J. M., and F. J. Ayala. 1983b. Effects of allozyme variation on fitness components in *Drosophila melanogaster*. *Genetica* 62:139–146.

Shaw, D., and R. W. Allard 1982. Isozyme heterozygosity in adult and open-pollinated embryo samples of Douglas-fir. *Silva. Fenn.* 16:115–121.

Shugart, H. H., and G. B. Blaylock. 1973. The niche-variation hypothesis: An experimental study with *Drosophila* populations. *Amer. Nat.* 107:575–579.

Shugart, H. H., and B. C. Patten. 1972. Niche quantification and the concept of niche pattern. Pp. 283–327 in B. C. Patten (ed.) *Systems Analysis and Simulation in Ecology*. Vol. 2. Academic Press, New York.

Sick, K. 1961. Haemoglobin polymorphism in fishes. *Nature* 192:894–896.

Sick, K. 1965a. Haemoglobin polymorphism of cod in the Baltic and the Danish Belt sea. *Hereditas* 54:19–48.

Sick, K. 1965b. Haemoglobin polymorphism of cod in the North Sea and the North Atlantic Ocean. *Hereditas* 54:49–69.

Siebenaller, J. F., and G. N. Somero. 1978. Pressure adaptive differences in lactate dehydrogenases of congeneric fishes living at different depths. *Science* 201:255–257.

Silva, P. J. N., R. K. Koehn, W. J. Diehl III, R. P. Ertl, E. B. Winshell, and M. Santos. 1989. The effect of glucose-6-phosphate isomeraase genotype in in vitro specific activity and in vivo flux in *Mytilus edulis*. *Biochem. Genet.* 27:451–467.

Simpson, G. G. 1953. *The Major Features of Evolution*. Columbia University Press, New York.

Singh, S. M. 1982. Enzyme heterozygosity associated with growth at different developmental stages in oysters. *Can. J. Genet. Cytol.* 24:451–458.

Singh, R. S., and L. R. Hale. 1990. Are mitochondrial DNA variants selectively non-neutral? *Genetics* 124:995–997.

Singh, R. S., J. L. Hubby, and L. H. Throckmorton. 1975. The study of genetic variation by electrophoretic and heat denaturation techniques at the octanol dehydrogenase locus in members of the *Drosophila virilis* group. *Genetics* 80:637–650.

Singh, R. S., R. C. Lewontin, and A. A. Felton. 1976. Genetic heterogeneity within electrophoretic "alleles" of xanthine dehydrogenase in *Drosophila pseudoobscura*. *Genetics* 84:609–629.

Singh, S. M., and R. H. Green. 1984. Excess of allozyme homozygosity in marine molluscs and its possible biological significance. *Malacologia* 25:569–581.

Singh, S. M., and E. Zouros. 1978. Genetic variation associated with growth rate in the American oyster (*Crassostrea virginica*). *Evolution* 32:342–353.

Skibinski, D. O. F., T. F. Cross, and M. Ahmad. 1980. Electrophoretic investigation of systematic relationships in the marine mussels *Modiolus modiolus* L., *Mytilus edulis* L., and *Mytilus galloprovincialis* Lmk. (Mytilidae: Mollusca). *Biol. J. Linn. Soc.* 13:65–73.

Skibinski, D. O. F., and R. D. Ward. 1981. Relationship between allozyme heterozygosity and rates of divergence. *Genet. Res. Camb.* 38:71–92.

Skibinski, D. O. F., and R. D. Ward. 1982. Correlations between heterozygosity and evolutionary rate of proteins. *Nature* 298:490–492.

Skibinski, D. O. F., M. Woodwark, and R. D. Ward. 1993. A quantitative test of the neutral theory using pooled allozyme data. *Genetics* 135:233–248.

Smith, D. A. S. 1975. Sexual selection in a wild population of the butterfly *Danaus chrysippus* L. *Science* 187:664–665.

Smith, D. A. S. 1980. Heterosis, epistasis and linkage disequilibrium in a wild population of the polymorphic butterfly *Danaus chrysippus* (L.). *Zool. J. Linn. Soc.* 69:87–109.

Smith, D. A. S. 1981. Heterozygous advantage experessed through sexual selection in a polymorphic African butterfly. *Nature* 289:174–175.

Smith, M. H., R. K. Chesser, E. G. Cothran, and P. E. Johns. 1982. Genetic variability and antler growth in a natural population of white-tailed deer. Pp. 365–387 in R. D. Brown (ed.) *Antler Development in Cervidae*. Caesar Kleberg Wildlife Research Institute, Kingsville, TX.

Smouse, P. E. 1986. The fitness consequences of multiple-locus heterozygosity under the multiplicative overdominance and inbreeding depression models. *Evolution* 40:946–957.

Snyder, L. R. G. 1978a. Genetics of hemoglobin in the deer mouse, *Peromyscus maniculatus*. 1. Multiple α- and β-globin structural loci. *Genetics* 89:511–530.

Snyder, L. R. G. 1978b. Genetics of hemoglobin in the deer mouse, *Peromyscus maniculatus*. 2. Multiple alleles at regulatory loci. *Genetics* 89:531–550.

Snyder, L. R. G. 1979. Strong gametic phase disequilibrium between two closely linked a-type globin loci in the deer mouse, *Peromyscus maniculatus*. *Genetics* 91:121.

Snyder, L. R. G. 1981. Deer mouse hemoglobins: Is there genetic adaptation to high altitude? *Bioscience* 31:299–304.

Sober, E. 1984. *The Nature of Selection*. MIT Press, Cambridge, MA.

Sober, E., and R. C. Lewontin. 1982. Artifact, cause, and genetic selection. *Philos. Sci.* 49:157–180.

Sokal, R. R. 1965. Statistical methods in systematics. *Biol. Rev.* 40:337–391.

Sokal, R. R. 1976. The Kluge–Kerfoot phenomenon reexamined. *Amer. Nat.* 110:1077–1091.

Solignac, M., J. Generemont, M. Monnerot, and J. C. Mounolou. 1984. Genetics of mitochondria in *Drosophila*: Inheritance in heteroplasmic strains of *D. mauritiana*. *Mol. Gen. Genet.* 197:183–188.

Somero, G. N. 1978. Temperature adaptation of enzymes: Biological optimization through structure–function compromises. *Annu. Rev. Ecol. Syst.* 9:1–19.

Somero, G. N., and M. Soulé. 1974. Genetic variation in marine fishes as a test of the niche-variation hypothesis. *Nature* 249:670–672.

Soulé, M. E. 1967. Phenetics of natural populations. 2. Asymmetry and evolution in a lizard. *Amer. Nat.* 101:141–160.

Soulé, M. E. 1971. The variation problem: The gene-flow-variation hypothesis. *Taxonomy* 20:37–50.

Soulé, M. 1973. The epistasis cycle: A theory of marginal populations. *Annu. Rev. Ecol. Syst.* 4:165–187.

Soulé, M. 1976. Allozyme variation: Its determinants in space and time. Pp. 60–77 in F. J. Ayala (ed.) *Molecular Evolution*. Sinauer Associates, Sunderland, MA.

Soulé, M. E. 1979. Heterozygosity and developmental stability: Another look. *Evolution* 33:396–401.

Soulé, M. E. 1982. Allomeric variation. 1. The theory and some consequences. *Amer. Nat.* 120:751–764.

Spassky, B., Th. Dobzhansky, and W. W. Anderson. 1965. Genetics of natural populations. 36. Epistatic interactions of the components of the genetic load in *Drosophila pseudoobscura*. *Genetics* 52:623–664.

Spencer, N., D. A. Hopkinson, and H. Harris. 1964. Phosphoglucomutase polymorphism in man. *Nature* 204:742–745.

Spiess, E. B. 1977. *Genes in Populations*. Wiley, New York.

Stebbins, G. L., and R. C. Lewontin. 1972. Comparative evolution at the levels of molecules, organisms, and populations. Pp. 23–42 in L. M. Le Cam, J. Neyman, and E. L. Scott (eds.)

Proceedings of the Sixth Berkeley Symposium on Mathematical Statistics and Probability. Vol. 5. University of California Press, Berkeley and Los Angeles.

Stiven, A. E. 1995. Genetic heterozygosity and growth rate in the southern Appalachian land snail *Mesodon normalis* (Pilsbry 1900): The effects of laboratory stress. *Malacologia* 36:171–184.

Strauss, S. H. 1986. Heterosis at allozyme loci under inbreeding and crossbreeding in *Pinus attenuata*. *Genetics* 113:115–134.

Strauss, S. H., Y.-P. Hong, and V. D. Hipkins. 1993. High levels of population differentiation for mitochondrial DNA haplotypes in *Pinus radiata*, *muricata*, and *attenuata*. *Theor. Appl. Genet.* 85:6065–6071.

Sved, J. A. 1971. An estimate of heterosis in *Drosophila melanogaster*. *Genet Res.* 18:97–105.

Sved, J. A., and F. J. Ayala. 1970. A population cage test for heterosis in *Drosophila pseudoobscura*. *Genetics* 66:97–113.

Sved, J. A., T. E. Reed, and W. F. Bodmer. 1967. The number of balanced polymorphisms that can be maintained in a natural population. *Genetics* 55:469–481.

Swaddle, J. P., and I. C. Cuthill. 1994a. Female zebra finches prefer males with symmetrically manipulated chest plummage. *Proc. Roy. Soc.* (Lond.) ser. B 258:267–271.

Swaddle, J. P., and I. C. Cuthill. 1994b. Preference for symmetric males by female zebra finches. *Nature* 367:165–166.

Templeton, A. R. 1979. The unit of selection in *Drosophila mercatorum*. 2. Genetic revolution and the origin of coadapted genomes in parthenogenetic strains. *Genetics* 92:1265–1282.

Templeton, A. R. 1980a. Modes of speciation and inferences based on genetic distances. *Evolution* 34:719–729.

Templeton, A. R. 1980b. The theory of speciation via the founder principle. *Genetics* 94:1011–1038.

Templeton, A. R. 1981. Mechanisms of speciation—A population genetic approach. *Annu. Rev. Ecol. Syst.* 12:23–48.

Templeton, A. R. 1982. Adaptation and the integration of evolutionary forces. Pp. 15–31 in R. Milkman (ed.) *Perspectives on Evolution*. Sinauer Associates, Sunderland, MA.

Templeton, A. R., C. F. Singh, and B. Brokaw. 1976. The unit of selection in *Drosophila mercatorum*. 1. The interaction of selection and meiosis in parthenogenetic strains. *Genetics* 82:249–376.

Teska, W. R., M. H. Smith, and J. M. Novak. 1990. Food quality, heterozygosity, and fitness correlates in *Peromyscus polionotus*. *Evolution* 44:1318–1325.

Thornhill, R. 1992. Fluctuating asymmetry and the mating system of the Japanese scorpionfly, *panorpa japonica*. *Anim. Behav.* 44:867–869.

Thornhill, R., and J. Alcock. 1983. *The Evolution of Insect Mating Systems*. Harvard University Press, Cambridge, MA.

Thorpe, J. P. 1982. The molecular clock hypothesis: Biochemical evolution, genetic differentiation, and systematics. *Annu. Rev. Ecol. Syst.* 13:139–168.

Tobias, J. W., T. E. Shrader, B. Rocap, and A. Varshavsky. 1991. The N-end rule in bacteria. *Science* 254:1374–1377.

Tomekpe, D., and R. Lumaret. 1991. Association between quantitative traits and allozyme heterozygosity in a tetrasomic species: *Dactylis glomerata*. *Evolution* 45:359–370.

Tosic, M., and F. J. Ayala. 1981. Density- and frequency-dependent selection at the Mdh-2 locus in *Drosophila pseudoobscura*. *Genetics* 97:679–701.

Tracey, M. L., and F. J. Ayala. 1974. Genetic load in natural populations: Is it compatible with the hypothesis that many polymorphisms are maintained by natural selection? *Genetics* 77:569–589.

Tracey, M., N. F. Bellet, and C. D. Gravem. 1975. Excess allozyme homozygosity and breeding population structure in mussels, *Mytilus californianus*. *Marine Biol.* 32:303–311.

Turelli, M., and L. Ginzburg. 1983. Should individual fitness increase with heterozygosity? *Genetics* 104:191–209.

Valentine, J. W. 1976. Genetic strategies of adaptation. Pp. 78–94 in F. J. Ayala (ed.) *Molecular Evolution*. Sinauer Associates, Sunderland, MA.

Valentine, J. W., and F. J. Ayala. 1975. Genetic variation in *Frieleia halli*, a deep sea brachiopod. *Deep-Sea Res.* 22:37–44.

Valentine, J. W., and F. J. Ayala. 1976. Genetic variability in krill. *Proc. Natl. Acad. Sci. USA* 73:658–660.

Valentine, J. W., and D. Jablonski 1983. Speciation in the shallow sea: General patterns and biogeographic controls. Pp. 201–226 in R. W. Sims, J. H. Price, and P. E. S. Whalley (eds.) *Evolution, Time and Space: The Emergence of the Biosphere*. Academic Press, New York.

Van Delden, W. 1982. The alcohol dehydrogenase polymorphism in *Drosophila melanogaster*: Selection at an enzyme locus. Pp. 187–222 in M. K. Hecht, B. Wallace, and G. T. Prance (eds.) *Evolutionary Biology*. Vol. 15.

Van Valen, L. 1962. A study of fluctuating asymmetry. *Evolution* 16:125–142.

Vrijenhoek, R. C., and P. L. Leberg. 1991. Let's not throw the baby out with the bathwater: A comment on the management for MHC diversity in captive populations. *Conserv. Biol.* 5:252–254.

Vrijenhoek, R. C., and S. Lerman. 1982. Heterozygosity and developmental stability under sexual and asexual breeding systems. *Evolution* 36:768–776.

Vrijenhoek, R. C., E. Pfeiler, and J. D. Wetherington. 1992. Balancing selection in a stream-dwelling fish, *Poeciliopsis monacha*. *Evolution* 46:1642–1657.

Wade, M. J. 1978. A critical review of the models of group selection. *Quart. Rev. Biol.* 53:101–114.

Wallace, B. 1975. Hard and soft selection revisited. *Evolution* 29:465–473.

Wallis, G. P., and J. A. Beardmore. 1984. Genetic variation and environmental heterogeneity in some closely related goby species. *Genetica* 62:223–237.

Waples, R. S. 1991. Heterozygosity and life-history variation in bony fishes. *Evolution* 45:1275–1280.

Ward, R. D. 1977. Relationship between enzyme heterozygosity and quaternary structure. *Biochem. Genet.* 15:123–135.

Ward, R. D. 1978. Subunit size of enzymes and genetic heterozygosity in vertebrates. *Biochem. Genet.* 16:799–810.

Ward, R. D., M. Sarfarazi, C. Azimigarakani, and J. A. Beardmore. 1985. Population genetics of polymorphisms in Cardiff newborn: Relationship between blood group and allozyme heterozygosity and birth weight. *Hum. Hered.* 35:171–177.

Ward, R. D., D. O. F. Skibinski, and M. Woodwark. 1992. Protein heterozygosity, protein structure, and taxonomic differentiation. *Evol. Biol.* 26:73–159.

Watt, W. B. 1968. Adaptive significance of pigment polymorphisms in *Colias* butterflies. 1. Variation of melanin pigment in relation to thermoregulation. *Evolution* 22:437–458.

Watt, W. B. 1977. Adaptation at specific loci. 1. Natural selection in phosphoglucose isomerase of *Colias* butterflies: Biochemical and population aspects. *Genetics* 87:177–194.

Watt, W. B. 1983. Adaptation at specific loci. 2. Demographic and biochemical elements in the maintenance of the *Colias* PGI polymorphism. *Genetics* 103:691–724.

Watt, W. B. 1985. Bioenergetics and evolutionary genetics: Opportunities for new synthesis. *Amer. Nat.* 125:118–143.

Watt, W. B. 1986. Power and efficiency as indexes of fitness in metabolic organization. *Amer. Nat.* 127:629–653.

Watt, W. B. 1991. Biochemistry, physiological ecology, and population genetics—The mechanistic tools of evolutionary biology. *Func. Ecol.* 5:145–154.

Watt, W. B. 1992. Eggs, enzymes, and evolution—Natural genetic variants change insect fecundity. *Proc. Natl. Acad. Sci. USA* 89:10608–10612.

Watt, W. B., P. A. Carter, and S. M. Blower. 1985. Adaptation at specific loci. 4. Differential mating success among glycolytic allozyme genotypes of *Colias* butterflies. *Genetics* 109:157–175.

Watt, W. B., P. A. Carter, and K. Donohue. 1986. Females' choice of "good genotypes" as mates is promoted by an insect mating system. *Science* 233:1187–1190.

Watt, W. B., R. C. Cassin, and M. S. Swan. 1983. Adaptation at specific loci. 3. Field behavior and survivorship differences among *Colias* PGI genotypes are predictable from in vitro biochemistry. *Genetics* 103:725–739.

Weir, B. S. 1990. *Genetic Data Analysis*. Sinauer Associates, Sunderland, MA.

Weir, B. S., A. H. D. Brown, and D. R. Marshall. 1976. Testing for selective neutrality of electrophoretically detectable protein polymorphism. *Genetics* 84:639–659.

Westfall, R. D., and M. T. Conkle. 1992. Allozyme markers in breeding zone designation. *New Forests* 6:279–309. Also pp. 279–309 in W. T. Adams, S. H. Strauss, D. L. Copes, and A. R. Griffin (eds.) *Population Genetics of Forest Trees*. Kluwer Academic, Dordrecht, 1992.

Wetton, J. H., R. E. Carter, D. T. Parkin, and D. Walters. 1987. Demographic study of a wild house sparrow population by DNA fingerprinting. *Nature* 327:147–149.

Wheeler N. C., and R. P. Guries. 1982. Population structure, genic diversity and morphological variation in *Pinus contorta* Dougl. *Can. J. For. Res.* 12:595–606.

White, M. J. D. 1978. *Modes of Speciation*. Freeman, San Francisco.

Whitehurst, P. H., and B. A. Pierce. 1991. The relationship between allozyme variation and life-history traits of the spotted chorus frog, *Pseudacris clarkii*. *Copeia* 1991:1032–1039.

Whitham, T. G., and S. Mopper. 1985. Chronic herbivory: Impacts on architecture and sex expression of pinyon pine. *Science* 228:1089–1091.

Wilkins, N. P. 1975. Phosphoglucose isomerase in marine molluscs. Pp. 931–943 in C. L. Markert (ed.) *Isozymes 4. Genetics and Evolution*. Academic Press, New York.

Williams, G. C. 1966. *Adaptation and Natural Selection*. Princeton University Press, Princeton, NJ.

Williams, G. C. 1975. *Sex and Evolution*. Princeton University Press, Princeton, NJ.

Williams, G. C. 1985. A defense of reductionism in evolutionary biology. Pp. 1–27 in R. Dawkins and M. Ridley (eds.) *Oxford Surveys in Evolutionary Biology*. Vol. 2. Oxford University Press, Oxford.

Williams, G. C., R. K. Koehn, and J. B. Mitton. 1973. Genetic differentiation without isolation in the American eel, *Anguilla rostrata*. *Evolution* 27:192–204.

Williams, G. C., and J. B. Mitton. 1973. Why reproduce sexually? *J. Theor. Biol.* 39:545–554.

Wills, C. 1978. Rank-order selection is capable of maintaining all genetic polymorphisms. *Genetics* 89:403–414.

Wills, C. 1981. *Genetic Variability*. Clarendon Press, Oxford.

Wilson, D. 1975. Variation in leaf respiration in relation to growth and photosynthesis of *Lolium*. *Ann. Appl. Biol.* 80:323–338.

Wilson, D. 1981. Response to selection for dark respiration rate of mature leaves in *Lolium perenne* and its effects on growth of young plants and similar simulated swards. *Ann. Bot.* 49:303–312.

Wilson, D., and J. G. Jones. 1982. Effect of selection for dark respiration rate of mature leaves on crop yields of *Lolium perenne* cv s23. *Ann. Bot.* 49:313–320.

Wolff, K., and J. Haeck, J. 1990. Genetic analysis of ecologically relevant morphological variability in *Plantago lanceolata* L. 6. The relation between allozyme heterozygosity and some fitness components. *J. Evol. Biol.* 3:243–255.

Wooten, M. C., and M. H. Smith. 1985. Large mammals are genetically less variable? *Evolution* 39:210–212.

Wright S. 1931. Evolution in Mendelian populations. *Genetics* 16:97–159.

Wright, S. 1978. *Evolution and the Genetics of Populations. 4. Variability Within and Among Populations*. University of Chicago Press, Chicago.

Wu, C.-I., and W.-H. Li. 1985. Evidence for higher rates of nucleotide substitution in rodents than in man. *Proc. Natl. Acad. Sci. USA* 82:1741–1745.

Yeh, F. C. 1988. Isozyme variation of *Thuja plicata* (Cupressaceae) in British Columbia. *Biochem. Syst. Ecol.* 16:373–377.

Yeh, F. C., and Y. A. El-Kassaby. 1980. Enzyme genetic variation in natural populations of Sitka spruce (*Picea sitchensis* [Bong.] Carr.). 1. Genetic variation patterns among trees from ten IUFRO provenances. *Can. J. For. Res.* 10:415–422.

Yeh, F. C., M. A. K. Khalil, Y. A. El-Kassaby, and D. C. Trust. 1986. Allozyme variation in *Picea mariana* from Newfoundland: Genetic diversity, populations structure, and analysis of differentiation. *Can. J. For. Res.* 16:713–720.

Yeh, F. C., and C. Layton. 1979. The organization of genetic variability in central and marginal populations of lodgepole pine, *Pinus contorta* ssp. *latifolia*. *Can. J. Genet. Cytol.* 21:487–503.

Yeh, F. C., and D. O'Malley. 1980. Enzyme variations in natural populations of Douglas-fir *Pseudotsuga menziesii* (Mirb.) 1. Genetic variation patterns in coastal populations. *Silvae Genet.* 29:83–92.

Yezerinac, S. M., S. C. Lougheed, and P. Handford. 1992. Morphological variability and enzyme heterozygosity: Individual and population level correlations. *Evolution* 46:1959–1964.

Young, J. P. W., R. K. Koehn, and N. Arnheim. 1979. Biochemical characterization of "Lap," a polymorphic aminopeptidase from the blue mussel, *Mytilus edulsis*. *Biochem. Genet.* 17:305–323.

Zamer, W. E., and R. J. Hoffmann. 1989. Allozymes of glucose-6-phosphate isomearse differentially modulate pentose-shunt metabolism in the sea anemone *Metridium senile*. *Proc. Natl. Acad. Sci. USA* 86:2737–2741.

Zapata, C., G. Gajardo, and J. A. Beardmore. 1990. Multilocus heterozygosity and sexual selection in the brine shrimp *Artemia franciscana*. *Mar. Ecol. Prog. Ser.* 62:211–217.

Zera, A. J., R. K. Koehn, and J. G. Hall. 1983. Allozymes and biochemical adaptation. Pp. 633–674 in G. A. Kerkut and L. I. Gilbert (eds.) *Comprehensive Insect Physiology, Biochemistry and Pharmacology*. Pergamon, New York.

Zouros, E. 1975. Electrophoretic variation in allelozymes related to function or structure. *Nature* 154:446–448.

Zouros, E., and D. W. Foltz. 1984. Possible explanations of heterozygote deficiency in bivalve molluscs. *Malacologia* 25:583– 591.

Zouros, E., and D. W. Foltz. 1987. The use of allelic isozyme variation for the study of heterosis. Pp. 2–59 in M. C. Rattazzi, J. G. Scandalios, and G. S. Whitt (eds.) *Isozymes: Current Topics in Biological and Medical Research*. Vol. 13. Alan R. Liss, New York.

Zouros, E., and G. H. Pogson. 1993. The present status of the relationship between heterozygosity and heterosis. Pp. 135–146 in A. Beaumont (ed.) *Genetics and Evolution of Aquatic Organisms*. Chapman and Hall, London.

Zouros, E., S. M. Singh, D. W. Foltz, and A. L. Mallet. 1983. Post-settlement viability in the American oyster (*Crassostrea virginica*): An overdominant phenotype. *Genet. Res. Camb.* 41:259–270.

Zouros, E., S. M. Singh, and H. E. Miles. 1980. Growth rate in oysters: An overdominant phenotype and possible explanations. *Evolution* 34:856–867.

Zuk, M. 1984. A charming resistance to parasites. *Nat. Hist.* 93:28–34.

Zuk, M., K. Johnson, R. Thornhill, and J. D. Ligon. 1990. Mechanisms of female choice in red jungle fowl. *Evolution* 44:472–485.

Index